NUCLEAR SPECTROSCOPY ON CHARGE DENSITY WAVE SYSTEMS

Physics and Chemistry of Materials with Low-Dimensional Structures

VOLUME 15

The titles published in this series are listed at the end of this volume.

NUCLEAR SPECTROSCOPY ON CHARGE DENSITY WAVE SYSTEMS

Edited by

Tilman Butz

Physics Department,
Technische Universität München,
Garching, Germany

KLUWER ACADEMIC PUBLISHERS
DORDRECHT / BOSTON / LONDON

Library of Congress Cataloging-in-Publication Data

Nuclear spectroscopy on charge density wave systems / edited by Tilman Butz.
p. cm. -- (Physics and chemistry of materials with low-dimensional structures ; v. 15)
Papers from a workshop held in Bavaria in 1989.
Includes bibliographical references and index.
ISBN 978-90-481-4165-4 (alk. paper)
1. Charge density waves. 2. Solid state physics. 3. Nuclear spectroscopy. I. Butz, Tilman. II. Series.
QC176.8.E4N83 1992
530.4'1--dc20 92-12827

ISBN 978-90-481-4165-4 ISBN 978-94-015-1299-2 (eBook)
DOI 10.1007/978-94-015-1299-2

Published by Kluwer Academic Publishers,
P.O. Box 17, 3300 AA Dordrecht, The Netherlands.

Kluwer Academic Publishers incorporates
the publishing programmes of
D. Reidel, Martinus Nijhoff, Dr W. Junk and MTP Press.

Sold and distributed in the U.S.A. and Canada
by Kluwer Academic Publishers,
101 Philip Drive, Norwell, MA 02061, U.S.A.

In all other countries, sold and distributed
by Kluwer Academic Publishers Group,
P.O. Box 322, 3300 AH Dordrecht, The Netherlands.

Printed on acid-free paper

Table of Contents

PREFACE

This book is the first coherent presentation of investigations of charge density wave (CDW) systems by nuclear spectroscopic techniques. It belongs to the series "Physics and Chemistry of Materials with Low-Dimensional Structures" published by Kluwer Academic Publishers. It addresses the following readership: graduate students and elder scientists with an education in physics and chemistry who are interested in modern aspects of solid state physics and want to acquire a broader knowledge of nuclear spectroscopy techniques applied to CDW systems; experts in the field of CDW's who are familiar with a dozen of modern techniques but never bestowed all their affections on nuclear spectroscopy techniques; active researchers in the field of *one* of the nuclear spectroscopy techniques applied to CDW systems who wish to learn that even this highly specialized topic – which certainly does not deserve a separate entry in the physics and astronomy classification scheme – is full of surprises "next door".

The limited space of this volume necessitated a number of restrictions: first, a selection of systems had to be made which was guided by the concept of discussing a few systems thorougly studied by nuclear spectroscopy techniques in greater detail rather than to present an incoherent collection of data; secondly the reader is advised to make extensive use of previous books in this series because it would not have made sense to repeat the content here.

Chapter 1 gives a short introduction to CDW's in general and to the question what can be learned about CDW's by nuclear spectroscopy techniques. Chapter 2 gives a Landau theory description of CDW formation in chain-like tetrachalcogenides. This chapter is intended to elucidate the essential ingredients into a theoretical description of a prototypical system and should stimulate a rigorous research activity on these substances by nuclear spectroscopy techniques. For CDW theories on layered transition metal compounds, the reader is referred to the book "Structural Phase Transitions in Layered Transition Metal Compounds", K. Motizuki (ed.) in the same series. Chapter 3 treats experimental resuls on layered transition metal compounds. A short introduction to nuclear magnetic resonance (NMR), nuclear quadrupole resonance (NQR), time differential perturbed angular correlation (TDPAC), and the Mössbauer effect (ME) is included in this chapter because all three techniques have been extensively applied to Ta-dichalcogenides which are prominent members of this family of materials. Whereas chapter 3 in mainly concerned with static aspects of CDW's, chapter 4 and 5 treat in great detail CDW dynamics and CDW transport in chain-like transition metal chalcogenides and molybdenum bronzes, respectively. The books "Electronic Properties of Inorganic Quasi-One Dimensional Compounds I + II", E. Monceau (ed.), and "Low-Dimensional Electronic Properties of Molybdenum Bronzes and Oxides", C. Schlenker (ed.), of the same series are indispensable. Chapter 6 treats the one-dimensional inorganic complex salt $K_2Pt(CN)_4Br_{0.3}3.2H_2O$ (KCP) studied by NMR on a variety of nuclei. Chapter 7 demonstrates the tremendous advances of high resolution NMR in yielding spin density maps of organic radical salts and is intended to stimulate the application of this powerful technique more directly to CDW phenomena. Despite the broad variety of materials treated in this book the leitmotif should be clear: the crystalline state of the solid is by no means dull and dead but rather alive.

The idea to edit this book matured during a workshop with the same title held at the beautiful former monastery Irsee in upper Bavaria in 1989 which was generously sponsored by the Stiftung Volkswagenwerk. I am indebted to Dr. A. Lerf for his organisational contribution to this workshop and to the edition of this book in its early stage. It is a pleasure to thank all authors who contributed to this volume. Last but not least, I am grateful to B. and M. Weber who tirelessly fought the battle of floppies, formats, fonts and alike *i.e.* the blessings of the computer age.

T. Butz

Garching, February 1992

INTRODUCTION

TILMAN BUTZ
Physik-Department, Technische Universität München
8046 Garching
Germany

Our knowledge of the structure of the solid state of matter has increased incredibly with the discovery of X-ray diffraction, followed by neutron and electron diffraction, techniques which derive their power form the translational invariance inherent in crystalline matter. In fact, concepts like the reciprocal lattice and k-space as well as classifications in terms of Bravais-lattices and group theory have proven immensely powerful. The desire for simplicity and harmony even tempts to consider the crystalline state with translational invariance to be superior, *i.e.* energetically more favored, to the amorphous state. Of course, the real world is not perfect and crystals have defects, point defects or extented defects, as well as there is short-range order in amorphous systems.

However, this view of the structure of solids is far too limited. Nature has much more fantasy! It suffices to mention the element carbon: there is graphite, diamond, and now the Buckminster-Fullerite. What an impressive architecture! Moreover, phase transitions accompanied by superstructure formation and symmetry breaking are Nature's popular tools to keep scientists busy. Let us have a closer look to superstructure formation in a monatomic one-dimensional chain. In the normal state, all atoms are equidistant. For a half-filled conduction band, however, dimerization could take place with a doubling of the unit cell and a gap opening up at the Brillonin zone boundary. This transition from a conducting state to an insulating state was first described by Peierls in 1955 [1]. Obviously, the elastic energy required for dimerization has to be overcompensated by electronic energy gain should the transition take place. Since electronic charge should accumulate on the dimer with a depletion in the region between dimers, a socalled charge density wave (CDW) forms. Of course, other band fillings would lead to larger unit cells. Examples for this Peierls transition are partially oxidized potassium tetracyanoplatinate $K_2Pt(CN)_4Cl(Br)_{0.33}H_2O$ and the purely organic conductor TTF-TCNQ, both to be discussed in detail in chapters 6 and 7, respectively. The extension of this concept to two-dimensional systems leads to two-dimensional clusters. As an example consider a hexagonal lattice and a charge density distribution described by three

T. Butz (ed.), Nuclear Spectroscopy on Charge Density Wave Systems, 1–5.

wavevectors of equal magnitude at 120° intervals. If the phasing of the CDW is hexagonal, *i.e.* the density maxima are located at atomic positions, seven-atom clusters, 13-atom clusters *etc.* can be formed eventually with additional isolated atoms. Other phasings lead to new types of clusters. Much more complicated structures arise for CDW's with lower symmetry. Examples for such two-dimensional clusters are several transition metal dichalcogenides as discussed in chapter 3.

Thus far we have discussed superstructures which are commensurate with the lattice only. All these systems are perfectly describable by conventional methods, *i.e.* by a unit cell and the associated basis of atoms. The complexity might eventually jeopardize modern *ab initio* band structure calculations, but basically no new concepts are required. An extreme case is the consecutive formation of various commensurate superstructures, the socalled devils staircase [2]. However, the possibility to cover an area completely with two types of tiles *without* translational invariance, *i.e.* Penrose tiling [3], by far exceeding M.C. Escher's beautiful "regelmatige vlakverdeling", and the discovery of quasi-crystals [4] severely shakes our traditional view of a crystalline solid. There is yet another rather subtile attack to the concept of a unit cell with a basis: there is no reason whatsoever for the CDW-wavevector to be commensurate with the lattice. An incommensurate modulation of the atomic positions invalidates the concept of a unit cell alltogether. Whereas the diffraction techniques apparently still cope with these new problems, P. Bak hit the nail on the head by his desperate exclamation: "Where are the atoms?" [5]. Whereas incommensurate structures can still be described by projections from a higher dimensional space – what an uncomfortable abstraction for an experimentalist! – the dismantling of "conventional order" culminates in Ruelle's "turbulent crystal" [6].

Interestingly enough, just in time a "real space" technique with atomic resolution was invented: the scanning tunneling microscope (STM) and the related atomic force microscope (AFM). In fact, the STM achieved a great triumph in the field of CDW's within a few years only [7]. However, it does not seem obvious how to translate our familiar macroscopic concepts like mechanical properties or electronic transport properties to objects of an atomic size, *i.e.* to an atom itself. What a mess: the atom at the tip of the STM senses the atom of the sample – or vice versa? Who is the master, who is the slave? Who raises questions and who answers?

Thus far, we have discussed structural aspects concerning the average atomic positions in a CDW-modulated state only, *i.e.* a static situation. However, for an incommensurate CDW there is no energy required to slide the CDW over the lattice. This offers the possibility of CDW-transport, first recognized by Fröhlich [8] in 1955, a sort of "cost-free transport" like superconductivity or ballistic transport. Field-dependent non-ohmic conductivity and narrow-band noise are but two catchwords in this fascinating field which has been reviewed recently by Monceau [9] and Grüner [10]. Chapters 4 and 5 of this book are devoted to sliding CDW's in quasi-one-dimensional conductors. What an effervescent source of "-ons": phonons, phasons, amplitudons, let alone excitons, solitons and so ons, a never-ending list of exciting excitations. I can't help thinking of the following metaphor: what absurdity to reduce real life with all emotions, irrationalism, hate and love, common sense, excitements – and what have you – to a military parade with primitive cubic symmetry!

Now let us turn to nuclear spectroscopy. There are quite a few techniques: nuclear magnetic resonance (NMR), nuclear quadrupole resonance (NQR), time differential perturbed angular correlations (TDPAC), Mössbauer effect (ME), nuclear orientation (NO), and muon spin rotation (μSR). The first four techniques have been widely applied in the field of CDW systems and a brief introduction will be given in chapter 3. NO has been applied to 2H-TaS_2 but has no resolving power because it is an integral technique. Thus far, no NMR on oriented nuclei has been reported. Similary, μSR has not given any information on CDW systems thus far. Therefore these last two techniques will not be discussed any further.

The nuclei are coupled to the "extranuclear world" via the hyperfine interaction. There are basically three types of hyperfine parameters: first, there are energy shifts like Knight-shift, chemical shift, or isomer shift (ME) which in some way are related to electron densities at the nuclei; secondly, there are energy splittings like magnetic dipole interactions (nuclear Zeeman effect) or electric quadrupole interactions; third, there are relaxation times: T_1, the spin-lattice or longitudinal relaxation time and T_2, the spin-spin or transverse relaxation time. In the context of CDW systems the various shifts and the quadrupole splittings are the key quantities: they allow the detection of very subtle site inequivalencies with a precision often exceeding by far the diffraction results. Of particular use is the nuclear quadrupole interaction because it yields under suitable circumstances the sign, magnitude, symmetry, and orientation of the electric field gradient tensor all of which characterize a single site. In order to illustrate the complementary nature of hyperfine data and diffraction data let us consider a simple example: the periodicity of a commensurate CDW is easily determined by electron diffraction, say $3a_0 \times 3a_0$ like in 2H-$TaSe_2$, but it is rather difficult to extract the phasing of the CDW onto the lattice. Depending on this phasing the number and population of inequivalent Ta-sites (and Se-sites) is completely different. For a hexagonal phasing, a 1:6:2 population of three inequivalent Ta sites should be observed (the center of a seven-atom cluster and its 6 nearest neighbours plus two isolated atoms). For an orthorhombic phasing with the CDW maxima located between three Ta atoms, a 3:3:3 population of three inequivalent Ta sites should be observed, which is indeed the case. Moreover, the center atom in the case of hexagonal phasing should possess axial symmetry whereas the center triangle of the orthorhombic phasing should deviate strongly from axial symmetry. This shows immediately how easy it is to determine the phasing form hyperfine data, provided the periodicity of the CDW is known. Moreover, the stacking of the CDW in adjacent layers has a subtle effect on the cluster architecture eventually rendering initially equivalent lattice sites inequivalent. Another example pertaining to incommensurate CDW's: a simple plane wave incommensurate CDW yields characteristic charge density profiles which translate into shift profiles of profiles of the electric field gradient tensor components. Nature generously provided such simple examples, as will be shown in chapters 3 to 5. The book on "Incommensurate Phases in Dielectrics" [11] is strongly recommended. A final example illustrates the power of hyperfine techniques: the motional narrowing of resonance lines observed for sliding CDW's is of crucial importance because it unambiguously proves that the CDW-current affects the bulk of the material and is not of filamentary nature (see chapters 4 and 5). What a rich source of information we get by simply asking a nucleus about his "feelings" in the solid!

There are a few victims to my enthusiasm: the history of CDW's and a sound bibliography. The latter problem can be cured: a list of recommended books and review articles not yet cited previously is given at the end of the reference list. The former task exceeds the possibilities of this volume and is left to more competent experts. Finally, I regret that such hot and strongly debated topics like CDW's in potassium or in a variety of other systems like ω-Zr, CuS_2 and many more could not be included in this volume.

References

[1] R.E. Peierls, *Quantum Theory of Solids*, University Press, Oxford (1955).

[2] S. Aubry in: "*Solitons and Condensed Matter Physics*", A.R. Bishop and T. Schneider (eds.), Springer, New York (1978), p. 264.

[3] see the beautiful tiles on page 32 of the article by D.R. Nelson, *Scientific American*, August (1986).

[4] D. Shechtman, I. Blech, D. Gratias, and J.W. Cahn, *Phys. Rev. Lett.* **53** (1985) 1951.

[5] P. Bak, *Phys. Rev. Lett.* **56** (1986) 861.

[6] R. Ruelle, *Physica* **113**A (1982) 619.

[7] R.V. Coleman, B. Giambattista, P.K. Hansma, A. Johnson, W.W. McNairy, and C.G. Slough, *Adv. Phys.* **37** (1988) 559.

[8] H. Fröhlich, *Proc. Roy. Soc.* A **223** (1955) 296.

[9] "*Electronic Properties of Inorganic Quasi-One-Dimensional Compounds*", vol. I and II, P. Monceau (ed.), Reidel, Dordrecht (1985).

[10] G. Grüner, *Rev. Mod. Phys.* **60** (1988) 1129.

[11] "*Incommensurate phases in Dielectrics*", Modern Problems in Condensend Matter Science Vol. 14, R. Blinc and A.P. Levanyuk (eds.), North-Holland, Amsterdam (1986).

Recommended books and review articles

J.T. Devreese, R.P. Evrard, and V.E. Doren (eds.), *Highly Conducting One-Dimensional Solids*, Plenum Press, New York (1979).

C. Schlenker (ed.), *Low-Dimensional Electronic Properties of Molybdenum Bronzes and Oxides*, Kluwer Academic Press, Dordrecht (1986).

H.J. Keller (ed.), *Low-Dimensional Cooperative Phenomena*, NATO ASI Series B7, Plenum Press, New York (1975).

L. Pal, G. Grüner, A. Janossy, and J. Solyom (eds.), *Organic Conductors and Semiconductors*, Lecture Notes in Physics, Springer, Berlin (1977).

S. Barisic, A. Bjelis, J.R. Cooper, and B. Leontic (eds.), *Quasi-One-Dimensional Conductors*, Lecture Notes in Physics, Springer, Berlin (1979).

L. Alcacer (ed.), *The Physics and Chemistry of Low-Dimensional Solids*, NATO ASI Series C56, Reidel, Dordrecht (1980).

P. Delhaes and M. Drillon (eds.), *Organic and Inorganic Low-Dimensional Crystalline Materials*, NATO ASI Series B168, Plenum Press, New York (1986).

D. Jérome and L.G. Caron (eds.), *Low-Dimensional Conductors and Superconductors*, NATO ASI Series B, Plenum Press, New York (1987).

J. Rouxel (ed.), *Crystal Chemistry and Properties of Materials with Quasi-One-Dimensional Structures*, Reidel, Dordrecht (1986).

J.A. Wilson, F.J. DiSalvo, and S. Mahajan, *Adv. Phys.* 24 (1975) 117.

G.Y. Hutiray and Y. Solyom (eds.), *Charge Density Waves in Solids*, Lecture Notes in Physics, Springer, Berlin (1985).

K. Motizuki (ed.), *Structural Phase Transitions in Layered Transition Metal Compounds*, Reidel, Dordrecht (1986).

R.H. Friend and A.D. Yoffe, *Adv. Phys.* **36** (1987) 1.

PHENOMENOLOGICAL THEORY OF CHARGE-DENSITY-WAVE PHASE TRANSITIONS IN THE $NbTe_4$ - $TaTe_4$ SERIES OF COMPOUNDS

MICHAEL B. WALKER
Department of Physics
University of Toronto
Toronto, Ontario
Canada M5S 1A7

1. Introduction

This article reviews work on the development of a phenomenological theory of phase transitions accounting for the various modulated phases (and the phase transitions between these phases) which occur in the compounds $NbTe_4$, $TaTe_4$ and $(Nb_xTa_{1-x})Te_4$ $(0 \leq x \leq 1)$. Although this may appear to be a rather restricted selection of compounds, the selection is appropriate for a general review for several reasons. Firstly, the model appropriate to charge-density-wave formation on a single column of the $NbTe_4$ is one of the most widely studied prototypical models of charge-density-wave formation. Secondly, $NbTe_4$ and $TaTe_4$ are structurally among the simplest materials in which charge-density waves have been studied: the appropriate theoretical model is therefore correspondingly simple, contains relatively few parameters, and can be analyzed in detail. Thirdly, $NbTe_4$ and $TaTe_4$ possess a fascinating variety of modulated phases and phase transitions between them; accounting for this varied behavior within the framework of a single model provides a nontrivial test of the theory.

The general approach of this article was previously developed and found to be successful in understanding the properties of certain layered transition-metal dichalcogenides by Jacobs and Walker [1], following the initial work of Moncton *et al.* [2] and McMillan [3]. Although a number of studies described later in this book were performed on the transition metal dichalcogenides, the theory of charge-density waves in these materials has been so well reviewed by Shiba and Nakanishi [4] in a previous book in this series that there is no need for further discussion here.

Since other articles in this book deal primarily with various magnetic resonance studies, the present article discusses as an example an appropriate form for the scalar contribution to the hyperfine interaction for both commensurate and incommensurate phases. The role of symmetry in determining the form of the hyperfine

T. Butz (ed.), Nuclear Spectroscopy on Charge Density Wave Systems, 7–33.

interaction is particularly stressed. No experimental magnetic resonance studies have yet been reported for $NbTe_4$ or $TaTe_4$, but such would be of considerable interest since these materials have been among the more intensively studied theoretically. Magnetic resonance studies on a number of other compounds are reviewed elsewhere in this book, and in the review article by Blinc [5].

2. A Prototypical Model of Charge-Density-Wave Formation

This section describes a model which has become one of the more important prototypical models of charge-density-wave formation. The model was introduced by McMillan [3], who studied it in an approximation in which the charge-density-wave amplitude was assumed to be independent of position. In this approximation, the model is equivalent to a continuum approximation to the Frenkel-Kontorova model [6], which was analysed in detail by Franck and Van der Merwe [7]. Particularly complete discussions of this model (which do not make the above mentioned approximation) can be found in the article by Jacobs and Walker [1], and in the review of Shiba and Nakanishi [4]. The model has a peculiar critical point of infinite order [8] which can be removed by adding certain terms to the free energy [9].

2.1. THE ORDER PARAMETER, SYMMETRY CONSIDERATIONS, AND THE FREE ENERGY

Different ways of arriving at what is called here the prototypical model can be found in Refs. [1,3, and 4]. In order to illustrate in detail the role played by symmetry in constructing the free energy, a simple one-dimensional model having the desired properties is introduced. This one-dimensional model is a monatomic lattice of evenly spaced ions at high temperatures which is assumed to have a low-temperature commensurate phase in which the ions form trimers as shown in Figure 1(a). (It will be seen below that this is precisely what occurs in the case of a chain of Nb ions in $NbTe_4$.) The displacements of the ions in the commensurate state (as shown in Figure 1) from their normal-state positions are given by the expression

$$u_n = A\sin(Qz_n) \tag{2.1}$$

where $Q = \frac{1}{3}c'^*$ with $c'^* = 2\pi/c'$ and $z_n = (n+\frac{1}{2})c'$ (the lattice constant is written c' with a prime to distinguish it from the parameter c occurring in section 3). To allow for the possibility of incommensurate states, the displacements are written as

$$u_n = Im[\psi(z_n)e^{iQz_n}] \tag{2.2}$$

where $\psi(z) = A(z)\exp[i\phi(z)]$ is a slowly varying function of z. Notice that if the phase $\phi(z) = 0$ and the amplitude $A(z)$ is independent of z, the expression of Eq. (2.1) is recovered.

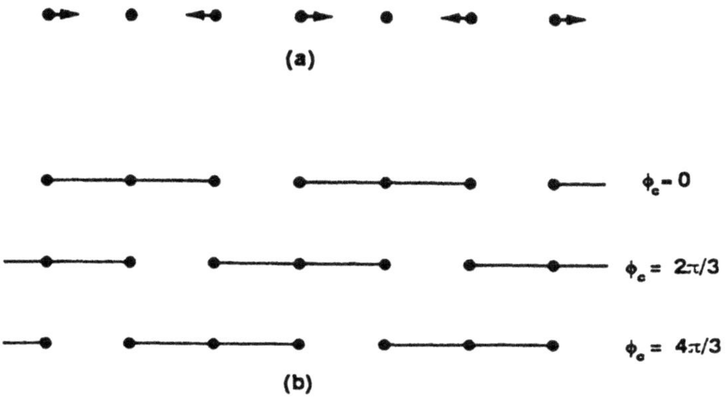

Fig. 1(a) The solid circles represent a one-dimensional lattice of evenly spaced ions. The arrows represent the displacements of the ions in the low temperature commensurate phase of the prototypical model. Note that the ions form trimers, each trimer being composed of a central undisplaced ion and two neighboring ions which are displaced towards it.

Fig. 1(b) The three distinct domains of the commensurate phase of the prototypical model are characterized by the relative positions of the trimers. The domains characterized by the phases $\phi_c = 2\pi/3$ and $\phi_c = 4\pi/3$ have the trimers shifted one and two lattice constants to the left respectively, relative to the domain characterized by $\phi_c = 0$. Note that here each trimer is represented by three ions connected by a solid line.

The state of our one-dimensional crystal is specified by specifying $\psi(z)$, which is called the order parameter. Since the free energy depends on the state of the crystal, and the state of the crystal is determined by the order parameter $\psi(z)$, the free energy of the crystal will be a functional of $\psi(z)$. What form this functional takes is determined by the symmetry of the problem, and it will therefore be necessary to determine how the order parameter transforms in symmetry operations.

Suppose we have a crystal in which the displacements of the ions u_n are given by Eq. (2.2) in terms of some arbitrary but fixed order parameter $\psi(z)$. Now translate the crystal rigidly to the right by a distance of one lattice constant c'. The displacements of the translated crystal are given by

$$\tilde{u}_n = u_{n-1} = Im\left[\tilde{\psi}(z_n)e^{iQz_n}\right] \tag{2.3}$$

where

$$\tilde{\psi}(z) = e^{-i2\pi/3}\psi(z - c')\,. \tag{2.4}$$

Since the free energy of a crystal is unaffected by a rigid translation, it must be unchanged if the order parameter $\psi(z)$ is replaced by $\tilde{\psi}(z)$ as defined by Eq. (2.4). Similarly, since the free energy is unchanged by an inversion of the crystal through the origin it must be invariant with respect to the transformation

$$\psi(z) \rightarrow \psi^*(-z) \; . \tag{2.5}$$

It will be assumed that the displacements u_n, and hence the order parameter $\psi(z)$, are relatively small for the states of interest, and the free energy will therefore be expanded in powers of the order parameter $\psi(z)$. The free energy can thus be written

$$F = F_N + \int f(z) dz \tag{2.6}$$

where F_N is the free energy of the normal state (i.e. for $\psi = 0$) and

$$f(z) = (T - T_i)|\psi|^2 + \gamma \left| \left(i\frac{d}{dz} + \delta \right) \psi \right|^2 - Re(\psi^3) + \frac{1}{2} \left| \psi \right|^4 \; . \tag{2.7}$$

Note that the free energy is invariant with respect to the transformations indicated by Eqs. (2.4) and (2.5). For example, in the transformation of Eq. (2.4), ψ is multiplied by $\exp(-i2\pi/3)$; thus a term in ψ^3 (for which the three factors of $\exp(-i2\pi/3)$ give unity) is allowed. In Eq. (2.7), T is the temperature, T_i is the normal-state to incommensurate-state transition temperature, and γ is positive. At any given temperature, the optimal value of ψ is determined by minimizing the free energy with respect to $\psi(z)$. The fact that this minimization process is unaffected if both F and ψ are multiplied by constants is used to set the coefficients of the terms in $Re(\psi^3)$ and $|\psi|^4$ equal to -1 and $\frac{1}{2}$, respectively (see Ref. 1 for further details).

A way of checking that the various terms in Eq. (2.7) are consistent with the requirements of translational invariance (with respect to translations by Bravais-lattice vectors of the underlying crystal lattice) is as follows. Note that the order parameter ψ is associated with the wave vector $Q = c'^*/3$, and ψ^* is associated with wave vector $-Q$. The sum of the wave vectors of each of the factors of a given term in Eq. (2.7) must be a reciprocal lattice vector (including zero). For example, the sum of the wave vectors of the term $\psi\psi^*$ is $Q - Q = 0$, whereas the sum of the wave vectors of the term ψ^3 is $3Q = c'^*$. Those terms for which the sum of the wave vectors is a nonzero reciprocal lattice vector are called lock-in terms and tend to favour a commensurate phase.

Although the problem studied in this section concerns a one-dimensional column of ions, the real problem of interest studied later is that of a three-dimensional system of coupled chains. Thus, the mean field approach which is described in this article is expected to give good qualitative results.

2.2. COMMENSURATE AND INCOMMENSURATE PHASES

The phase diagram as determined [1] by minimizing the free energy of Eqs. (2.6) and (2.7) with respect to variations in the order parameter is shown in Figure 2. In the normal phase, the order parameter is zero, whereas in the commensurate phase the order parameter is nonzero and independent of position, and in the incommensurate phase the order parameter is a spatially varying periodic function of position. The commensurate and incommensurate phases are now described in more detail.

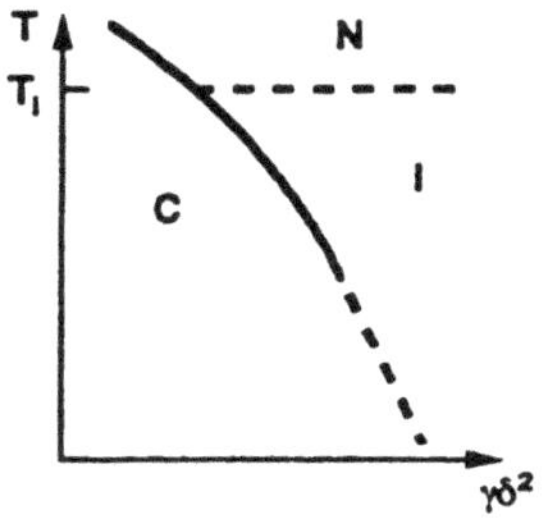

Fig. 2 The phase diagram determined by Jacobs and Walker [1] by a mean field analysis of the prototypical model showing the regions of stability of the normal (N), commensurate (C) and incommensurate (I) phases. The solid line indicates a phase transition which is first order, whereas the broken lines indicate phase transitions which are continuous. Note that the incommensurate to commensurate phase transition is first order at relatively high temperatures and continuous at relatively low temperatures.

In the commensurate phase the order parameter is independent of z and will be written $\psi = a_c \exp(i\phi_c)$. The free energy then becomes

$$F = F_N + L\left[(T - T_i + \gamma\delta^2)a_c^2 - a_c^3 \cos 3\phi_c + \frac{1}{2}a_c^4\right] . \tag{2.8}$$

The optimal value of the amplitude is

$$a_c = \frac{3}{4} + \frac{1}{4}[1 - 16(T - T_i + \gamma\delta^2 - \frac{1}{2})]^{\frac{1}{2}} , \tag{2.9}$$

and the optimal value of the phase is three-fold degenerate:

$$\phi_c = 0, \frac{2}{3}\pi, \frac{4}{3}\pi, \ (\text{mod } 2\pi) . \tag{2.10}$$

The ion displacements for the phase $\phi_c = 0$ have already been shown in Figure 1(a). The displacements for $\phi_c = 2\pi/3$ and $4\pi/3$ are found by shifting the $\phi_c = 0$ pattern rigidly to the left by one and two lattice constants, respectively. These three degenerate states are said to be different domains, and are shown in Figure 1(b).

The symmetry of the commensurate phase is easily deduced given the result of Eq. (2.1) for the displacements (corresponding to the solution $\phi_c = 0$ of Eq. (2.10)). The new unit cell has dimension $3c'$ and the origin is a centre of inversion symmetry.

It is the term linear in γ in the free energy of Eq. (2.7) which stabilizes the incommensurate phase. Notice that this term has its minimum value (i.e. zero) for an order parameter of the form $\psi(z) = a\exp(i\delta z)$; this order parameter describes an incommensurate phase since, according to Eq. (2.3), the corresponding displacements are characterized by a wavevector $Q + \delta$ (and the ratio of the corresponding period $\Lambda = 2\pi/(Q+\delta)$ to the lattice constant c' is presumed to have no particular rational value). Notice also that this choice of order parameter gives a rather poor value for the ψ^3 contribution to the free energy (which averages to zero on integrating over z). It is clear that there is a competition between the term in γ in the free energy, which stabilizes the incommensurate phase, and the term in ψ^3 in the free energy, which stabilizes the commensurate phase. The term in γ is quadratic in ψ and is relatively more important at high temperatures; the incommensurate phase thus tends to be the stable phase at high temperatures. On the other hand, as the temperature is lowered, ψ grows in magnitude, the cubic term becomes relatively more important, and a phase transition to the commensurate phase occurs as shown in Figure 2.

The assumption $\psi(z) = a\exp(i\delta z)$ is an approximation to the order parameter which is valid only at temperatures close to the normal to incommensurate transition temperature T_i. A somewhat better approximation would be to take the first two terms of the infinite series representing $\psi(z)$, namely

$$\psi(z) = a_1 e^{i\delta' z} + a_2 e^{-2i\delta' z} \tag{2.11}$$

where a_1, a_2 and δ' are determined by minimizing the free energy to be

$$\begin{aligned} a_1 &= \left[\frac{\gamma\delta^2(T_i - T)}{\gamma\delta^2 - \frac{1}{2}}\right]^{\frac{1}{2}}, \\ a_2 &= \frac{T_i - T}{6(\gamma\delta^2 - \frac{1}{2})}, \\ \delta' &= \delta\left[1 - \frac{T_i - T}{6\gamma\delta^2(\gamma\delta^2 - \frac{1}{2})}\right]. \end{aligned} \tag{2.12}$$

Note that the observed wave vectors, which depend linearly on δ', will be temperature dependent. Also note that the ratio u_2/u_1 is small only for $|T_i - T|$ relatively small; thus as T is lowered further and further below T_i, contributions to $\psi(z)$ in addition to those shown in Eq. (2.12) become important, and a numerical solution of the problem becomes necessary.

In general, $\psi(z)$ can be shown to have the following properties:

$$\psi[z + 2\pi/(3\delta')] = e^{i2\pi/3}\psi(z) \ ,$$
$$\psi(-z) = \psi^*(z) \ , \tag{2.13}$$

and can be written as

$$\psi(z) = e^{i\delta' z}\chi(z) = e^{i\delta' z}\left(\chi_0 + \chi_1 e^{i3\delta' z} + \chi_{-1}e^{-i3\delta' z} + \chi_2 e^{i6\delta' z} + \ldots\right) \tag{2.14}$$

where the coefficients χ_i are real. This expression will be useful in a later discussion of the parametrization of the nuclear hyperfine interaction in the incommensurate phase. Furthermore, it can be shown that if the order parameter $\psi(z)$ gives a minimum of the free energy, $\psi(z-\Delta)$ does also, where Δ is arbitrary. In other words, the state of lowest free energy has a continuous degeneracy which corresponds to the fact that translating the modulation pattern by an arbitrary amount Δ in the z direction gives another state of the same energy. It is this continuous degeneracy of the ground state which results in phason modes having a frequency which goes to zero when the phason wave vector goes to zero (see below).

It is sometimes useful to write $\psi(z)$ in the form

$$\psi(z) = a(z)\exp[i\phi(z)] \ . \tag{2.15}$$

Eq. (2.11) can be regarded as the first two terms in an expansion of the order parameter in powers of a_1, a_2 being of order a_1^2. From Eq. (2.11) it follows that the first two terms in the expansions of $a(z)$ and $\phi(z)$ in powers of a_1, are given by

$$a(z) = a_1 + a_2\cos(3\delta' z)$$
$$\phi(z) = \delta' z - (a_2/a_1)\sin(3\delta' z) \ . \tag{2.16}$$

The spatial variation in $a(z)$ and $\phi(z)$ is shown qualitatively in Figure 3 for three different temperatures as determined by Nakanishi and Shiba [4]. Recall that in the commensurate state, the phase ϕ can have only the values $0, 2\pi/3$, and $4\pi/3$ (mod 2π). In the lowest approximation, the phase ϕ in the incommensurate phase varies linearly with z, but the correction proportional to $\sin(3\delta' z)$ can be seen to have the effect making the phase vary relatively slowly in the neighborhood of the values $0, 2\pi/3, 4\pi/3$ (mod 2π) so that it "spends more time" close to these values which are characteristic of the commensurate phase. In a region where the phase ϕ has a value close to $2\pi n/3$ for some integer n, the local structure is close to that of one of the commensurate domains (i.e. the ions form trimers). Thus the incommensurate phase can be viewed as being formed from a periodic sequence of commensurate domains, each pair of commensurate domains being separated by a domain wall. The length of a given domain is $\lambda = 2\pi/(3\delta')$, as can be seen from Eq. (2.13). It should be noted, however, that the variation in local structure as one goes from one commensurate domain to another is continuous. Also, the distinction between which regions of the crystal should be considered as belonging to a given commensurate domain, and which should be considered as belonging to a domain wall, is somewhat blurred; this is particularly true at temperatures close to the normal-to-incommensurate transition temperature T_i, as can be seen from Figure 3.

The interpretation in terms of domain walls can be pursued by defining the energy of a domain wall to be the energy of the state containing a single domain wall minus the energy of the state containing no domain walls. In the commensurate phase the energy of a domain wall is positive and the state of lowest energy contains no domain walls. As the temperature is raised in the commensurate phase, the domain wall energy decreases and eventually becomes negative. For negative domains wall energies, the lowest energy state contains domain walls, which arrange themselves in a periodic array, and the crystal is now in an incommensurate phase. If the interaction between domain walls is repulsive, the transition from the commensurate to the incommensurate phase is thermodynamically continuous (but may be hysteretic due to the problem of nucleating incommensurate regions in the commensurate phase) and occurs when the domain wall energy is zero. If the interaction between domain walls is attractive, the transition from the commensurate to the incommensurate phase is first order, and occurs at a temperature above that for which the domain wall energy becomes negative. The incommensurate to commensurate phase transition occurring in the prototypical model can be either first order or second order, as can be seen from Figure 2.

A convenient way to investigate the symmetry of the incommensurate phase is to consider the symmetry properties of the ion density defined by

$$\rho(z,\Delta) = \sum_n \delta\left(z - z_n - Im[\psi(z_n - \Delta)e^{iQz_n}]\right) \ . \tag{2.17}$$

As noted above, the ground state is continuously degenerate, with different ground states being characterized by different values of Δ. It is convenient to write in a single expression, Eq. (2.17), the ion density for all possible ground states, and to show explicitly the dependence of the density on the parameter Δ labelling the different ground states. Now introduce the following two-dimensional vectors:

$$\mathbf{r} = \begin{pmatrix} z \\ \Delta \end{pmatrix}, \quad \mathbf{a}_1 = \begin{pmatrix} c' \\ c' + \lambda \end{pmatrix}, \quad \mathbf{a}_2 = \begin{pmatrix} 0 \\ 3\lambda \end{pmatrix} \ . \tag{2.18}$$

The ion density can be shown by using Eq. (2.13) to be a periodic function of the two-dimensional vector $\mathbf{r}$. More explicitly, it satisfies

$$\rho(\mathbf{r} + n_1\mathbf{a}_1 + n_2\mathbf{a}_2) = \rho(\mathbf{r}) \tag{2.19}$$

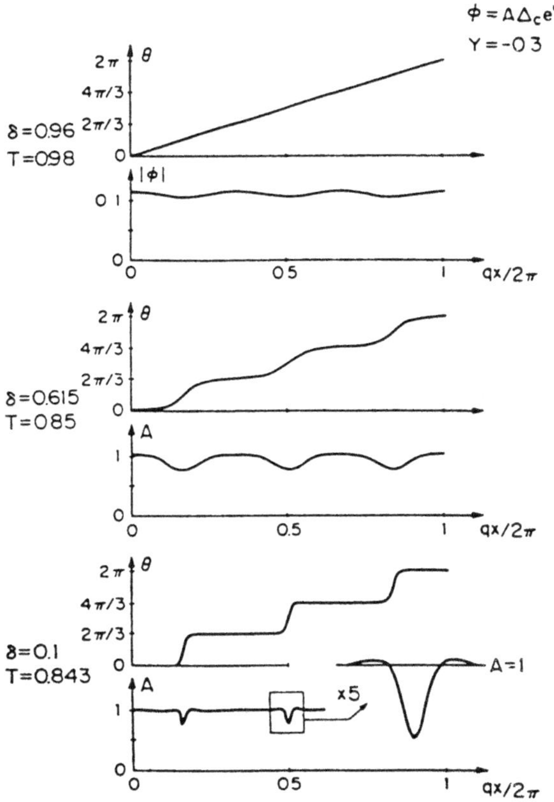

Fig. 3 The spatial variation of the amplitude and phase of the order parameter for the prototypical model for three different temperatures as determined by Nakanishi and Shiba [15]. The upper two curves for $T = 0.98$ are for a temperature just below the normal to incommensurate transition temperature T_i, the lower two curves for $T = 0.843$ are for a temperature equal to the incommensurate to commensurate transition temperature, and the intermediate two curves are for an intermediate temperature.

where n_1 and n_2 are integers. Furthermore, the density can be shown [using Eq. (2.13)] to have the inversion symmetry property

$$\rho(\mathbf{r}) = \rho(-\mathbf{r}) \ . \tag{2.20}$$

The two-dimensional space in which the density $\rho(\mathbf{r})$ is defined is called superspace, the two-dimensional density is said to be that of the supercrystal, and the density of the physical crystal is given by a constant-Δ section in the two-dimensional superspace. The two symmetry operations implied by Eqs. (2.19) and (2.20) are sufficient to generate the full superspace-group symmetry of the incommensurate phase. The ideas of superspace symmetry have been developed by de Wolff, Janner and Janssen [10].

2.3. THE NUCLEAR HYPERFINE INTERACTION

Suppose that the ions of this example have a nuclear spin. There will then be an interaction between the nuclear spin of the n-th ion, $\mathbf{I}_n$, and the electron spin density in the neighborhood of the nth ion. The electron spin density fluctuates in time; it is the average value of this fluctuating quantity which determines the nuclear-spin-resonance frequency, whereas fluctuations about the average value give rise to nuclear spin relaxation. Since the average value of the electron spin density is proportional to the external magnetic field, the interaction between the nuclear spin and the electron spin density becomes an interaction between the nuclear spin and the external magnetic field. The interaction Hamiltonian for the n-th nucleus will be written in the form

$$H_n = B_n \mathbf{I}_n \cdot \mathbf{H} \ . \tag{2.21}$$

Here, the interaction strength is assumed to be well represented by a single scalar quantity B_n. This is not generally correct. The following arguments can, however, be extended to the more general case by replacing the scalar quantity B_n by a tensor.

In the normal state of our example, there is only one ion in the unit cell. Thus B_n has the same value for all nuclei. In the commensurate phase, there are three ions in the unit cell. However, as a result of the centre of inversion symmetry, two of these nuclei are equivalent. Thus, in the commensurate phase there are two inequivalent nuclear sites with a 2:1 intensity ratio.

The form of B_n appropriate to the incommensurate phase must be consistent with the superspace symmetry of the incommensurate phase. The expression

$$B_n = \bar{B} + Re[\phi(z_n - \Delta)e^{iQz_n}] \ , \tag{2.22}$$

where $\phi(z)$ is a function satisfying the same conditions (Eq. (2.13)) as the order parameter, can be seen to be an appropriate representation of the hyperfine interaction parameter; here $\bar{B}$ is independent of n. The fact that the real part is taken of the expression in square brackets in Eq. (2.22), rather than the imaginary part, as in Eq. (2.3) for the displacement u_n, reflects the fact that B_n is a scalar and does not change sign under inversion, whereas u_n, being a component of a vector, does change sign under inversion.

It should be emphasized that although symmetry arguments constrain $\phi(z)$ to have the same symmetry properties as the order parameter $\psi(z)$, these two functions are not in general the same. Furthermore, $\phi(z)$ can not be obtained theoretically by minimizing some free energy functional, as was the case for $\psi(z)$. However, $\phi(z)$ might be expected to have a somewhat similar qualitative behavior to $\psi(z)$. For example, at temperatures just below T_i, the contribution proportional to $\exp(i\delta' z)$ should give the dominant contribution to $\phi(z)$. Also, at low temperatures, where $\psi(z)$ is well approximated by a sequence of approximately commensurate regions, $\phi(z)$ should be also.

Although $\phi(z)$ is not in general equal to $\psi(z)$, $\phi(z)$ will be a functional of $\psi(z)$ because the state of the system is assumed to be completely specified by a knowledge of the order parameter $\psi(z)$. Since the order parameter is assumed to be a relatively small quantity, $\phi(z)$ can be expanded in powers of $\psi(z)$; this approximation will be particularly good close to the normal to incommensurate transition temperature. Furthermore, since the order parameter $\psi(z)$ is a slowly varying function of z, the local approximation, in which $\phi(z)$ depends only on the value of ψ at z, is expected to be a good one. Thus

$$\phi(z) = c_1\psi(z) + ic_2\frac{d\psi(z)}{dz} + c_3\psi^*(z)\psi^*(z) + \ldots . \tag{2.23}$$

The constants c_1, c_2 and c_3 must be real, since both ψ and ϕ must satisfy the symmetry relations specified by Eq. (2.13); the signs of c_1, c_2 and c_3 are however undetermined by symmetry arguments. It should be emphasized that the fact that the constants c_1, c_2 and c_3 are real, and hence that the phase of the nuclear hyperfine interaction bears a definite relationship to that of the order parameter, is a consequence of the inversion symmetry present in this particular case.

In an approximation often used in the literature (e.g. see Ref. [5] and references therein) the nuclear hyperfine interaction B_n is written as a local function of the ion displacement u_n. Since a rigid translation of the crystal as a whole can not change the nuclear hyperfine interaction at a given ion, however, this interaction can not be a local function of the displacement of the ion itself, but depends rather on the displacement of the ion relative to its neighbours. The expansion of Eq. (2.23) is consistent with these remarks.

2.4. Dynamics of the Modulation Wave

In the discussion following Eq. (2.14) it was noted that the incommensurate state of lowest free energy has a continuous degeneracy which corresponds to the fact that translating the modulation pattern by an arbitrary amount Δ in the z direction gives a state of the same energy. As a result of this degeneracy an incommensurately modulated state will have a branch of excitations, called phasons, whose frequency goes to zero as the wave vector of the excitation goes to zero. Phasons were initially studied by Overhauser [11]. There is also an amplitudon branch of the vibrational spectrum which corresponds roughly to vibrations of the amplitude of the modulation wave, and which was initially studied by Lee *et al.* [12]. Phasons and amplitudons make up a part of the overall lattice vibrational spectrum. Early theoretical studies of the ways in which phasons and amplitudons could be observed in scattering experiments were carried out by Axe [13] and by Walker [14], while Currat and Janssen [15] have reviewed these topics recently. A detailed numerical study of the order-parameter vibrations for the prototypical charge-density wave model described in this section has been carried out by Nakanishi and Shiba [16].

More recently, a different formulation of the problem was given by Goff *et al.* [17], who studied the order-parameter vibrations of the prototypical model as a preliminary to an analysis of the order-parameter vibrations of $NbTe_4$. In the remainder of this section, we follow the approach of Goff *et al.* [17].

The order parameter is written in the form

$$\psi(z,t) = e^{i\delta' z}\left[\chi(z) + \xi(z,t)\right] \tag{2.24}$$

where $\chi(z)$ describes the equilibrium state (as above in Eq. (2.14)) and $\xi(z,t)$ is the vibrating part. The order parameter vibrations are determined by the Lagrangian

$$L = \int dz \left(\frac{1}{2}\rho\dot{\Xi}_M^+\dot{\Xi}_M - \frac{1}{2}\Xi_M^+ H \Xi_M\right) \tag{2.25}$$

where Ξ_M is the two component vector

$$\Xi = \begin{bmatrix} \xi(z,t) \\ \xi^*(z,t) \end{bmatrix} \ . \tag{2.26}$$

The subscript M attached to Ξ indicates that the two components of Ξ are not independent, but are complex conjugates of one another. Also, a dot above Ξ_M indicates differentiation with respect to time. The Hamiltonian operator is given by

$$H = \gamma\left(i\sigma_3\frac{d}{dz} + \delta - \delta'\right)^2 + V(z) \tag{2.27}$$

where

$$V(z) = \begin{bmatrix} T - T_i + 2|\chi|^2 & \chi^2 - 3e^{-3i\delta' z}\chi^* \\ \chi^{*2} - 3e^{i3\delta' z}\chi & T - T_i + 2|\chi|^2 \end{bmatrix} \ , \tag{2.28}$$

and σ_1, σ_2 and σ_3 are the Pauli matrices. The potential energy contribution to the Lagrangian, i.e. the term in Eq. (2.25) involving H, is obtained by substituting the expression (2.24) for the order parameter $\psi(z,t)$ into Eqs. (2.5) and (2.6) for the free energy and collecting terms which are quadratic in the vibrating part of the order parameter $\xi(z,t)$. The equation of motion which follows from the Lagrangian is

$$\rho\ddot{\Xi}_M = -H\Xi_M \ . \tag{2.29}$$

Note that because of the special form of Ξ_M given in Eq. (2.26), it must satisfy the condition

$$\sigma_1 K \Xi_M = \Xi_M \tag{2.30}$$

where K is the complex conjugation operator. Because of this, the assumption of a time dependence $exp(-i\omega t)$ for $\Xi_M(z,t)$ is not allowed. However, if $\Xi(z)$ is a solution of the time-independent equation

$$H\Xi(z) = E\Xi(z) \tag{2.31}$$

then

$$\Xi_M(z,t) = \frac{1}{2}(1+\sigma_1 K)e^{-i\omega t}\Xi(z) \tag{2.32}$$

is a solution of the equation of motion (2.29) satisfying the condition (2.30) provided $E = \rho\omega^2$. The problem is thus reduced to finding solutions $\Xi(z)$ of the time-independent Eq. (2.31) (where $\Xi(z)$ has two independent components and is not required to satisfy Eq. (2.30)).

The order parameter vibrations of the normal state (for which the present formalism is unnecessarily complicated) can be studied by putting $\chi(z) = 0$ and $\delta' = \delta$ in the above equations. A solution of Eq. (2.31) is

$$\Xi(z) = \begin{bmatrix} e^{ikz} \\ 0 \end{bmatrix} \tag{2.33}$$

with frequency

$$\omega^2 = (T - T_i + \gamma k^2)/\rho \ . \tag{2.34}$$

From Eqs. (2.2), (2.24), (2.32) and (2.33), the corresponding displacement field for this solution is

$$u_n = a\sin[(Q+\delta+k)z_n - \omega t] \ . \tag{2.35}$$

It is of interest that in this mean field treatment of the excitations of the normal phase, the phonon with wave vector $Q+\delta$ (i.e. $k=0$ in Eqs. (2.35) and (2.36)) has a frequency which goes to zero as the temperature T approaches the normal-incommensurate transition temperature T_i.

The order-parameter vibrations for the commensurate phase (corresponding to a trimerization of the one-dimensional lattice) can be found by taking $\delta' = 0$ and $\chi = A$ in Eqs. (2.24) - (2.31); here A is a real constant which is the amplitude of the equilibrium-state order parameter. There are two modes for each phonon wave vector $Q+k$, and these have frequencies given by

$$\rho\omega_\pm^2 = 2A^2 + T - T_i + \gamma(k^2+\delta^2) \pm \sqrt{(A^2-3A)^2 + (2\gamma k\delta)^2} \ . \tag{2.36}$$

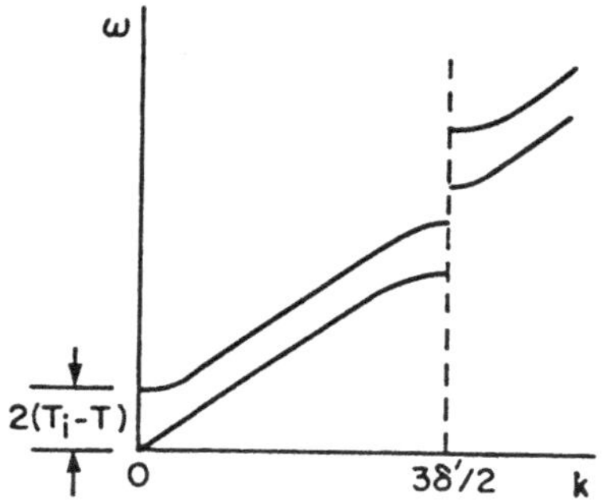

Fig. 4 The frequencies of the order-parameter vibrations of the incommensurate phase of the prototypical model as a function of the wave vector k of the eigenvector $\Xi(z)$; the observed phonon wave vector will be $Q + \delta' + k$. The frequency of the phason branch goes linearly to zero with k whereas the amplitudon branch has a gap $2(T_i - T)$ at $k = 0$.

For the $k = 0$ modes, the displacement fields corresponding, respectively, to frequencies ω_+ and ω_- are

$$\begin{aligned} u_n &= a \sin(Qz) \sin(\omega_+ t) \\ u_n &= a \cos(Qz) \sin(\omega_- t) \ . \end{aligned} \tag{2.37}$$

It can thus be seen that the $k = 0$ modes with frequencies ω_+ and ω_- correspond, respectively, to vibrations of the amplitude and phase of the equilibrium displacement field $u_n = A \sin Qz_n$.

Since, in the incommensurate phase, $\chi(z)$ is periodic with period $2\pi/(3\delta')$, Eq. (2.31) for $\Xi(z)$ is equivalent to a time-independent Schrödinger equation for a particle (described by a two-component wave function) in a periodic potential. Its solutions can therefore be taken to be Bloch functions characterized by a wave vector k, and the energy spectrum will have gaps at the Brillouin zone boundaries $k = \pm 3\delta'/2$. The frequency as a function of wave vector is shown qualitatively in Figure 4 (the quantitative details depend on the parameters T, γ and δ). From Eqs. (2.2), (2.24), (2.26) and (2.32) it follows that a solution $\Xi(z)$ characterized by a wave vector k corresponds to a lattice vibration of wave vector $Q + \delta' + k$. At small wave vectors, the phason branch has the dispersion relation $\omega = vk$ with the phason velocity given by

$$v^2 = \frac{\gamma}{\rho} \left\{ 1 - \left[\frac{4}{9} + \frac{\gamma}{18(\gamma - \frac{1}{2})} \right] \frac{(T_i - T)^2}{\gamma(\gamma - \frac{1}{2})} \right\} \ . \tag{2.38}$$

For further details, see Refs. [16 and 17]; for example, the viscous damping of the phason mode is included in Ref. [17].

3. Modulated Structures of the $NbTe_4$ Series of Compounds

3.1. OVERVIEW

The basic structures of $NbTe_4$ and $TaTe_4$ were determined by Selte and Kjekshus [18] and by Bjerkelund and Kjekshus [19], respectively, and are shown in Figure 5. (The basal plane is shown as a mirror plane in this Figure, but there is some question as to whether or not this is exactly correct [20]). The present interest in the various modulated structures of this series of compounds blossomed with the discovery by Boswell *et al.* [21], and by Böhm and von Schnering [22] that the room temperature phase of $NbTe_4$ is incommensurate. These results were soon confirmed by Mahy *et al.* [23]. Since then, a number of different modulated phases and phase transitions between these phases have been found, as shown in Figure 6.

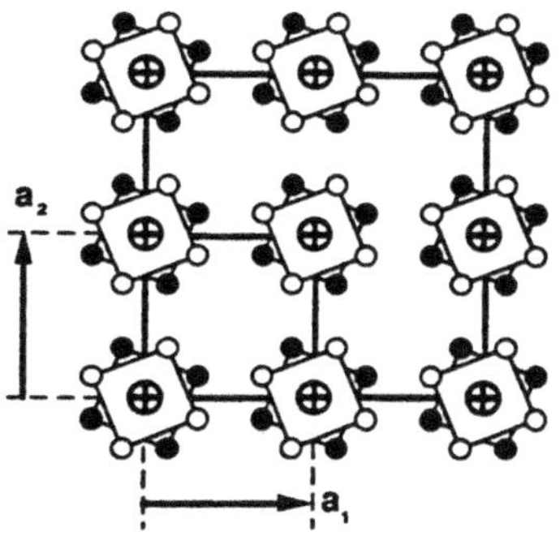

Fig. 5 The presumed structure of the normal phase of $TaTe_4$ projected onto the basal plane. The structure is made up of columns parallel to the c-axis with Ta ions, indicated by open circles containing crosses, at positions $z = \pm c/4$ along the axis of each column. Squares of Te ions at $z = 0$ and $z = c/2$ are indicated by open circles and closed circles, respectively. The vectors $\mathbf{a}_1$ and $\mathbf{a}_2$ both have length a.

A detailed structural study by Böhm and von Schnering [24], a theoretical study in terms of a Landau type model by Walker [25], an electron microscopy study by Eaglesham *et al.* [26], and a further structural study by Van Smaalen *et al.* [27], have all contributed to producing a detailed and consistent picture of the $\sqrt{2}a \times \sqrt{2}a$ incommensurate phase of $NbTe_4$. It is of interest that the detailed structural work of Böhm and von Schnering [24] describes the distortions of the ions on a given column in terms a longitudinal modulation of the Nb ions (which produces primarily Nb triplets), and a breathing, rotation and longitudinal translation of the Te squares.

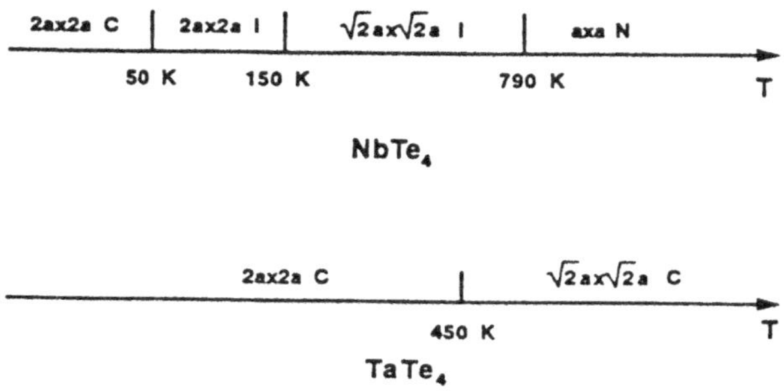

Fig. 6 The temperature range over which each of the known phases of $NbTe_4$ and $TaTe_4$ is stable is shown. Each phase is characterized by giving the size of the basal plane unit cell and whether it is commensurate (C) or incommensurate (I).

Precisely the same description of the displacements was arrived at by a group theoretical analysis [28] of the symmetry modes of a column of the $NbTe_4$ structure which deduced that the incommensurate phase was due to distortions transforming according to the Γ_2 irreducible representation of the space group of a column. Furthermore, both the theoretical and the experimental studies mentioned above are in agreement on the relative phasing of the modulations on the different columns. Thus, the $\sqrt{2}a \times \sqrt{2}a$ incommensurate phase of $NbTe_4$ appears at present to be well understood.

The nature of the $2a \times 2a$ commensurate state exhibited by $TaTe_4$ at room temperature and by $NbTe_4$ at temperatures below 50K has been more difficult to establish firmly. The theoretical study of Walker [25] had shown that the only way in which the theoretical model could arrive at a $2a \times 2a$ basal-plane unit cell for the commensurate state would be to phase the modulation waves on the different columns in such a way as to arrive at a structure with space group P4/ncc. However, the convergent beam electron microscopy studies of Eaglesham *et al.* [26], and Bird *et al.* [29], determined the space group of the commensurate phase to be P4/mcc (i.e. with the basal plane being a simple mirror plane m and not a diagonal glide plane n). Subsequent attempts were made by Sahu and Walker [28] and by Morelli, Sahu and Walker [30] to develop a theoretical model which contained a $\sqrt{2}a \times \sqrt{2}a$ incommensurate phase and a phase transition to a $2a \times 2a$ commensurate phase with P4/mcc symmetry, but such behaviour was not found to be possible within the context of the models studied. These difficulties have apparently been resolved by a detailed structural study by Bronsema *et al.* [31], and by a high-resolution electron microscopy study by Corbett *et al.* [32], which find that the space group of the $2a \times 2a$ commensurate phase is not P4/mcc, but instead P4/ncc as predicted by the theory. Given that the original convergent beam determination [26,29] of the space group was unusually difficult and necessitated the use of an unusually high-index superlattice diffraction spot, the combined weight of the theoretical and experimental evidence appears to favor the P4/ncc space group determination.

Boswell *et al.* [33] have found that heating $TaTe_4$ causes a phase transition from the $2a \times 2a$ commensurate phase to a $\sqrt{2}a \times \sqrt{2}a$ commensurate phase at a temperature of 450K. Chen *et al.* [34] have proposed that the high temperature phase is the $\sqrt{2}a \times \sqrt{2}a$ phase with P4/ncc symmetry found in earlier theoretical analyses [25,28] and have described a mechanism which would drive such a transition. There are no experimental structural or space group determinations for this high temperature commensurate phase.

A puzzling aspect of the earlier [21,23,26] electron microscopy studies of $NbTe_4$ was the appearance of certain "streaked" reflections. The origin of these streaks was clarified considerably by Mahy *et al.* [35], who proposed that a $2a \times 2a$ incommensurate phase existed between the lock-in temperature of 50K and 150K (above which the $\sqrt{2}a \times \sqrt{2}a$ room-temperature incommensurate phase exists).

Transitions between two incommensurate phases are relatively rare in comparison with incommensurate to commensurate (or lock-in) transitions. One example is the incommensurate to incommensurate phase transition which occurs in tetrathiafulvalene-tetracyanoquinodimethane (TTF-TCNQ), a theoretical description of which has been given by Bak and coworkers [36]. The incommensurate to incommensurate phase transition which occurs at 150K in $NbTe_4$ is thus of considerable interest and has been studied in some detail theoretically by Morelli and Walker [37-39]. It is possible in principle for this transition to be continuous [37] and the results of a theoretical analysis [38] indicate that it is continuous and propose a mechanism driving the transition.

The superspace symmetry of the $\sqrt{2}a \times \sqrt{2}a$ incommensurate phase of $NbTe_4$ has been discussed by Van Smaalen *et al.* [27], and by Prodan and Boswell [40]. Also, the relation of the superspace symmetry modes to the incommensurate to incommensurate phase transition, as well as the consequences of the superspace symmetry of both incommensurate phases for the resulting systematic absences expected in the diffraction patterns, has been discussed by Chen and Walker [41].

One observation which is not accounted for by any of the theoretical work is the observation by Boswell and Prodan [42] of diffraction spots with a c-axis wave vector of $(2c^*/3) - \delta/2$.

An interesting aspect of the electron microscopy study of $NbTe_4$ by Eaglesham *et al.* [26], was the observation by satellite dark-field imaging of the lock-in transition. The lock-in transition was found to be extremely slow, taking nearly an hour to complete.

A more or less continuous series of mixed crystals $(Nb_{1-x}Ta_x)Te_4$, $(0 \leq x \leq 1)$, have been prepared by Boswell and Prodan and the wave vector of the modulation has been shown to vary linearly with x [43].

The electronic band structure of $NbTe_4$ has been investigated by Bullet [44], and by Whangbo and Gressier [44].

The domain walls or discommensurations in $NbTe_4$ have been studied by electron microscopy by Eaglesham *et al.* [26], Bird *et al.* [29], and Mahy *et al.* [45]. Because the incommensurateness is one dimensional, there is not as much variety in these domain wall configurations as has been observed in the case of 2H-$TaSe_2$. Domain walls in this latter material have been studied in considerable detail experimentally by Fung *et al.* [46], and Chen *et al.* [47], and theoretically by Walker and Jacobs [48] and by Walker [49].

3.2. THE FREE ENERGY OF A CHARGE-DENSITY WAVE ON A SINGLE COLUMN

In the basic structure, shown in Figure 5, a periodic array of equally spaced Nb ions lies along the axis of each column. Since I will principally be interested in describing each of the modulated phases which occurs in terms of its symmetry, and since this symmetry is reflected in the statement of the Nb ion positions, I will focus from now on solely on the Nb ions. (It is understood, however, that Te ion distortions accompany the Nb ion distortions.) Note that there are two Nb ions in each unit cell, and that the z coordinates of the Nb ions on a given column are $z_n = (2n+1)c/4$ where n is an integer. Since in the commensurate phases of $NbTe_4$ and $TaTe_4$ the Nb ions on a given column are known to form trimers, the same arguments as those given in section 2 indicate that the displacements of the ions on a particular column can be written in terms of an order parameter ψ as in Eq. (2.2).

The theoretical model reviewed in this article was introduced by Walker [25] and considers first the formation of the modulation wave on a given column of the structure, and later takes account of the coupling between the columns. Both strong and weak coupling limits have been considered and it is the results of the limit of weak coupling between the columns which are in agreement with experiment. The case of strong coupling between columns gives the wrong size of the basal plane unit cell in the commensurate phase, and also gives a pattern of displacements for the commensurate phase which is not in agreement with that determined experimentally [31].

The c-axis glide plane and the centre of inversion symmetry which exist for a single column of $NbTe_4$ are equivalent to the translation and inversion symmetries of Eqs. (2.4) and (2.5) in so far as their effects on the order parameter of the present problem are concerned. (Note that the c-axis glide plane symmetry operator is $\{\sigma_x|\frac{1}{2}\mathbf{c}\}$ where σ_x is a reflection in a plane normal to the x axis and the magnitude of the translation vector $\mathbf{c}/2$ is equal to the Nb-Nb spacing, called c' in section 2.) Thus, the free energy describing charge-density-wave formation on a single columns of $NbTe_4$ is precisely that of the prototypical model described in section 2! The results reviewed in that section thus form the starting point for the investigation of charge-density waves in the $NbTe_4$-$TaTe_4$ series of compounds.

The existence of a centre of inversion symmetry for a single column of the basic structure of $NbTe_4$ has been questioned by Böhm [20]. Walker [25] has given a transformation which reduces the theoretical problem for the case where the inversion centre is absent, to one where it is present. Thus, it is sufficient here to review the results for the case where the inversion centre is present.

3.3. COMMENSURATE PHASES OF $TaTe_4$

As shown in section 2, the symmetry conditions built into the free energy require the phase ϕ_c of a commensurate charge-density wave on a single column to be $0, 2\pi/3$ or $4\pi/3$ corresponding to the three distinct domains of trimerized Nb ions as shown in Figure 1(b). A strong interaction between charge-density waves on different columns would cause the phase of the wave on a given column to deviate significantly from one of the three above-mentioned values (eg. see Ref. [24]). Experimentally, however, no such deviation is observed (e.g. see Ref. [30]) and it is therefore possible to treat the intercolumn interaction in perturbation theory. Such a perturbation theory treatment of the commensurate phases of the model is given in Ref. [33]. The results of this treatment are important because they allow the determination of the signs of the interaction between columns, and because the model makes some non-trivial predictions which can be tested by experiment.

The interaction free energy between charge-density waves on columns i and j is written

$$F_{int} = G \int dz Re[\psi_i(z)\psi_j^*(z)] \quad . \tag{3.1}$$

The interaction between columns has the same form for columns which are nearest neighbors, next-nearest neighbors, or third-nearest neighbors; however, the real constant G characterizing the magnitude of this interaction is different for the three cases just mentioned and is called $G_n(n = 1,2,3)$ for the n-th nearest neighbor case. The article assumes $|G_1| >> |G_2| >> |G_3|$.

The interaction between charge-density waves on different columns can be either repulsive (G>0), in which case the two columns must contain different domains (i.e. domains characterized by different values of ϕ_c) as shown in Figure 7(a), or attractive ($G < 0$), in which case the two columns must contain the same domain, as shown in Figure 7(b). It can be shown that in order to obtain the correct size for the basal plane unit cell (which has dimension $2a \times 2a$) the intercolumn interactions must be repulsive for first-and second-neighbor columns, and attractive for third-neighbor columns. The theoretically determined structure which is consistent with a $2a \times 2a$ basal plane unit cell is shown in Figure 8(a).

A non-trivial prediction of the model is that the space-group symmetry of the commensurate phase shown in Figure 8(a) is P4/ncc. The initial experimental determination [26,29] of the symmetry by convergent beam electron diffraction arrived at a space group P4/mcc, which differed from the theoretical prediction. However,

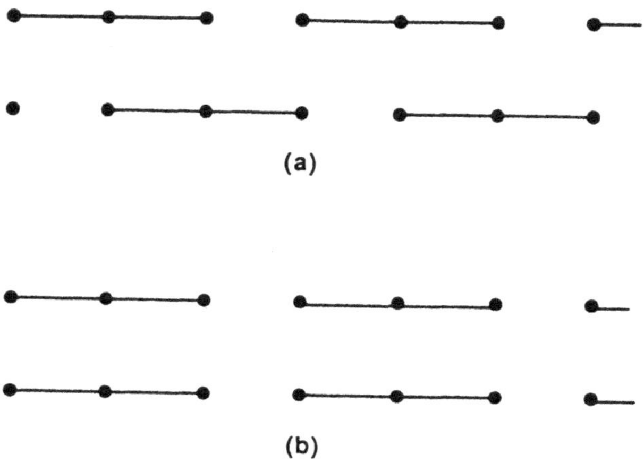

Fig. 7 A repulsive interaction between nearest-neighbor columns leads to unlike domains on the two columns as in (a), whereas an attractive interaction between two columns leads to like domains on the columns as in (b).

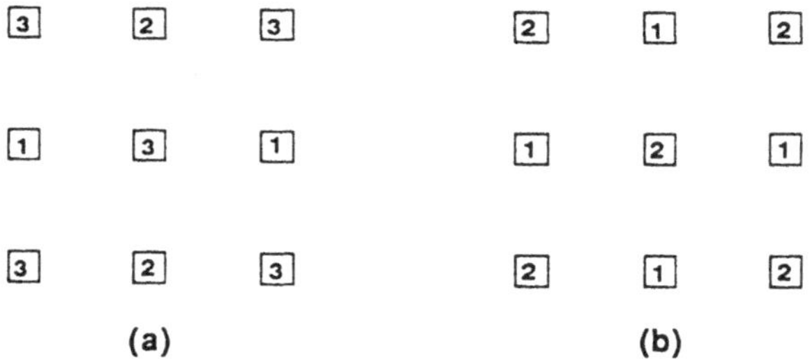

Fig. 8 Basal plane projections of the two commensurate phases predicted by theory for $TaTe_4$. The basal plane projection of each column is represented by a square box. The integers 1,2 and 3 in the square boxes refer to the three possible commensurate domains of a single column and can be taken to be the integer n in the expression $\phi_c = 2\pi n/3$ for the phase of the order parameter associated with that particular column.

a more recent detailed structural determination [31], as well as high resolution microscopy studies [32], indicate that the space group is P4/ncc, as predicted.

Boswell *et al.* [33] have observed experimentally a phase transition between two commensurate phases in $TaTe_4$ at 450K. Below 450K, the commensurate phase has a $2a \times 2a$ basal-plane unit cell and P4/ncc symmetry. The commensurate phase which exists above 450K has a $\sqrt{2}a \times \sqrt{2}a$ basal-plane unit cell. A plausible explanation of this transition has been given by Chen *et al.* [34], who show that a phase transition between the two phases shown in (a) and (b) of Figure 8, occurs when the effective second-neighbor interaction changes sign. The effective second-neighbor interaction is

$$G_{2eff} = G_2 - \frac{2G_1^2}{9|\psi|} \tag{3.2}$$

where $|\psi|$ is the magnitude of the order parameter. The first term G_2 represents the direct interaction between two second-neighbor columns, whereas the second term represents an indirect contribution resulting from two successive first-neighbor interactions. The term G_2 is positive and larger than the second term at room temperature. As the temperature increases, the magnitude of the order parameter decreases, and ultimately G_{2eff} becomes negative as a result of the contribution of the second term in Eq. (3.2). When the effective second-neighbor interaction becomes attractive (i.e. $G_{2eff} < 0$), a phase transition to the structure shown in Figure 8(b) occurs. For $G_{2eff} < 0$, the energetically favored state is one in which second-neighbor columns have the same domain, as shown in Figure 8(b). The predicted [25,34] space group [which follows from Figure 8(b)] for this structure is P4/ncc. There is at present no experimental determination of either the space group, or the detailed structure, in the high-temperature commensurate state of $TaTe_4$.

The commensurate phase of $NbTe_4$ is believed to have the same symmetry as the room-temperature commensurate phase of $TaTe_4$. However, there has at present been no experimental structural determination or space-group symmetry determination for the commensurate phase of $NbTe_4$.

3.4. INCOMMENSURATE PHASES OF $NbTe_4$

As in the discussion of the commensurate phase, in the lowest approximation the interactions between the charge-density waves on the different columns are neglected, and the order parameter describing the charge-density wave on the i-th column can be written

$$\psi_i(z) = \psi(z - \Delta_i) \ . \tag{3.3}$$

The properties and the general form of $\psi(z)$ are given by Eqs. (2.13) and (2.14). Changing Δ_i corresponds to rigidly sliding the charge-density in the z-direction by the amount of the change, but does not change the energy of the modulation wave in the approximation that intercolumn interactions are absent. The problem of determining the three-dimensional structure is equivalent to the problem of determining the relative values of the Δ_i on the different columns so as to minimize the intercolumn interaction energy.

The interaction energy between nearest-neighbor columns 1 and 2 has been evaluated [38] as a function of the relative displacement, $\Delta = \Delta_1 - \Delta_2$, of the modulation waves, and the result is shown in Figure 9. This nearest-neighbor interaction energy has its minimum at $\Delta = 3\lambda/2$ where, as above, λ is the length of a commensurate domain on a given column. In the study of the commensurate phase above, it was found that neighboring columns prefer to contain different domains. Here, the minimum of the intercolumn interaction energy of two neighboring columns occurs at a point where equivalent commensurate domains on the neighboring columns are as far as possible from one another. The three-dimensional structure which results is shown in Figure 10(a), and has a $\sqrt{2}a \times \sqrt{2}a$ basal-plane unit cell.

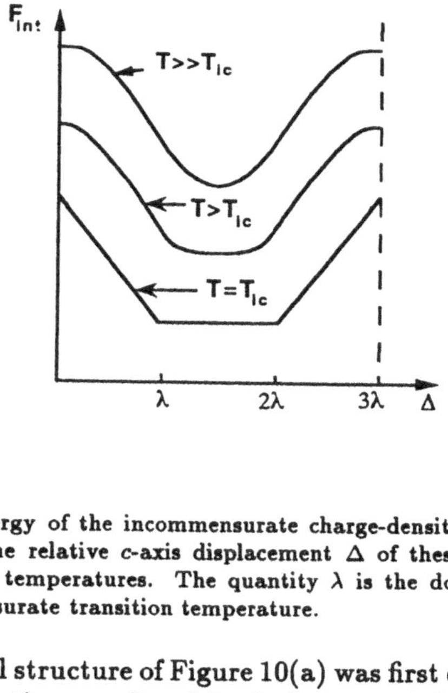

Fig. 9 The interaction energy of the incommensurate charge-density waves on two neighboring columns as a function of the relative *c*-axis displacement Δ of these two charge density waves is shown for three different temperatures. The quantity λ is the domain length and T_{ic} is the incommensurate to commensurate transition temperature.

The three-dimensional structure of Figure 10(a) was first determined theoretically in Ref. [25]. The diffraction results of Eaglesham *et al.* [26] were consistent with this structure, and the structural determination of Refs. [22 and 27] confirmed the correctness of the structure.

To investigate the phase transition from the high-temperature incommensurate state to the low-temperature incommensurate state it is of interest to probe the stability of the high-temperature incommensurate state with respect to the departures of the displacements Δ_i from their equilibrium values Δ_{io} shown in Figure

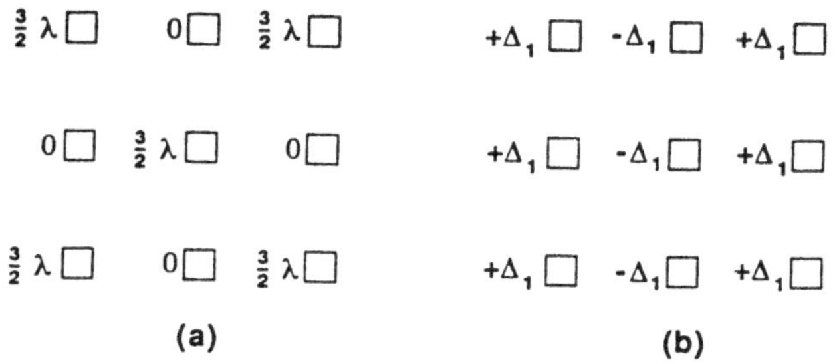

Fig. 10(a) Each square box represents the basal-plane projection of a column of the NbTe$_4$ structure in the $\sqrt{2}a \times \sqrt{2}a$ incommensurate phase. The quantities $3\lambda/2$ or 0 beside each box represent the c-axis displacement of the charge-density wave on that particular column relative to some fixed reference position common to all columns.
Fig. 10(b) In the $2a \times 2a$ incommensurate phase, the charge-density wave on each column has a c-axis displacement of $+\Delta_1$ or $-\Delta_1$ as indicated in addition to the displacements shown in (a).

10(a). Thus, write

$$\Delta_i = \Delta_{io} + \delta\Delta_i \tag{3.4}$$

where

$$\delta\Delta_i = \sum_{\mathbf{q}} \delta\Delta_{\mathbf{q}} e^{i\mathbf{q}\cdot\mathbf{r}_i} \tag{3.5}$$

and

$$\mathbf{r}_i = n_1\mathbf{a}_1 + n_2\mathbf{a}_2 \ . \tag{3.6}$$

Also, $\mathbf{q}$ is a two-dimensional vector with components q_x, q_y such that $-\pi/a < q_x, q_y \leq \pi/a$. Expanding the free energy in powers of the $\delta\Delta_q$ up to second order one finds

$$F = F_0 + \sum_{\mathbf{q}} D(q)|\delta\Delta_q|^2 \ . \tag{3.7}$$

The quantity $D(q)$ has been evaluated numerically in Ref. [39], where it was found that the modes with $(q_x, q_y) = (\pi/a, 0)$ and $(0, \pi/a)$ are soft modes which become unstable at the critical temperature of the phase transition.

The physical reason for the mode softening leading to the phase transition between the two incommensurate phases is that the sharpening up of the domain walls as the temperature is lowered causes the effective first-neighbor contribution to the force constant to be very weak (e.g. see Figure 8). When the first-neighbor force constant becomes sufficiently weak, the second-neighbor interaction begins to play a relatively more prominent role, and ultimately causes the phase transition. This is described in more detail in Refs. [38 and 39]. A detailed dynamical theory of the order parameter vibrations for $NbTe_4$, and of the soft modes, can be found in Ref. [17].

A Landau theory of this phase transition has been developed in Ref. [37]. The free energy relative to that of the $\sqrt{2}a \times \sqrt{2}a$ incommensurate phase is

$$\delta F = A(\Delta_1^2 + \Delta_2^2) + B(\Delta_1^2 + \Delta_2^2)^2 + C\Delta_1^2\Delta_2^2 \tag{3.8}$$

where Δ_1 and Δ_2 are $\delta\Delta_q$ for $\mathbf{q}$ equal to $(\pi/a, 0)$ and $(0, \pi/a)$, respectively. The $\sqrt{2}a \times \sqrt{2}a$ incommensurate phase becomes unstable relative to the $2a \times 2a$ incommensurate phase when A becomes negative. If the constant C is positive, a single-$\mathbf{q}$ state (e.g. $\Delta_1 \neq 0, \Delta_2 = 0$) which is shown in Figure 10(b) will be the stable phase, whereas if C is negative, a double-q state (i.e. $\Delta_1 = \Delta_2 \neq 0$) will be stable. The experiments of Mahy *et al.* [45], can best be interpreted in terms of the single-$\mathbf{q}$ state. Furthermore, a numerical analysis of the theoretical model suggests that the single-$\mathbf{q}$ state is stable (i.e. that $C > 0$).

A related but somewhat different approach to the incommensurate to incommensurate phase transition is taken in an article by Chen and Walker [41] who classify the possible soft modes giving rise to the phase transition according to irreducible representations of the superspace symmetry group of the high temperature incommensurate phase. The consequences of the superspace symmetries of both the high-temperature and low-temperature incommensurate phases in producing systematic absences in the diffraction pattern are also given in that article.

Much interesting experimental work remains to be done, including investigating whether or not the incommensurate to incommensurate phase transition is continuous, and whether or not an appropriate phason softens at the transition, and also in investigating the detailed temperature dependent structure in the low temperature incommensurate phase.

3.5. THE NUCLEAR HYPERFINE INTERACTION

Consider a scalar nuclear hyperfine interaction of the form shown in Eq. (2.21). In general, the hyperfine interaction constant for each nucleus in the crystal will be different, except where the nuclei are related by symmetry.

Consider first the commensurate phase which occurs at room temperature in $TaTe_4$ and which is presumed to occur below 50K in $NbTe_4$. This structure has a $2a \times 2a$ basal-plane unit cell with $P4/ncc$ space group symmetry and is illustrated in Figure 8(a). The structure contains three distinct columns labelled 1,2 and 3. The most general form for the hyperfine interaction at the n-th Ta (or Nb) nucleus belonging to column 1, 2 or 3 is

$$\begin{aligned} B_{1n} &= b_1 + c_1 \ \cos(Qz_n + \phi) \ , \\ B_{2n} &= b_1 + c_1 \ \cos(Qz_n - \phi) \ , \\ B_{3n} &= b_3 + c_3 \ \cos(Qz_n) \ , \end{aligned} \tag{3.9}$$

respectively. Thus, column 3 yields 2 inequivalent sites with intensity ratio 2:1. Columns 1 and 2 both yield the same set of three inequivalent sites with intensity ratio 1:1:1. Thus, for the crystal as a whole, there are predicted to be five inequivalent sites with intensity ratio 2:1:1:1:1. In the limit of negligible intercolumn interactions, $\phi = 2\pi/3$, $b_1 = b_3$ and $c_1 = c_3$; in this case only two distinct lines would be observed with intensity ratios 2:1.

The high temperature commensurate phase contains only two inequivalent columns labelled 1 and 2 in Figure 8(b). The predicted hyperfine interactions for these columns are B_{1n} and B_{2n} as given by Eq. 3.9, above. Thus, in this case there are three inequivalent sites with intensity ratios 1:1:1. Of course, the parameters b_1, c_1 and ϕ may be different in the two commensurate phases and may vary with temperature.

In the room temperature incommensurate phase of $NbTe_4$ there are two columns in the $\sqrt{2}a \times \sqrt{2}a$ basal plane unit cell shown in Figure 10(a). However, both are equivalent from the point of view of magnetic resonance. Symmetry considerations show that the form of the scalar contribution to the hyperfine interaction is the same as in the prototypical example, and is given by Eq. (2.22). An expansion of $\phi(z)$ in powers of $\psi(z)$ up to first order which is local in so far as the z coordinate is concerned yields

$$\phi(z) = c\psi(z) + d\psi(z - 3\lambda/2) + \ldots, \tag{3.10}$$

where c and d are real constants and λ is the length of a domain as above. The term in $\psi(z - 3\lambda/2)$ takes into account the contribution to the hyperfine interaction on a given column of the charge-density waves on neighboring columns. Further terms could be added to Eq. (3.10) analogous to those given in Eq. (2.23).

4. Conclusions

The structural simplicity of the compounds $NbTe_4$, $TaTe_4$ and $(Nb_xTa_{1-x})Te_4$ allows the development of a detailed phenomenological theory of the fascinating variety of phases and phase transitions which exist in these materials. Some of these phases and phase transitions are better understood than others. The $2a \times 2a$ commensurate phase is likely to be the phase with P4/ncc symmetry predicted by theory [25,34] and confirmed by experiment [31], although there remains the puzzle of why the convergent beam electron diffraction studies [26,29] disagree with these results. The theory yields a structural model [25, 34] with P4/ncc symmetry for the $\sqrt{2}a \times \sqrt{2}a$ commensurate phase of $TaTe_4$ which occurs [33] at temperatures greater than 450K; the detailed structure and the symmetry of this phase have yet to be determined experimentally. A plausible mechanism driving the transition between the two commensurate phases has been identified [38, 29].

The $\sqrt{2}a \times \sqrt{2}a$ incommensurate phase of $NbTe_4$ appears to be well understood. Although the $2a \times 2a$ incommensurate phase of $NbTe_4$ and the phase transition between the two incommensurate phases have been studied extensively theoretically [37-39], relatively few (except for the work of Mahy *et al.* [35]) experimental studies have been made on these topics. Experimental work relating to the order of the phase transition, to the possibility of a soft optical phason [37] driving the transition, and to the structure and symmetry (e.g. see Ref. [41]) of the $2a \times 2a$ incommensurate phase would be of interest.

Experimental magnetic resonance studies of the various phases and phase transitions occurring in $NbTe_4$ and $TaTe_4$ would be of interest. This article has described how symmetry can be used in arriving at a general form for the scalar contribution to the hyperfine interaction. This approach can be extended to other interactions of interest in magnetic resonance experiments.

Acknowledgements

I have much appreciated collaborations with Z.Y. Chen, W.E. Goff, A.E. Jacobs, Rose Morelli and D. Sahu on topics related to this review. This work was supported by the Natural Sciences and Engineering Research Council of Canada.

References

1 A.E. Jacobs and M.B. Walker, *Phys. Rev.* **B21**, 4132 (1980); M.B. Walker and A.E. Jacobs; *Phys. Rev.* **B25**, 4856 (1982), *Phys. Rev.* **B25**, 3424 (1982).

2 D.E. Moncton, J.D. Axe, and F.J. DiSalvo, *Phys. Rev. Lett.* **34**, 734 (1975); *Phys. Rev.* **B16**, 801 (1977).

3 W.L. McMillan, *Phys. Rev.* **B14**, 1496 (1976).

4 H. Shiba and K. Nakanishi in *Structural Phase Transitions in Layered Transition Metal Compounds*, K. Motizuki ed., Reidel, Dordrecht, 1986, p. 175.

5 R. Blinc, P. Prelovŝek, V. Rutar, J. Seliger, and S. Žumar in *Incommensurate Phases in Dielectrics-Fundamentals*, R. Blinc and A.P. Levanyuk eds., North Holland, Amsterdam, 1986, p. 143.

6 Y.I. Frenkel and T. Kontorova, *Zh. Eksp. Teor. Fiz.* **8**, 1340 (1938).

7 F.C. Frank and J.H. Van der Merve, *Proc. R. Soc. London*, **A198**, 205 (1949).

8 A.E. Jacobs, David Mukamel, and M.B. Walker, *Phys. Rev.* **B24**, 5195 (1981).

9 S.A. Jackson, P.A. Lee, and T.M. Rice, *Phys. Rev.* **B17**, 3611 (1978).

10 P.M. de Wolff, *Acta Crystallogr.* **A30**, 777 (1974); A. Janner and T. Janssen, *Phys. Rev.* B15, 643 (1977); A. Janner and T. Janssen, *Physica* **A99**, 47 (1979).

11 A.W. Overhauser, *Phys. Rev.* **B3**, 3173 (1971).

12 P.A. Lee, T.M. Rice, and P.W. Anderson, *Solid State Commun.* **14**, 703 (1974).

13 J.D. Axe in: Proc. Gatlinsberg Neutron Scattering Conference, R.M. Moon, ed. (1976), p.353.

14 M.B. Walker, *Can. J. Phys.* **56**, 127 (1978).

15 R. Currat and T. Janssen, *Solid State Physics* **41**, 201 (1988).

[16] K. Nakanishi and H. Shiba, *J. Phys. Soc. Japan* **45**, 1147 (1978).

[17] W.E. Goff, M.B. Walker, and Z.Y. Chen, *Phys. Rev.* **B43**, 655 (1991).

[18] K. Selte and A. Kjekshus, *Acta. Chem. Scand.* **18**, 690 (1964).

[19] E. Bjerkelund and A. Kjekshus, *J. Less Common Metals* **7**, 231 (1964).

[20] H. Böhm, *Z. Kristallogr.* **180**, 113 (1987).

[21] F.W. Boswell, A. Prodan, and J.K. Brandon, *J. Phys.* **C16**, 1067 (1987).

[22] H. Böhm and H.G. von Schnering, *Z. Kristallogr.* **162**, 26 (1983).

[23] J. Mahy, J. Van Landuyt, and S. Amelinckx, *Phys. Stat. Sol.* (a) **77**, K1 (1983).

[24] H. Böhm and H.G. von Schnering, *Z. Kristallogr.* **171**, 41 (1985).

[25] M.B. Walker, *Can. J. Phys.* **63**, 46 (1985).

[26] D.J. Eaglesham, D. Bird, R.L. Withers, and J.W. Steeds, *J. Phys.* **C18**, 1 (1985).

[27] S. van Smaalen, K.D. Bronsema, and J. Mahy, *Acta Cryst.* **B42**, 43 (1986).

[28] D. Sahu and M.B. Walker, *Phys. Rev.* **B32**, 1643 (1985).

[29] D.M. Bird, D.J. Eaglesham, R.L. Withers, S. McKernan, and J.W. Steeds in *Proceedings of the International Conference on Charge Density Waves in Solids*, Gy. Hutiray and J. Sólyom eds., Springer-Verlag, Berlin, 1985.

[30] Rose Morelli, D. Sahu, and M.B. Walker, *Phys. Rev.* **B33**, 4843 (1986).

[31] K.D. Bronsema, S. van Smaalen, J.L. de Boer, G.A. Wiegers, and F. Jellinek, *Acta Cryst.* **B43**, 305 (1987).

[32] J.M. Corbett, L.G. Hiltz, F.W. Boswell, J.C. Bennett, and A. Prodan, *Ultramicroscopy* **26**, 43 (1988).

[33] F.W. Boswell, A. Prodan, J.C. Bennett, J.M. Corbett, and L.G. Hiltz, *Phys. Stat. Sol.* **102**, 207 (1987).

[34] Z.Y. Chen, M.B. Walker, and R. Morelli, *Phys. Rev.* **B39**, 11742 (1989).

[35] J. Mahy, J. Van Landuyt, S. Amelinckx, K.D. Bronsema, and S. van Smaalen, *J. Phys.* **C19**, 5049 (1986).

[36] P. Bak and V.J. Emery, *Phys. Rev. Lett.* **36** 978 (1976); P. Bak, *Phys. Rev. Lett.* **37**, 1071 (1976); P. Bak and T. Jannssen, *Phys. Rev.* **B17**, 436 (1978).

[37] M.B. Walker and Rose Morelli, *Phys. Rev.* **B38**, 4836 (1988).

[38] Rose Morelli and M.B. Walker, *Phys. Rev. Lett.* **62**, 1520 (1989).

[39] Rose Morelli and M.B. Walker, *Phys. Rev.* **B40**, 7542 (1960).

[40] A. Prodan and F.W. Boswell, Acta. *Cryst.* **B43**, 165 (1987).

[41] Z.Y. Chen and M.B. Walker, *Phys. Rev.* **B40**, 8983 (1989).

[42] F.W. Boswell and A. Prodan, *Phys. Rev.* **B34**, 2979 (1986).

[43] F.W. Boswell and A. Prodan, *Mat. Res. Bull* **19**, 93 (1984).

[44] D.W. Bullett, *J. Phys.* **C17**, 253 (1984); M.H. Whangbo and P. Gressier, *Inorg. Chem.* **23**, 1228 (1984).

[45] J. Mahy, J. Van Landuyt, S. Amelinckx, Y. Uchida, K.D. Bronsema, and S. van Smaalen, *Phys. Rev. Lett.* **55**, 1188 (1985).

[46] K.K. Fung, S. McKernan, J.W. Steeds, and J.A. Wilson, *J. Phys.* **C14**, 5417 (1981).

[47] C.H. Chen, J.M. Gibson, and R.M. Fleming, *Phys. Rev.* **B26**, 184 (1982).

[48] M.B. Walker and A.E. Jacobs, *Phys. Rev.* **B25**, 3424 (1982).

[49] M.B. Walker, *Phys. Rev.* **B26**, 6208 (1982).

LAYERED TRANSITION METAL DICHALCOGENIDES

MICHIO NAITO
NTT Basic Research Laboratories
Tokyo, 180
Japan

HIRONORI NISHIHARA
Ryukoku University
Otsu, 520-21
Japan

TILMAN BUTZ
Physik-Department, Technische Universität München
8046 Garching
Germany

1. Hyperfine Techniques

Among the layered transition metal dichalcogenides, TaS_2 and $TaSe_2$ have attracted considerable attention because they exist in various polytypes and exhibit a rich variety of charge density wave (CDW) phenomena. Moreover, these materials are in some sense prototypical for nuclear spectroscopy on CDW systems because ^{181}Ta happens to be a suitable isotope for the different techniques NMR, NQR, TDPAC, and Mössbauer spectroscopy. All of these techniques have been applied extensively and complement each other.

Figure 1 summarizes the basic features of these techniques for nuclear quadrupole interactions. With NQR (and NMR) the hyperfine splitting of the nuclear ***ground-state*** is detected by resonant absorption of radiofrequency radiation; with Mössbauer spectroscopy the hyperfine splitting of the first excited nuclear state (or the groundstate, or both) is detected by resonant absorption of a γ-quantum and subsequent reemission (nuclear fluorescence); with TDPAC the hyperfine splitting of the *first excited state* – or more precisely, the ***intermediate nuclear state*** of a γ-γ-cascade – is detected as a quantum beat or interference phenomenon due to the coherent superposition of all possible population and depopulation pathways. This latter technique is thus of an extremely different character than the others: it does not require the absorption of quanta at all and therefore imposes far less restrictions

T. Butz (ed.), Nuclear Spectroscopy on Charge Density Wave Systems, 35–112.

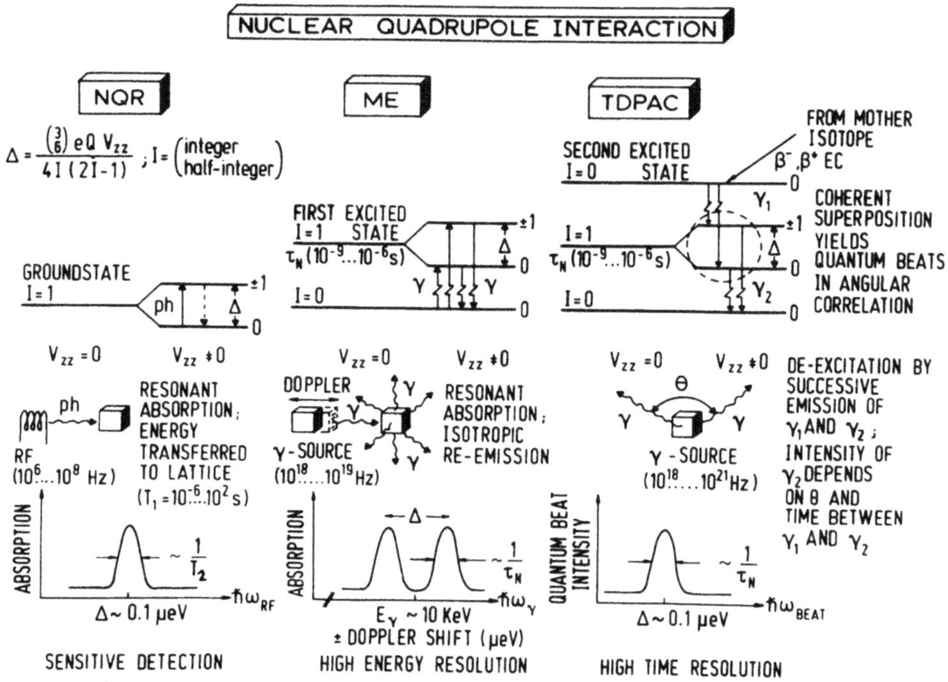

Fig. 1 Comparison between NQR, ME, and TDPAC (schematic).

on sample size and shape compared to NMR/NQR and Mössbauer spectroscopy. Moreover, the sensitivity is absolutely temperature independent.

Since this is the first chapter in this book which presents detailed experimental results on CDW systems obtained by various nuclear spectroscopy techniques a short introduction to the relevant aspects with special emphasis on the isotope ^{181}Ta will be given. The next paragraph surveys the structure of the materials to be presented in the following paragraphs. These will be divided into compounds with octahedral metal coordination, with trigonal prismatic metal coordination, and with both coordinations alternating in successive layers ("mixed coordination").

1.1. NMR

Nuclear magnetic resonance (NMR) and nuclear quadrupole resonance (NQR) are resonant absorption phenomena in the radio frequency range (1 – 1000MHz) accompanied by transitions between energy levels of nuclear spins in their ground-state due to magnetic and quadrupolar hyperfine interactions with surrounding electrons. Since the frequency is very low, or the energy of the relevant photons is very small, the resolution of the absorption spectrum is highest while the sensitivity of the detection is lowest compared to other hyperfine techniques. They are similar to other hyperfine techniques in that they probe microscopic or very local electronic properties at nuclear sites. Thus it is quite useful to use NMR and NQR for a study of local electronic properties in various CDW systems. Here, we give a short review on NMR and NQR for non-expert readers [1]. Detailed descriptions relevant to a particular section will be presented in each section.

A free nucleus with spin **I** and gyromagnetic ratio γ_N interacts with a static magnetic field **H** parallel to the z-axis as described by a Zeeman Hamiltonian

$$\mathcal{H} = -\gamma_N \hbar \mathbf{I} \cdot \mathbf{H} = -\gamma_n \hbar I_z H_0. \tag{1.1}$$

The degenerate $2I+1$ nuclear spin states split into Zeeman levels with equal spacing of $\gamma_N \hbar H_0$. An oscillating magnetic field with an angular frequency ω (or frequency f) induces transitions between adjacent Zeeman levels with the resonance condition

$$\omega = \omega_N = \gamma_N H_0$$

or

$$f = \frac{\gamma_N}{2\pi} H_0. \tag{1.2}$$

The nucleus in solids interacts with surrounding electrons with spin **s** and angular momentum **l** via the hyperfine interaction which consists of Fermi-contact, dipole-dipole and orbital terms:

$$\begin{aligned} \mathcal{H} = & \frac{8\pi}{3} \gamma_e \gamma_N \hbar^2 \mathbf{I} \cdot \mathbf{s} \delta(\mathbf{r}) \\ & - \gamma_e \gamma_N \hbar^2 \mathbf{I} \cdot \left[\frac{\mathbf{s}}{r^3} - \frac{3(\mathbf{s} \cdot \mathbf{r})\mathbf{r}}{r^5} \right] \\ & + \gamma_e \gamma_N \hbar^2 \frac{\mathbf{I} \mathbf{l}}{r^3}. \end{aligned} \tag{1.3}$$

This equation can be written as

$$\mathcal{H} = -\gamma_N \hbar \mathbf{I} \cdot \mathbf{H}_{loc} \tag{1.4}$$

with

$$\begin{aligned} \mathbf{H}_{loc} = & - \frac{8\pi}{3} \gamma_e \hbar \mathbf{s} \delta(\mathbf{r}) \\ & + \frac{\gamma_e \hbar}{r^3} \left[\mathbf{s} - \frac{3(\mathbf{s} \cdot \mathbf{r})\mathbf{r}}{r^2} \right] - \gamma_e \hbar \frac{\mathbf{l}}{r^3}. \end{aligned} \tag{1.5}$$

H_{loc} is an effective local field from electrons due to the hyperfine interaction. When a static magnetic field is applied to the sample in the paramagnetic state, the non-zero thermal average of H_{loc} appears in proportion to the applied field H_0, which shifts the resonance frequency in eq. (1.2) by an amount

$$K = \frac{< H^z_{loc} >}{H_0}. \tag{1.6}$$

This shift is called the "Knight shift" in metals and the "chemical shift" in insulating, diamagnetic compounds. In magnetically ordered materials such as ferromagnets or antiferromagnets, nonvanishing hyperfine fields may well exist even in zero applied field due to a non-zero thermal average of S_z in zero field. In a simple metal where the Fermi contact interaction with s-band conduction electrons is dominant, the Knight shift is expressed as

$$K = \frac{8\pi}{3} < |u_k(0)|^2 >_{\mathrm{FS}} \chi_s \tag{1.7}$$

with the Pauli paramagnetic susceptibility

$$\chi_s = \frac{1}{2}\gamma_e^2 \hbar^2 \rho(E_F) \tag{1.8}$$

where $< |u_k(0)|^2 >_{FS}$ is the amplitude of the conduction electron wave function at the nuclear site averaged over electronic states on the Fermi surface, and $\rho(E_F)$ is the electronic density of states at the Fermi surface. If a metallic material undergoes a transition to the CDW state with decreasing temperature, a gap may open or a part of the Fermi surface might be destroyed, and an anomaly may be observed in the temperature dependence of the Knight shift.

The nuclear spin-lattice relaxation rate T_1 is a characteristic time with which the nuclear spin system returns to the thermal equilibrium after the population of a nuclear spin system was disturbed, for example, to an "infinite spin temperature". This is usually measured by observing the two-pulse spin-echo signals after various delay times t from the saturation of the spin echo signals by a comb of RF pulses. The decay time of the spin echo signal when the time interval of the double pulse is increased is called spin-echo decay time T_2. The spin-lattice relaxation time T_1 is calculated from the transition probability W with which the nuclear spin flips by transferring its energy to the "lattice":

$$\frac{1}{T_1} = 2W. \tag{1.9}$$

The "lattice" for the nuclear spin system is usually the electron spin system coupled by the hyperfine interaction as described in eq. (1.3). The energy which is transferred is very small and only low-lying exited states (elementary excitations) of the electron spin system are relevant to the relaxation. In a simple metal where the Fermi-contact interaction with s-band conduction electrons is dominant, the relaxation rate T_1^{-1} is expressed as

$$\frac{1}{T_1} = \frac{64}{9}\pi^3\hbar^3\gamma_e^2\gamma_N^2 < |u_k(0)|^2 >_{FS} \rho^2(E_F)k_B T. \tag{1.10}$$

The relaxation rate is proportional to the square of the Fermi contact hyperfine coupling constant and the square of the density of states at the Fermi surface. If a metallic material undergoes a transition to the CDW state with decreasing temperature, an anomaly may also be observerd in the temperature dependence of the relaxation rate. The so-called "Korringa-relation" holds between the spin-lattice relaxation time T_1 and the Knight shift K,

$$K^2 T_1 T = \left(\frac{\gamma_e}{\gamma_N}\right)^2 \frac{\hbar}{4\pi k_B} \tag{1.11}$$

for the case of non-interacting electron spins. For a modified Korringa relation when the effect of electron-electron correlation is taken into account, see [5,6]. The simple Korringa relation (1.11) holds in the case of a two-dimensional metal even if the electron-electron correlation is taken into account within the framework of the random field approximation [7].

The local field in eq. (1.5) is considered to be the sum of the thermal average and a fluctuating component

$$\mathbf{H}_{loc} = <\mathbf{H}_{loc}> + \delta\mathbf{H}. \tag{1.12}$$

The nuclear spin-lattice relaxation rate is expressed in terms of the fluctuating component of the local field as

$$\frac{1}{T_1} = \frac{\gamma_N^2}{2}\int_{-\infty}^{\infty} < \{\delta H^-(t)\delta H^+(0)\} > e^{-i\omega_N t}dt \tag{1.13}$$

where $\{AB\}$ means a symmetrical product of the operators $(AB + BA)/2$. In the above simple case where the Fermi contact interaction with the s-band conduction electrons is dominant, this reduces to the Fourier transform of the spin correlation function

$$\frac{1}{T_1} = \frac{\gamma_N^2}{2}\sum_{\mathbf{q}} A_{\mathbf{q}}A_{-\mathbf{q}}\int_{-\infty}^{\infty} < \{S_{\mathbf{q}}^-(t)S_{-\mathbf{q}}^+(0)\} > \cos\omega_N t dt \tag{1.14}$$

where $A_{\mathbf{q}}$ is the expectation value of the Fermi contact interaction in eq. (1.3) versus the Bloch function of wave number $\mathbf{q}$. The nuclear spin-lattice relaxation rate is also expressed in terms of the generalized wave number and frequency dependent susceptibility $\chi(\mathbf{q},\omega)$. It is given in the same case as above by

$$\frac{1}{T_1}=\frac{2\gamma_N^2 k_B T}{(\gamma_e\hbar)^2}\sum_{\mathbf{q}} A_{\mathbf{q}}A_{-\mathbf{q}}\frac{\chi_\perp''(\mathbf{q},\omega_N)}{\omega_N}. \tag{1.15}$$

The nuclear spin-lattice relaxation data contain information on the generalized susceptibility on which a summation over $\mathbf{q}$ is taken at low frequency. For the equivalence of the expressions of T_1^{-1} with the transition probability eq.(1.9), to that employing the Fourier transform of the spin correlation function eq.(1.14), and to that with the q-summed generalized susceptibility eq.(1.15), see textbooks and references cited therein [1].

The spin-echo decay time T_2 contains information on both spin-lattice relaxation and spin-spin interactions. The decay rate is expressed as

$$\frac{1}{T_2}=\frac{1}{2T_1}+\frac{\gamma_N^2}{2}\int_{-\infty}^{\infty} <\{\delta H^z(t)\delta H^z(0)\}> dt \tag{1.16}$$

or

$$\frac{1}{T_2}=\frac{1}{2T_1}+\frac{\gamma_N^2 k_B T}{(\gamma_e\hbar)^2}\lim_{\omega\to 0}\sum_{\mathbf{q}} A_{\mathbf{q}}A_{-\mathbf{q}}\frac{\chi_\parallel''(\mathbf{q},\omega)}{\omega} \tag{1.17}$$

in correspondence to eq.(1.14) and eq.(1.15).

1.2. NQR AND QUADRUPOLE EFFECTS IN NMR

In addition to the magnetic interaction between the nucleus and the electrons, the electrostatic interaction of the nucleus with its surrounding electrons is relevant to the resonance phenomena in the radio-frequency range. When the electrostatic potential is expanded in a power series about the nuclear center of mass, the first important term is the interaction between the electric field gradient and the nuclear electric quadrupole moment. We will make here a short review on nuclear quadrupole resonance (NQR) [8], followed by a somewhat detailed description of low-field NMR, which is less well-known, with a special emphasis on the case of the nuclear spin $I = 7/2$ which prevails for the case of ^{181}Ta in layered tantalum dichalcogenides. Spectra in high magnetic fields are also reviewed.

The nuclear spin Hamiltonian which describes the quadrupole effects together with the Zeeman effects for a nucleus in the groundstate with spin $\mathbf{I}$, with gyromagnetic ratio γ_N and the quadrupole moment Q is given by

$$\mathcal{H}=\frac{e^2qQ}{4I(2I-1)}\left\{3I_z^2-I(I+1)+\frac{1}{2}\eta(I_+^2+I_-^2)\right\}-\gamma_N\hbar\mathbf{I}\cdot\mathbf{H}_{eff}. \tag{1.18}$$

In the case of a predominant quadrupole interaction, or in the case of a weak field, the quantization axes are conveniently taken to be the principal axis system for the electric field gradient (EFG) tensor such that

$$eq = V_{zz} \tag{1.19}$$

is the maximum field gradient component in the principal axis system and

$$\eta = \frac{V_{xx} - V_{yy}}{V_{zz}} \tag{1.20}$$

is the asymmetry parameter which measures the departure of the EFG from axial (or cylindrical) symmetry. The value of η has the property

$$0 \leq \eta \leq 1 \tag{1.21}$$

provided the x and y axes are chosen such that

$$|V_{xx}| \leq |V_{yy}| \leq |V_{zz}|. \tag{1.22}$$

The schematic energy level diagram for a strong quadrupole interaction is shown in Figure 2 for $I = 7/2$.

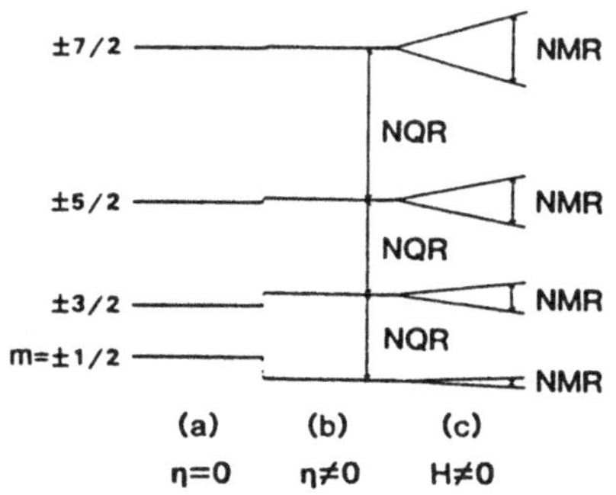

Fig. 2 Schematic energy levels of the nuclear spin system with $I = 7/2$. (a) $\eta = 0, H = 0$, (b) $\eta \neq 0$, (c) $H \neq 0$. NQR and low-field NMR transitions are shown.

1.2.1. *The case of zero applied magnetic field (H=0)* The Hamiltonian consists of the quadrupolar term only unless the system is magnetically ordered. The quadrupolar term does not vanish unless I equals 1/2 or the local symmetry around the nucleus is cubic. If the local symmetry around the nucleus is axial, the asymmetry parameter vanishes ($\eta = 0$). The resonance frequencies for the axially symmetric electric field gradient are, for example in the case of $I = 7/2$, one, two, or three times ν_Q, where

$$\nu_Q = \begin{pmatrix} 3 \\ 6 \end{pmatrix} \frac{e^2qQ}{4I(2I-1)h} \qquad \text{for } I = \begin{pmatrix} \text{integer} \\ \text{half} - \text{integer} \end{pmatrix} \tag{1.23}$$

is the so-called quadrupole frequency.[1] In the case of a non-axial electric field gradient, the resonance frequencies are no longer in the ratio 1:2:3, but depend on the value of η. The resonance frequencies in units of ν_Q as a function of η are given in Figure 3 for the case of $I = 7/2$ [9]. A single nuclear site creates three NQR transitions, and the quadrupole frequency ν_Q and the asymmetry parameter η can be determined if the proper assignment for the three transitions is made. However, the assignment becomes difficult if there are many inequivalent nuclear sites, especially in cases where not all transitions are observerd, and it is important to analyze the NQR data consistently with the data of NMR, Mössbauer effect, and TDPAC which are complementary to each other.

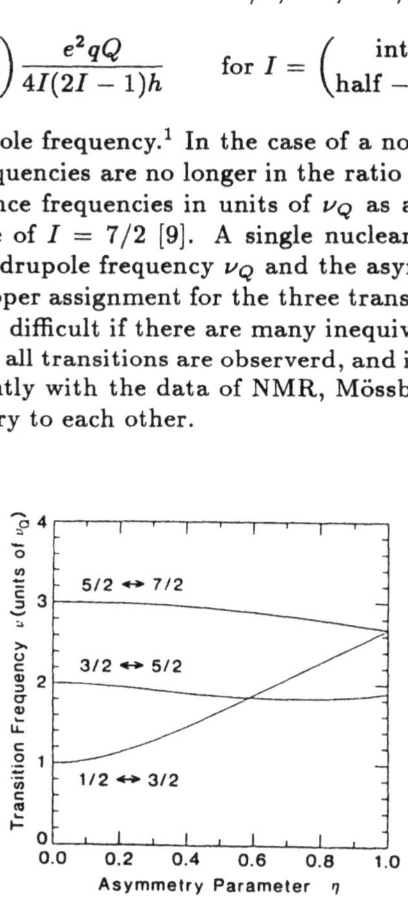

Fig. 3 NQR frequency in units of ν_Q as a function of η in the case of $I = 7/2$.

[1] In standard NMR textbooks one defines $\nu_Q = \frac{3e^2qQ}{2I(2I-1)h}$ for both integer and half-integer spins. In this case the smallest energy difference would be $h\nu_Q$ for half-integer and $h\nu_Q/2$ for integer spins. In standard TDPAC textbooks eq.(1.23) is used and we prefer to have a single definition throughout this chapter.

1.2.2. *The case of a weak magnetic field* A weak applied magnetic field removes the degeneracy of the $\pm m$ quadrupole levels as shown in Figure 2(c) and it becomes possible to observe low frequency transitions between the split levels [9]. For half-integer spins all $I + 1/2$ transitions should be observable unless the asymmetry parameter of the EFG in the principal axis system is very small. The resonance field of the transition between $\pm m$ levels, which is proportional to the operating frequency, is expressed as [9,10]

$$H = \frac{H_0}{\alpha_m} \tag{1.24}$$

where

$$\alpha_m = [4A\cos^2\theta + \sin^2\theta(B + 2C\cos 2\phi)]^{1/2}. \tag{1.25}$$

H_0 is the resonance field for a free nucleus, θ and ϕ are polar angles of the applied field with respect to the the principal axes, and A, B and C are functions of η as listed in [9]. The resonance field H is a function of θ and ϕ if the value of η is fixed at constant frequency, and one therefore gets a powder spectrum averaging over the whole range of angles. The intensity of the transitions as a function of H is calculated by taking into account a factor $W(\theta, \phi)$ for the angular dependence of the transition probability, as given in [10]

$$I(H) \propto \int_0^\pi W(\theta, \phi = f(H, \theta)) \left(\frac{\partial H}{\partial \phi}\right)^{-1}_{\phi = f} \sin\theta d\theta. \tag{1.26}$$

Some examples of the line shapes of the powder spectra as a function of the reduced magnetic field $h = H/H_0 = 1/\alpha_m$ are shown in Figure 4 for the $\pm 1/2$ and $\pm 3/2$ transitions for some values of η. These are qualitatively similar to the powder spectra with anisotropic Knight shifts for axially symmetric or non-axial cases, but the widths are very large. The peak reduced field h_p and the threshold reduced fields h_1, h_2 in which the signal appears or disappears in swept fields are shown as a function of η in Figure 5. The figure shows that the widths of the $\pm 1/2$ or $\pm 5/2$ transitions are rather large, the total width $|h_1 - h_2|$ is comparable or larger than H_0 itself, and the intensity of the signal is expected to be weak. The width of the $\pm 7/2$ transition is still larger (10 H_0 or more). The $\pm 3/2$ transitions is interesting; if η is less than about 0.15 the width is large, the transition probability is small, and the intensity of the signal is expected to be weak. However, if η is larger than about 0.15 the spectrum sharpens and the signal intensity could be rather strong. An enlarged figure for the $\pm 3/2$ transition is given in Figure 6, which shows that for $\eta = 0.35$ h_1, h_p, and h_2 coincide at 0.40 and the width vanishes even for a powder sample in a first order calculation. The signal intensity in this case should be very strong. The value of 0.35 for η is therefore called "magic asymmetry". It should be noticed that the powder spectra are determined only by the value of η and are not affected by the magnitude of the EFG. Thus this type of NMR experiment is a good method to get values of asymmetry parameters at various nuclear sites, independent of the magnitude of the EFG, because the NMR frequency is much smaller than ν_Q. This technique was found independently by Szeftel and Alloul for the case of $I = 3/2$ [11] and by Nishihara *et al.* for the case of $I = 7/2$ [10]. The broadening of the line shape is determined by the distribution of η at the nuclear site independent of a possible distribution of the magnitude of the EFG, whereas the NQR spectrum is affected by distributions of both ν_Q and η.

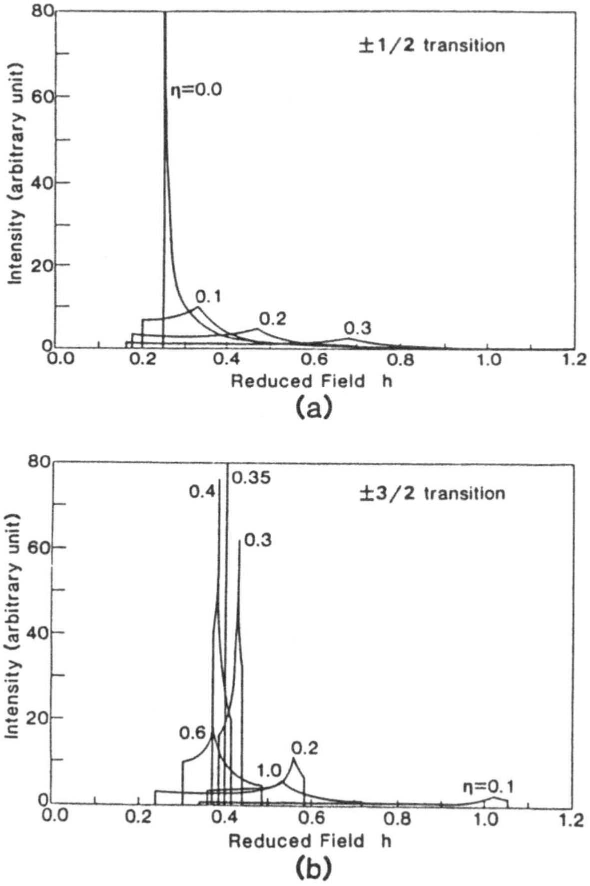

Fig. 4 Examples of line shapes of powder spectra as a function of the reduced magnetic field h for the $\pm 1/2$ and $\pm 3/2$ transitions.

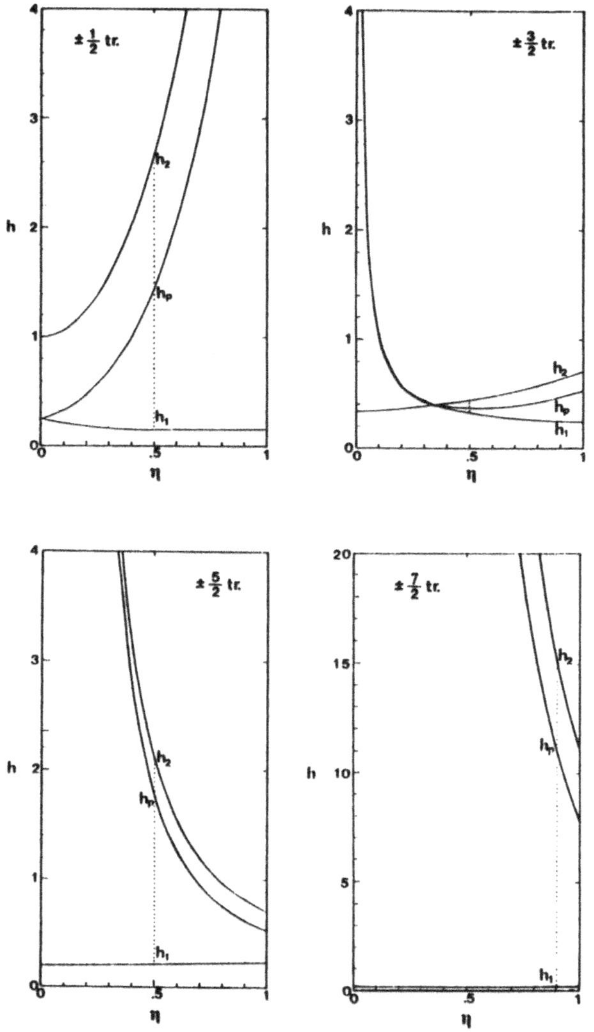

Fig. 5 The peak reduced field h_p and the threshold fields h_1 and h_2 at which the signal appears or disappears in swept magnetic fields are shown as a function of η for $\pm1/2, \pm3/2, \pm5/2$, and $\pm7/2$ transitions for $I = 7/2$ (low field case).

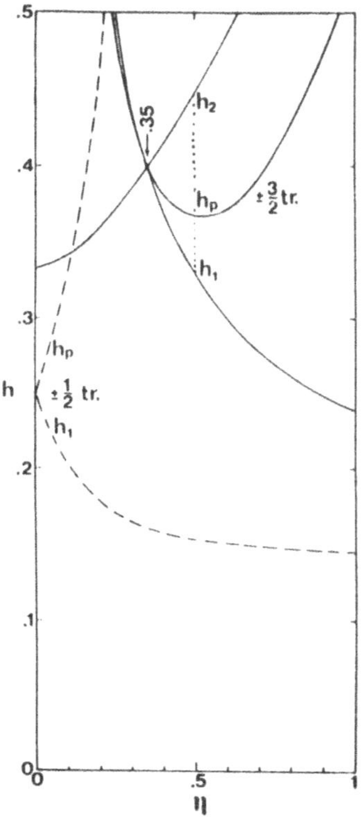

Fig. 6 An enlarged portion of Figure 5 of the ±3/2 transition.

1.2.3. *The case of a strong magnetic field* In the case of a strong magnetic field, the quantization axis z is taken to be the direction of H_{eff} which is the direction of the externally applied field H, or the direction of the vector sum of the applied field and the hyperfine fields from surrounding electrons in magneticallly ordered materials. The quadrupole coupling is treated as a perturbation to the Zeeman coupling. The $2I+1$ equidistant magnetic energy levels separated by $\hbar\omega$

$$E_m = -\gamma_N \hbar H_0 m \tag{1.27}$$

are perturbed in a first order calculation to

$$E_m = -\hbar\omega m + \frac{e^2 qQ}{8I(2I-1)}(3\cos^2\theta - 1)[3m^2 - I(I+1)] \tag{1.28}$$

in the case of axial symmetry. Since the quadrupole Hamiltonian is traceless, the quadrupole coupling does not shift the center of gravity of the resonance to first order. The shift of the $+m$ and $-m$ levels is identical. Therefore, for a half-integer spin, the central line is unaffected by the quadrupolar coupling to first order. The central line is affected in a second order calculation with the shift being of the order of $\nu_Q^2/\gamma_N \hbar H_0$. The effect of the quadrupole coupling is schematically shown in Figure 7 for the case of spin 3/2. The resonance frequencies depend on the relative orientation of the principal axes of the EFG with respect to the quantization axis; therefore one gets a powder spectrum by averaging the intensity over the whole range of angles (see textbooks for details of such "quadrupolar powder spectra" [8]). On can get information on the Knight shift, ν_Q, and η from such a powder spectrum in principle, but it may happen that the satellite transitions are not clear, or even wiped out due to the randomness of the local symmetries. The width of the central line due to the second order quadrupole effects is of the order $\nu_Q^2/\gamma_N \hbar H_0$ which is inversely proportional to the applied field, while the width due to magnetic origins is proportional to the applied field. Thus one can get information on the origin of the width by examining the field dependence of the width. The nuclear spin-lattice relaxation time T_1 is difficult to determine accurately in the case of "quadrupolar powder spectra" since the relaxation behaviour can no longer be described by a single time constant T_1 in the multilevel spin system. For the relaxation behaviour in multilevel spin systems, see references [12].

1.3. TDPAC

Time differential perturbed angular correlations (TDPAC) belongs to the group of hyperfine techniques. Here, we focus on γ-γ-angular correlations and nuclear quadrupole interactions (NQI's).

TDPAC requires a nucleus in an excited state which decays via the successive emission of two γ-quanta (γ-γ-cascade). The countrate in a detector for either the first (γ_1) or the second (γ_2) quantum will be isotropic for randomly oriented nuclear spins. However, for each nucleus there is an anisotropic emission probability for the γ-quantum because of angular momentum conservation: the initial nuclear spin vector, the final nuclear spin vector, and the vector characterizing the direction and angular momentum of the γ-quantum have to satisfy the "triangle rule" of quantum mechanical vector coupling. Therefore, by the mere observation of γ_1 in a given detector one selects an oriented – more precisely – an aligned subensemble of nuclei which were likely to have emitted γ_1. By detecting γ_2 *in coincidence* with γ_1, an anisotropic coincidence countrate results, because γ_2 was emitted from the selected aligned subensemble. The advantage of TDPAC compared to low temperature nuclear orientation is obvious: the anisotropy is entirely temperature independent. The disadvantage is that coincidence rather than single counts have to be measured, and hence, weak sources have to be used in order to guarantee that γ_2 was emitted from the *same* nucleus as γ_1.

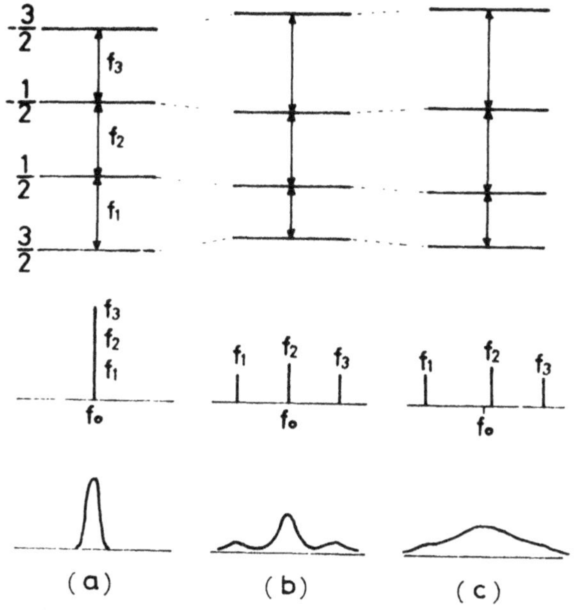

Fig. 7 The effect of the quadrupole coupling is schematically shown for the case of a strong magnetic field for $I = 3/2$. (a) No quadrupole coupling. (b) First order splitting of satellites. (c) Second order effect. Top: energy levels. Middle: spectrum of a single nucleus. Bottom: line shapes in an imperfect crystal.

This angular correlation can be perturbed by the interaction of nuclear moments and extranuclear fields, provided the intermediate nuclear state has a sufficiently long lifetime. In this case, the spin will precess between the population of the intermediate state (=emission of γ_1) and its depopulation (=emission of γ_2). Together with the spin precession there will also be a precession of the anisotropic emission probability distribution of γ_2. A prompt event will not perturb the angular correlation (γ_1 and γ_2 emitted at the same time), whereas a delayed coincidence (γ_2 later than γ_1) will lead to a rotated angular correlation. Rather than to integrate over all possible cascade events with varying delay time between γ_1 and γ_2 it is obviously advantageous to store these coincidences in a time differential manner. In this way, the nuclear spin precession signal appears as a modulation of the exponentially decaying lifetime curve.

The derivation of all relevant formulae for TDPAC is well documented in articles e.g. by Frauenfelder and Steffen [13]. Here, the most important aspects are summarized only, as far as they are relevant for the present purpose.

For random powder samples the coincidence countrate can be written as

$$W(\theta,t) \simeq \exp(-t/\tau_N)(1 + A_2(1)A_2(2)G_{22}(t)P_2(\cos\theta) + \ldots). \tag{1.29}$$

Here, τ_N denotes the lifetime of the intermediate state; $A_2(1)A_2(2)$ describes the anisotropy of the cascade and depends on spins, multipolarities and mixing ratios of the γ-transitions (each transition contributes a factor); $G_{22}(t)$ is the perturbation function which contains *all* information about the hyperfine interaction; $P_2(\cos\theta)$ is a Legendre polynomial with θ being the angle between γ_1 and γ_2, i.e. between the detectors. In principle, higher order terms like $A_4(1)A_4(2)G_{44}(t)P_4(\cos\theta)$ can occur, but frequently are negligibly small. From eq.(1.29) it is clear that θ will be chosen such as to give the maximum effect, i.e. $\theta = 180°$. For reasons of normalization, a further measurememt at $\theta = 90^o$ (the other extremum of P_2) is performed. In practice, a setup of four or more detectors is used and 180° coincidences as well as 90° coincidences are recorded simultaneously. From these data one usually forms the following ratio which eliminates the exponential decay factors as well as detector efficiency differences

$$R(t) = 2\frac{W(180^o) - W(90^o)}{W(180^o) + 2\cdot W(90^o)} \simeq A_2(1)A_2(2)G_{22}(t). \tag{1.30}$$

Under the influence of an axially symmetric electric field gradient the intermediate nuclear state splits energetically according to

$$E_m = \hbar\omega_Q(3m^2 - I(I+1)) \tag{1.31}$$

with

$$\omega_Q = -\frac{eQV_{zz}}{4I(2I-1)\hbar}.$$

In the following we focus on the $133keV$-$482keV$ cascade in ^{181}Ta which is fed via β^--decay from ^{181}Hf ($T_{1/2} = 42d$). (All samples therefore have to be doped with ^{181}Hf at a 50-200ppm level!) The intermediate state of this cascade has $I = 5/2$. The energies are in units of $\hbar\omega_Q$

$$E_{\pm 5/2} = 10, \quad E_{\pm 3/2} = -2, \quad E_{\pm 1/2} = -8. \tag{1.32}$$

The perturbation function $G_{22}(t)$ contains all possible energy differences in the following way

$$G_{22}(t) = 0.2 + 0.371\cos\omega t + 0.286\cos 2\omega t + 0.143\cos 3\omega t \tag{1.33}$$

with

$$\omega = \frac{1}{\hbar}(E_{\pm 3/2} - E_{\pm 1/2}) = 6\omega_Q = -\frac{3\pi}{10h} eQV_{zz}.$$

The intensities of the cosine functions result from summations over Clebsch-Gordon coefficients. For non-axially symmetric electric field gradients, both the precession frequencies as well as the intensities depend on the asymmetry parameter η. In this case the frequencies are no longer in the ratio 1:2:3. Figure 8 shows Fourier transformed perturbation functions for $I = 5/2$ versus η. It is clear that the ratio of the first harmonic to the fundamental unambiguously determines η while the second harmonic serves for peak assignments in multicomponent spectra since it is by definition the sum of the fundamental and its first harmonic. The relationship between this ratio R and η is as follows [14]

$$R(\eta) = \frac{1}{2}(\sqrt{3}\cot(\phi/3) - 1) \tag{1.34}$$

with

$$\begin{aligned}
\cos\phi &= q/r^3 \\
r &= \mathrm{sign}(q)\sqrt{|p|} \\
p &= -28\left(1 + \frac{\eta^2}{3}\right) \\
q &= -80(1 - \eta^2).
\end{aligned}$$

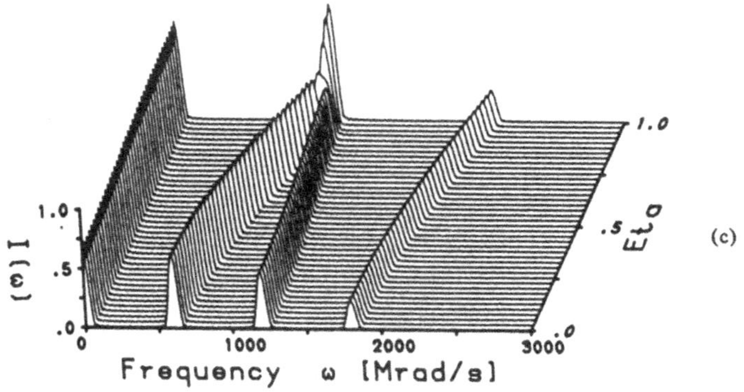

Fig. 8 Fourier transformed random powder perturbation functions verus asymmetry parameter η for pure quadrupole interaction and $I = 5/2$.

Analytic perturbation functions for single crystals and powder samples for static non-axially symmetric nuclear quadrupole interactions are given in [14].

Since the superposition of several spectral components in $G_{22}(t)$ often leads to very complicated patterns, a Fourier transformation of the time spectrum $R(t)$ is advantageous. Details concerning window functions, zero padding, and mirrowing are described in [15]. Any static distribution of NQI's will show up as linebroadening of the peaks. The peak at $\omega = 0$, the so-called hardcore, however, will not be broadened by static distributions of NQI's. The only way to obtain a broad $\omega = 0$ peak is reorientational motion. In this respect, TDPAC offers an unambiguous discrimination between static and dynamic broadening. We shall not discuss time-dependent hyperfine interactions in more detail because they are not relevant for the following paragraphs.

1.4. MÖSSBAUER EFFECT

Fluorescence is quite a common phenomenon in the visible or UV spectrum. However, resonance fluorescence of γ-rays emitted from radioactive nuclei was not discovered before 1957. The reason is rather simple: the recoil losses during γ-emission are many orders of magnitude larger than the natural linewidth $\Gamma = \hbar/\tau_N$ (τ_N = lifetime of the excited state). Similarly, a frequency shift to higher energies occurs for the absorption process for the very same reason. Thus, there is no overlap between the emitted spectrum and the absorption spectrum. However, in a solid there is a finite probability that the γ-emission occurs without recoil losses, i.e. the recoil is taken up by the entire solid rather than by a single atom. This recoil-free fraction is called Lamb-Mössbauer factor

$$f = e^{-k^2 <u^2>/3} \tag{1.35}$$

with $k = 2\pi/\lambda$ being the wave number of the γ-quantum and $< u^2 >$ being the mean squared displacement of the emitting atom. Via the f-factor information on the phonon density of states, i.e. on lattice dynamics is obtainable. Very often the Lamb-Mössbauer factor is appreciable at low temperatures only. Fortunately, this is not the case for the $6.2keV$ resonance in ^{181}Ta.

A typical Mössbauer setup consists of a single-line γ-source, which is movable, and an absorber followed by a γ-detector. By moving the source it is possible to Doppler-shift the γ-energy by very small amounts around the resonance frequency. The resonance effect can be detected in transmission geometry (loss of countrate at resonance because of isotropic reemission) or in backscattering geometry (detecting the fluorescence quanta). For further details the reader is referred to [16]. In the following we shall discuss two aspects only which are relevant to Mössbauer effect studies of CDW systems with special emphasis on ^{181}Ta.

1.4.1. *The isomer shift* For nuclei with a different mean squared radius of the excited and the groundstate there is a shift of the resonance frequency S which depends on the electron density difference at the nucleus between source and absorber

$$S = \frac{2}{5}\frac{\pi c}{E_\gamma} Z e^2 (<r^2>_{excited} - <r^2>_{ground}) \cdot (\rho_{absorber}(0) - \rho_{source}(0)). \quad (1.36)$$

This offers the unique possibility to deduce information on the valence state of the Mössbauer atom. However, an empirical calibration of S versus valence is required. For more details on the isomer shift the reader is referred to [17]. In that book the peculiarities of the $6.2keV$ resonance of ^{181}Ta can be found, too [18].

1.4.2. *Line splittings* The resonance line can be split by magnetic dipole and electric quadrupole interactions. A simple example for a quadrupole split line is ^{57}Fe: The first excited state has $I = 3/2$ and splits into $\pm 3/2$ and $\pm 1/2$ whereas the groundstate with $I = 1/2$ remains unsplit. Hence, there are two resonance transitions whose energy difference is given by eq.(1.23). The line intensities result from summations over Clebsch-Gordon coeffients. They depend – of course – on whether one is dealing with random powder samples or with single crystals. Usually, it is immediately possible to deduce the *sign* of the electric field gradient tensor component V_{zz} from the order of resonance lines (including line intensities), contrary to NMR and γ-γ-TDPAC.

If both the excited and the groundstate are split, the resonance line eventually splits into a large number of lines making the observation of the resonance effect difficult. An example for the ^{181}Ta NQI in rhenium metal is given in Figure 9 for two different orientations of the crystal with respect to the γ-ray emission direction. Note that in this example a split source ($=^{181}$Ta diffused into a Re single crystal) and a single-line absorber (Ta-foil) was used, contrary to the examples of TaS_2 discussed below. Hence, the ^{181}Ta$\underline{Re}$ and $^{181}TaS_2$ spectra are mirror images of one another despite the same sign of the electric field gradient. A deviation from axial symmetry of the electric field gradient tensor manifests itself in both a shift of the line position as well as in modified line intensities.

In priciple, the resonance lineshape is a Lorentzian. In the case of ^{181}Ta, the interference between photoelectric absorption and Mössbauer absorption followed by internal conversion leads to a dispersion-modified Lorentzian [19].

Relaxation effects due to fluctuating isomer shifts or hyperfine fields lead to characteristic modifications of the lineshapes. We shall not discuss these effects any further because they are not relevant for the following paragraphs. A severe drawback of Mössbauer spectroscopy and TDPAC which both use shortlived nuclear excited states is the fact that it is practically impossible to distort the system from equilibrium while being in the excited nuclear state by e.g. an external pulse in order to monitor the relaxation towards equilibrium, a feature which renders NMR so extremely powerful for dynamic studies.

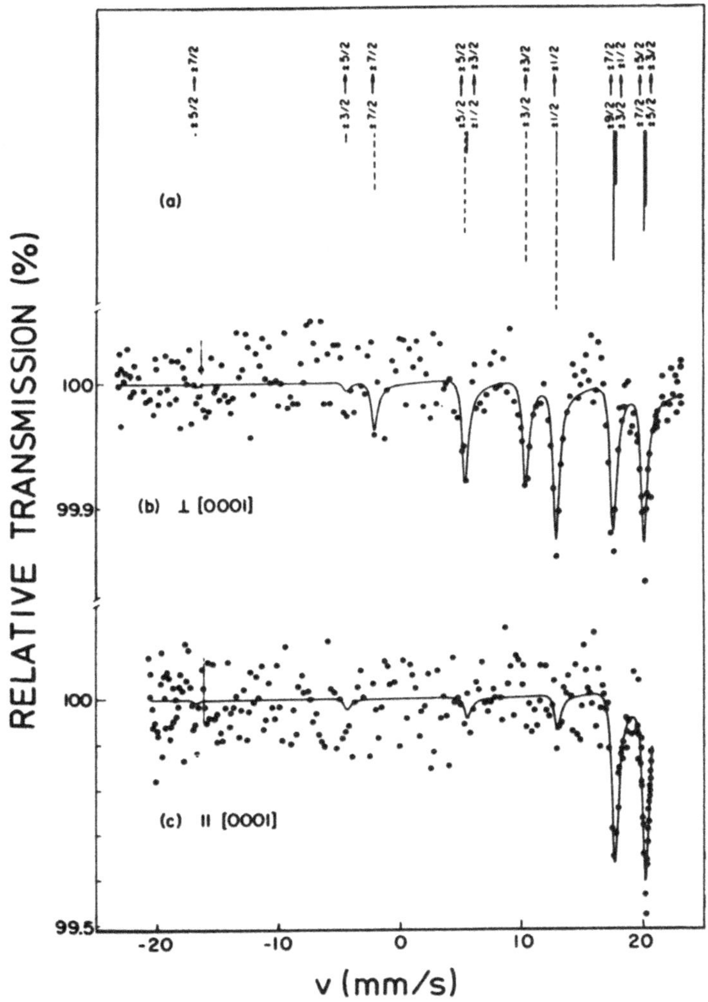

Fig. 9 ME spectrum for ^{181}Ta in a Re single crystal [Kaindl *et al.* [18]].
a) line positions and intensities: $\Delta m = 0$ dashed; $\Delta m = \pm 1$ solid.
b) γ-direction perpendicular to [0001] direction.
c) γ-direction parallel to [0001] direction.

Finally, Table 1.1 compares the features of the four kinds of nuclear spectroscopies decribed in this section, namely Mössbauer spectroscopy (ME), time differential perturbed angular correlations (TDPAC), nuclear quadrupole resonance (NQR), and nuclear magnetic resonance (NMR) as far as their application to ^{181}Ta is concerned. Each of the four spectroscopies has one weak point. ME and TDPAC do not have high enough resolution to resolve the fine structure due to the stacking ordering. NQR does not give quantitative information on the population ratios of inequivalent sites, and NMR does not give the magnitudes of the quadrupole interaction. Therefore the results from all of these spectroscopies have to be combined to obtain a complete set of data.

	ME	TDPAC	NQR	NMR
Quadrupole Frequency (ν_Q)	yes	yes	yes	no
Asymmetry Parameter (η)	yes	yes	yes	yes
Population	yes	yes	no	yes
Fine Structure due to Stacking	no	no	yes	yes

Table 1.1 Comparison of features of ME, TDPAC, NQR, and NMR for ^{181}Ta spectroscopy.

2. Structure of layered transition metal dichalcogenides

Dichalcogenides with layered structures are formed mainly by the early transition metals (group IVb, Vb and VIb) [20]. In these compounds the anions (S,Se) build up densely packed monolayers. Two of them are held together by the metal ions in octahedral (T-type) or trigonal prismatic holes (H-type) depending on the position of the chalcogen layers with respect to each other. In the first case, they are arranged in the same way as in the hexagonally close-packed structures; in the second case the sulfur atoms of the different layers lie on top of each other. The dichalcogenides of the group IV metals crystallize in the T-type structure (CdI_2-type). The compounds of Nb occur exclusively in a MoS_2-like structure (a stacking variant of the MoS_2-type), VSe_2 in the CdI_2-type. VS_2, which also exhibits the CdI_2-type structure, can only be prepared by deintercalation of $LiVS_2$ and is only metastable at room temperature [21]. At higher temperatures, only nonstoichiometric compounds can be prepared [22]. TaS_2 and $TaSe_2$ exist in both structural types. In addition, there are further polytypes (6R, $4H_b$), in which both structural elements alternate in subsequent layers. Among the tellurides only those of the IVb-metals (Ti, Zr, Hf) crystallize in the simple CdI_2-type structure. In the tellurides of the group Vb and VIb metals the Te-M-Te-layers are puckered due to metal-metal dimer or metal chain formation [23].

Due to their crystallographic and electronic structure [20] the compounds of the group IV metals, $TiSe_2$ and $TiTe_2$ exhibit a semimetallic or metallic character. With the exception of some tellurides and NbS_2 all of these metallic compounds show anomalies in the electronic transport properties, like conductivity, magnetic susceptibility, Hall-effect, Seebeck coefficient, and in the specific heat [24,25]. Below these temperatures superstructures appear. These superstructures can be commensurate (CCDW) or incommensurate (ICDW) with the underlying basic structure. In some examples a sequence of superstructures during cooldown is observed. The structural transitions are fully reversible, but sometimes with a large hysteresis. Apart from X-ray diffraction, these superstructures are more easily observable by electron diffraction. They are always observed with the incident electron beam perpendicular to the layer planes. In this way the size and symmmetry of the superstructure and the degree of commensurability can be deduced unambiguously. For the determination of the metal positions in the superstructure cell the intensity of the diffraction spots has to be taken into account. However, the intensities of the electron diffraction are difficult to evaluate due to multiple diffraction. Hence, only models of the metal distribution within the supercell can be given. For the construction of these models the charge density wave (CDW) theory serves as a guideline. Depending on the phasing of the charge density waves, different metal arrangements can be deduced which differ in the number of inequivalent sites and their population. In some cases an unambiguous decision between these models cannot be given from the electron diffraction pattern alone. The hyperfine techniques which are sensitive to the local electron distribution around the probe atom are suitable to determine the phasing unambiguously, provided the differences in the electron distribution at the different metal sites are large enough. Thus they complement the information obtained by the diffration method and allow a decision between the proposed models.

As far as incommensurate charge density waves are concerned, there are characteristic density distributions depending on the number of wave-vectors involved. An example for a single-**q** plane-wave model is the familiar $1/\sin x$-distribution. This density distribution translates into a distribution of the electric field gradient tensor. In a local picture the largest component of the electric field gradient tensor is proportional to the density and hence the line-profile is a direct measure of the density distribution. On the contrary, in a non-local picture both the largest component of the electric field gradient tensor and the asymmetry parameter will be distributed. In this case the actual wavelength of the ICDW as well as the CDW amplitude will affect the line-profiles. Finally, situations intermediate between commensurate and plane-wave incommensurate will lead to discrete spectral components superimposed onto the characteristic distributions. Thus a careful line-profile analysis is of crucial importance for a proper description of ICDW's or quasi CCDW's.

Finally, it should be possible to investigate CDW dynamics by NMR, NQR, TD-PAC, and Mössbauer spectroscopy. However, apart from spin-lattice relaxation time measurements in various compounds, no information on CDW dynamics has been obtained so far with these techniques in the layered transition metal dichalcogenides, contrary to $NbSe_3$ and the Mo-bronzes to be described in chapters 4 and 5.

3. Compounds with octahedral metal coordination

3.1. 1T-$TiSe_2$

Group IVb layered transition metal dichalcogenides are usually semiconductors in which there is an energy gap between the chalcogen p band and the transition metal d band. In 1T-$TiSe_2$, however, these p and d bands overlap, which leads to a semimetallic behaviour in this compound. 1T-$TiSe_2$ shows a phase transition at $T_0 = 202$K, below which a $2a_0 \times 2a_0 \times 2c_0$ commensurate CDW appears [26,27]. In the case of the CDW formation in 1T-$TiSe_2$, the nesting between the hole pocket in the Se_{2p} band and the electron pocket in the Ti_{3d} band is realized, and hence 1T-$TiSe_2$ may be called an excitonic insulator.

The Ti nucleus is not suitable for NMR measurements. Hence, only the ^{77}Se ($I = 1/2$, $\gamma_N/2\pi = 0.8118$MHz/kOe) NMR measurements of 1T-$TiSe_2$ have been performed. Figure 10 shows the Se NMR spectra taken at various temperatures on 1T-$TiSe_2$ powders by Dupree *et al.* [28]. Their sample was prepared using Se enriched to 94.4% in ^{77}Se in order to raise the NMR sensitivity. Above T_0, the line shape shows the characteristic powder pattern produced by an anisotropic Knight shift when the nuclei occupy sites of axial symmetry. As the sample is cooled below T_0, the line shape begins to change progressively, eventually reaching a new temperature independent profile below about 100K. The low temperature line shape is recognizable as characteristic of an anisotropic Knight shift in non-axial symmetry.

In general, an anisotropic Knight shift can be expressed in the following form:

$$K(\theta,\phi) = K_{iso} + K_{ax}(3\cos^2\theta - 1) + K_{asym}\sin^2\theta\cos 2\phi \tag{3.1}$$

where (θ,ϕ) describes the orientation of the principal axes of the shift tensor with respect to the magnetic field. The experimental line shapes can be parametrized in terms of three shift values K_1, K_2 and K_3 as indicated in the bottom spectrum of Figure 10. For axial symmetry there are two parameters $K_\parallel = K_3$ and $K_\perp = K_1 = K_2$. When eq.(3.1) is averaged over all values of θ and ϕ, as appropriate for a powder sample, the resulting line shape yields the following correspondences:

$$K_{iso} = \frac{1}{3}(K_1 + K_2 + K_3) \xrightarrow{\text{axial symmetry}} \frac{1}{3}(2K_\perp + K_\parallel)/3 \tag{3.2a}$$

$$K_{ax} = \frac{1}{3}(K_3 - \frac{K_2}{2} - \frac{K_1}{2}) \xrightarrow{\text{axial symmetry}} \frac{1}{3}(K_\parallel - K_\perp) \tag{3.2b}$$

$$K_{asym} = \frac{1}{2}(K_2 - K_1) \xrightarrow{\text{axial symmetry}} 0 \tag{3.2c}$$

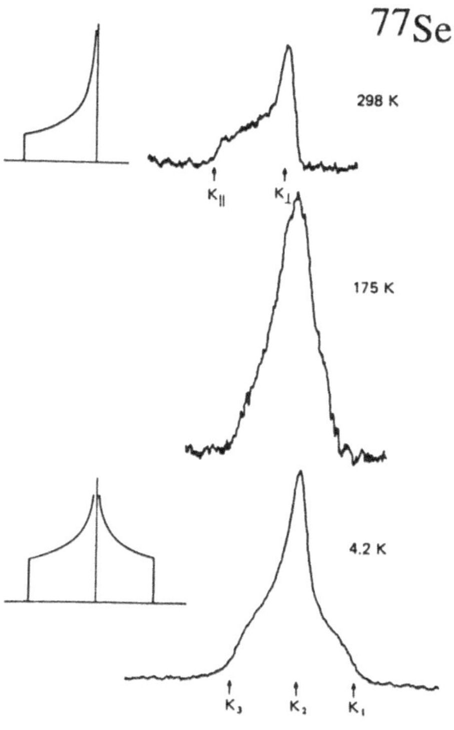

Fig. 10 ^{77}**Se NMR line shape of 1T-$TiSe_2$ at representative temperatures above and below the phase transition (T_0 = 202K). The high-temperature (axial) and low-temperature (nonaxial) Knight shift components are indicated by arrows. Theoretical line shapes in the absence of dipolar broadening are shown on the left-hand side for axial and non-axial symmetry [Dupree *et al.* [28]].**

The values of K_{iso}, K_{ax}, and K_{asym} are obtained by these equations from the values of K_1, K_2, and K_3. Their temperature dependences are shown in Figure 11.

In the low-temperature phase, the CDW formation leads to two inequivalent Se sites as shown in Figure 12. The majority (3/4) site is not axially symmetric, while the minority (1/4) site is axially symmetric. The axial resonance line of the minority site nuclei is evident only as an additional asymmetry in the low-temperature line shape. The sharp peak of the axial resonance profile must nearly coincide with the central peak of the non-axial majority resonance since only one main peak is observed. In the above-mentioned analysis for the low-temeperature spectrum, the presence of a minority-site resonance is disregarded and the values of K_{iso},

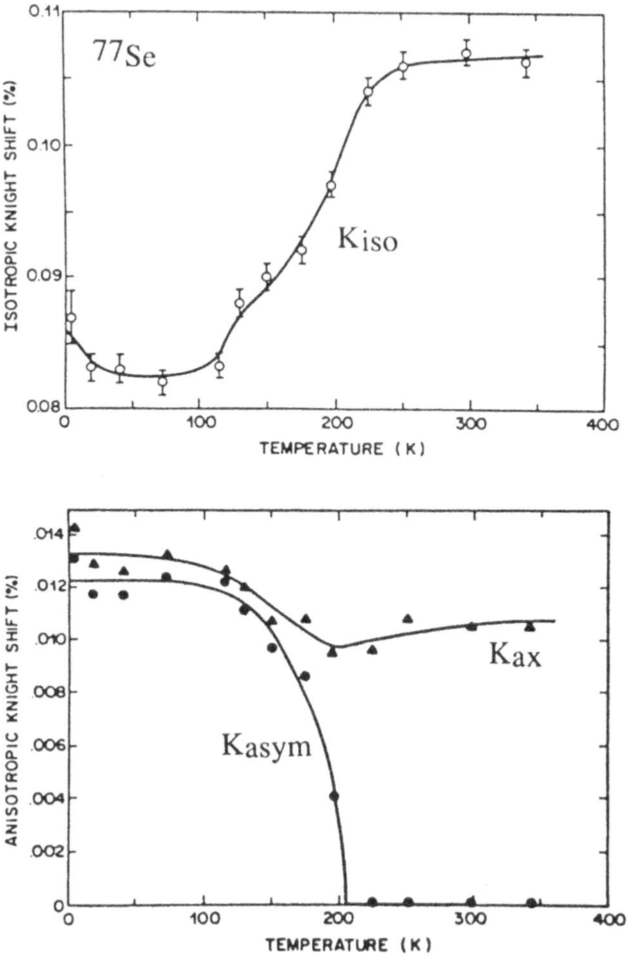

Fig. 11 Temperature dependences of K_{iso}, K_{ax} and K_{asym} of ^{77}Se in 1T-$TiSe_2$ [Dupree *et al.* [28]].

K_{ax}, and K_{asym} represent shifts for the majority sites. The isotropic Knight shift K_{iso} descreases rapidly below T_0. The axial shift K_{ax} exhibits a weak temperature dependence, and the non-axial component K_{asym} vanishes in the high-temperature phase, but increases rapidly below T_0. The temperature dependence of K_{asym} is similar to that of the order parameter for the second order transition.

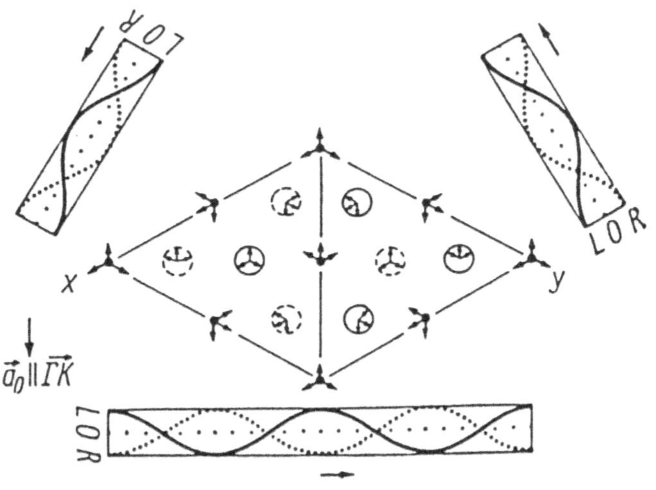

Fig. 12 Schematic representation of the low-temperature displacements of Ti and Se in the CDW state of 1T-$TiSe_2$. *Dots:* Ti, *solid circles:* upper Se, and *broken circles:* lower Se [Wilson [27]].

The temperature dependence of the nuclear spin-lattice relaxation time was also measured by Dupree *et al.* [28]. Figure 13 shows the plot of $(T_1T)^{-1}$ vs T. The relaxation rate $(T_1T)^{-1}$ decreases rapidly by a factor $\sim$ 25 between room temperature and 77K, and then rises sharply to a maximum between 4.2K and 10K. The observed rapid rise below 40K may be of magnetic origin, the detail of which remains an open question. Except for this low-temperature rise of $(T_1T)^{-1}$ the behaviour of both K_{iso} and $(T_1T)^{-1}$ is similar to that in 1T-VSe_2, indicating that the density of states at the Fermi level, at least of the s electron character at Se sites, seems to decrease below T_0 and almost vanishes in the liquid nitrogen temperature region due to CDW formation.

3.2. 1T-VSe_2

1T-VSe_2 exhibits a continuous charge density wave transition at T_0=110K accompanied by a small resistivity hump [29-31]. Although this compound is of the

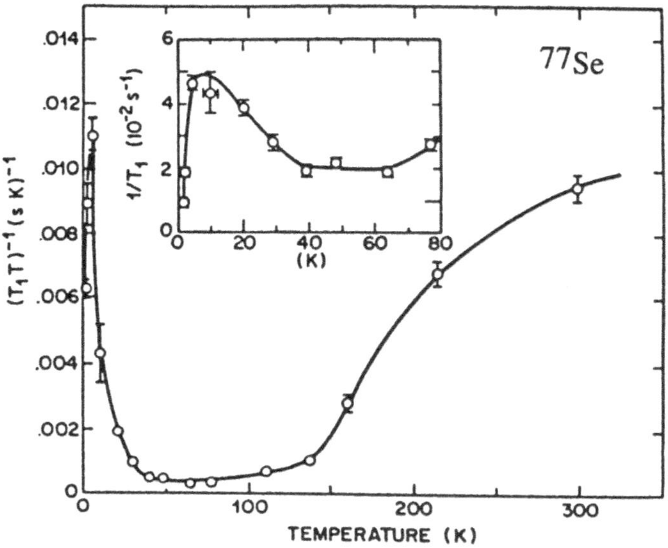

Fig. 13 Temperature dependence of $\frac{1}{T_1T}$ of ^{77}Se in 1T-$TiSe_2$ [Dupree *et al.* [28]].

1T polytype, not only the transition behaviour but also the CDW structure itself are quite different from those observed in 1T-TaX_2. The CDW wave vector is $0.25\mathbf{a}^* + 0.314\mathbf{c}^*$ above 85K, and $0.25\mathbf{a}^* + 0.307\mathbf{c}^*$ below 85K [32]. This means, the superlattice is commensurate (4×4) in the plane, but incommensurate along the c-axis. Although the c^* component of the CDW wave vector remains incommensurate down to the lowest temperature investigated, it shows a slight abrupt change in the vicinity of 85K.

The ^{51}V nucleus ($I = 7/2, \gamma_N/2\pi = 1.1193$MHz/kOe, and $Q = -0.04 \times 10^{-24}$ cm^2) has a good NMR sensitivity. The NMR measurement of the ^{77}Se nucleus ($I = 1/2, \gamma_N/2\pi = 0.8118$MHz/kOe) is also possible, although the natural abundance of ^{77}Se is low (7.5%). However, 1T-VSe_2 has received less attention compared to 1T-TaX_2, and hence there are not many NMR studies on 1T-VSe_2. This is partly due to the fact that the preparation of highly stoichiometric specimens of 1T-VSe_2 is very difficult. Thompson and Silbernagel first reported the V and Se NMR on

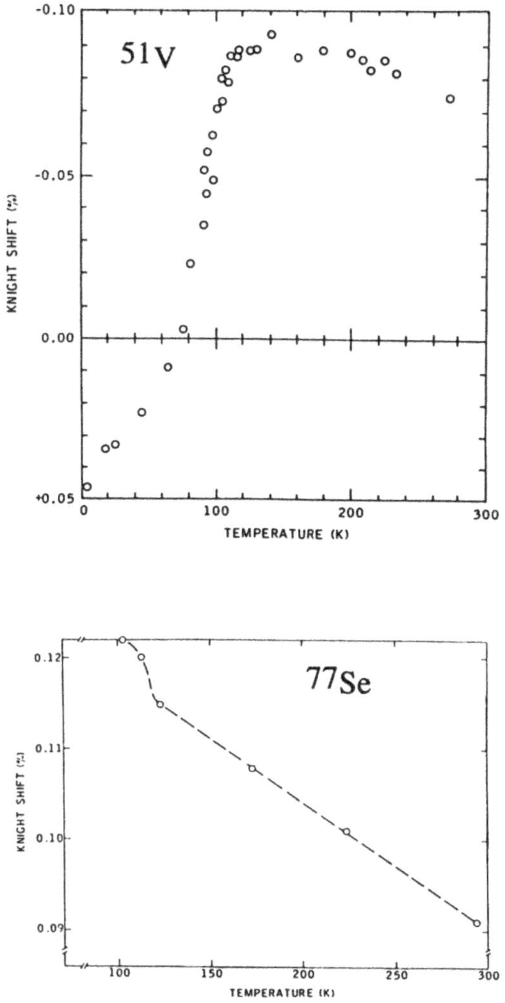

Fig. 14 Temperature dependences of the ^{51}V (a) and ^{77}Se (b) Knight shifts in 1T-VSe_2 [Thompson *et al.* [33]].

1T-VSe_2 [33]. They measured the temperature dependences of the Knight shift (K_{iso}: the isotropic part of the Knight shift) of ^{51}V and ^{77}Se for 1T-VSe_2 powders, and found a sharp drop below T_0 for V and a small break in the slope at T_0 for Se as shown in Figure 14. These results were essentially confirmed by the following studies by Tsuda *et al.* [34] using single crystals of 1T-VSe_2. Tsuda *et al.* also measured the temperature dependence of the nuclear spin-lattice relaxation time (T_1) in the "magic angle" configuration ($3\cos^2\theta = 1, \theta = 54.7°$) at which the first-order quadrupole splittings vanish and the spectrum becomes a single line. The result is plotted in the form of $(T_1 T)^{-1}$ *vs* T in Figure 15, which also shows a sharp drop below T_0. These observations in the Knight shift and the relaxation time can be understood systematically as a 30% reduction of the density of states due to the opening of partial gaps at the Fermi surface associated with the CDW formation.

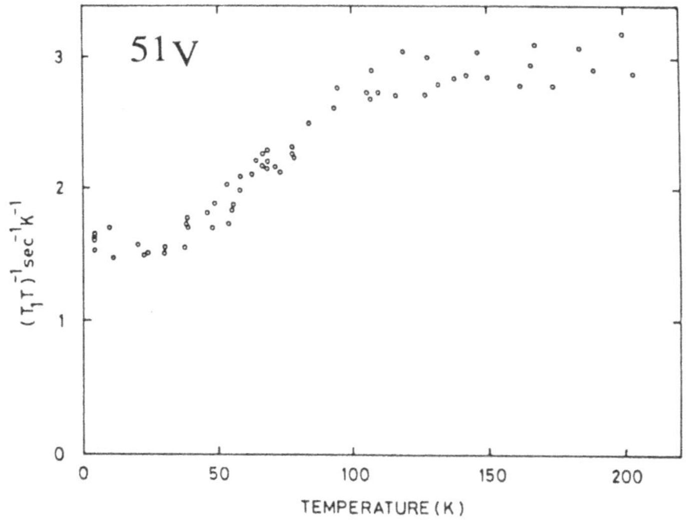

Fig. 15 Temperature dependence of $\frac{1}{T_1T}$ of ^{51}V in 1T-VSe_2 (Tsuda *et al.* [34]).

Below T_0, a significant broadening is observed in the central line and especially in quadrupole satellites for the ^{51}V NMR spectrum, indicating the existence of a Knight shift and an EFG distribution at vanadium sites due to a spatial modulation of charge density. The detailed analysis of the lineshape of the first quadrupole satellite, the evolution of which is shown in Figure 16, was performed by Skripov *et al.*[35]. The following distribution is assumed for the quadrupole coupling constant,

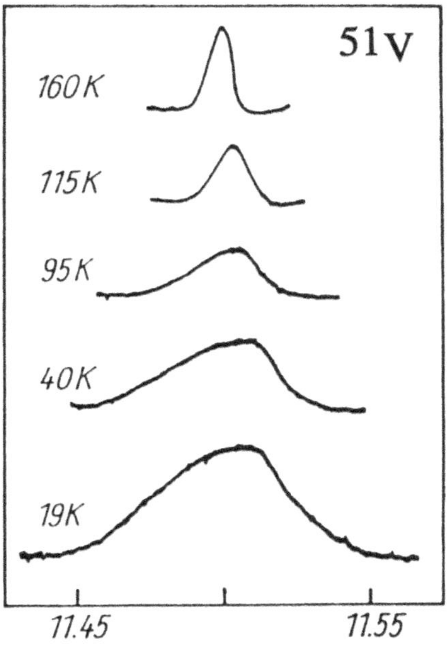

Fig. 16 Evolution of the lineshape of the first quadrupole satellite of ^{51}V NMR in 1T-VSe [Skripov *et al.* [35]].

$\nu_Q = e^2qQ/h$, in a single layer,

$$\nu_Q(r) = \nu_{Q^0} + \nu_{Q^1} \sum_{i=1}^{3} \cos(\mathbf{q}_i \cdot \mathbf{r} + \phi_i) \tag{3.3}$$

where ν_{Q^0} is the quadrupole coupling constant in the absence of the CDW, ν_{Q^1} is the ν_Q modulation amplitude proportional to the CDW amplitude, $\mathbf{q}_i(|\mathbf{q}_i| = |\mathbf{a}^*|/4)$ are the CDW in-plane wave vectors and ϕ_i are the CDW phases. The CDW of 1T-VSe_2 is incommensurate in the c-axis direction. In this case, each of the three phases ϕ_i changes from layer to layer scanning the interval $(0,\ 2\pi)$ continuously. The resulting ν_Q distribution strongly depends on the relative phasing of the three CDW's within a layer. Figure 17(a) shows the ν_Q distribution calculated for the hexagonal symmetry of the CDW ($\phi_1 = \phi_2 = \phi_3$). This distribution clearly does not agree with the experimental one which has no pronounced structure. On the other hand, the orthorhombic symmetry ($\phi_2 = \phi_3, \phi_1 = \phi_2 + \Delta\phi$) of the CDW is also possible as will be discussed for 2H-TaX_2 in sections 4.1.1., 4.1.2.. Figure 17(b) shows the ν_Q distribution calculated for orthorhombic symmetry with $\Delta\phi = 2\pi/3$. If the background broadening is taken into account, this distribution is in better agreement with the spectrum in Figure 16 than the hexagonal one. Furthermore, the CDW amplitude can be estimated from the linewidth of the observed spectrum as $\Delta n/n(= 2\nu_{Q^1}/\nu_{Q^0}) \sim 0.12$.

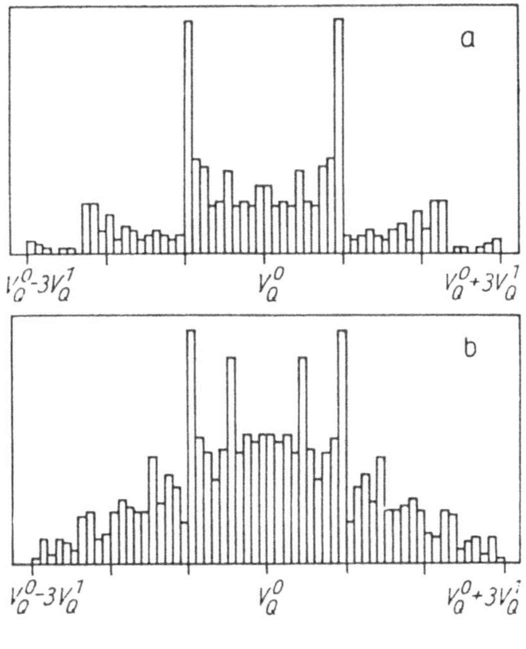

Fig. 17 ν_Q distribution calculated for the CDW state for 1T-VSe_2: (a) hexagonal symmetry ($\phi_1 = \phi_2 = \phi_3$), orthorhombic symmetry ($\phi_2 = \phi_3, \phi_1 = \phi_2 + 2\pi/3$) [Skripov *et al.* [35]].

3.3. TANTALUM DICHALCOGENIDES

Of the layered transition-metal dichalcogenides, Ta-dichalcogenides are most extensively investigated with respect to charge density waves. This section describes the nuclear spectroscopy on the charge density waves of 1T-modifications of Ta-dichalcogenides.

The layered compounds 1T-$TaSe_2$ and 1T-TaS_2 exhibit successive charge density wave transitions and attain a commensurate $\sqrt{13} \times \sqrt{13}$ superstructure at low temperatures. This 3q-superstructure leads to a charge density pattern shown in Figure 18, both of which show the formation of a so-called "13-atom star-of-David cluster" [36]. Ignoring the stacking ordering, the undistored star-of-David cluster consists of three inequivalent Ta atoms with populations $\alpha : \beta : \gamma = 1 : 6 : 6$, that is, a central site (α), an inner ring of six sites (β) moving towards α, and an outer ring of six next-nearest-neighbour sites (γ) which also shift inwards. The stacking ordering of the CDW makes the β and γ sites further split and yields the intra-cluster fine structure.

Figure 19 summarizes the behaviour of the resistivity and the susceptibility due to the successive CDW transitions in 1T-$TaSe_2$ and 1T-TaS_2 [24]. Both compounds show discontinuous jumps with hysteresis in the resistivity and the susceptibility at the first order phase transitions of incommensurate to commensurate (IC-C) in 1T-$TaSe_2$, or incommensurate to nearly commensurate (IC-NC) and nearly commensurate to commensurate (NC-C) in 1T-TaS_2. The behaviour at these transitions, namely the rise in the resistivity and the fall in the susceptibility as the temperature is lowered, seem to correspond to the decrease in the density of states of conduction electrons at the Fermi level. With the exception that 1T-TaS_2 shows more complicated successive phase transitions (the existence of NC and T phases) than 1T-$TaSe_2$, the charge density waves of 1T-$TaSe_2$ and 1T-TaS_2 are similar to each other in many aspects such as the CDW wave vectors, the in-plane charge density pattern and the lattice distortion pattern (the formation of 13-atom star-of-David clusters).

In spite of these similarities, there exists a marked difference between the two compounds in the low-temperature resistivity behaviour as seen in Figure 19. Namely, 1T-$TaSe_2$ is metallic while 1T-TaS_2 is semiconducting and has a low-temperature resistivity three orders of magnitude higher than 1T-$TaSe_2$. The semiconductive behaviour of 1T-TaS_2 has been one of the most puzzling problems in the charge density waves of layered transition metal dichalcogenides for a long time [38]. It is because generally the charge densitity wave formulation in two-dimensional (2D) metals does not destroy the Fermi surface completely, and hence does not give rise to a metal-insulator transition. This general rule is true for almost all the transition metal dichalcogenides. The only exception is the 1T-modification of TaS_2. This problem will be discussed in section 3.3.2..

Historically, in the initial stage, the structural studies of the charge density waves in these compounds were made for the most part by diffraction experiments such as electron [24,37,39], X-ray [36] and neutron diffractions [40], from which the above star-of-David cluster formation has been deduced. A complementary approach is to look microscopically and directly at the spatial variations of the local environ-

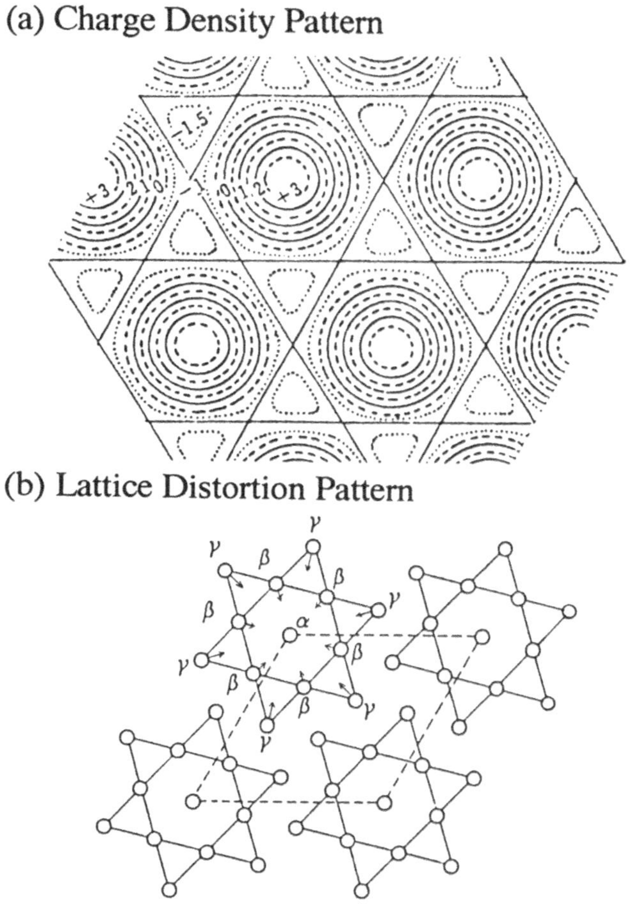

Fig. 18 Charge density pattern and lattice distortion pattern for 1T-TaX_2.

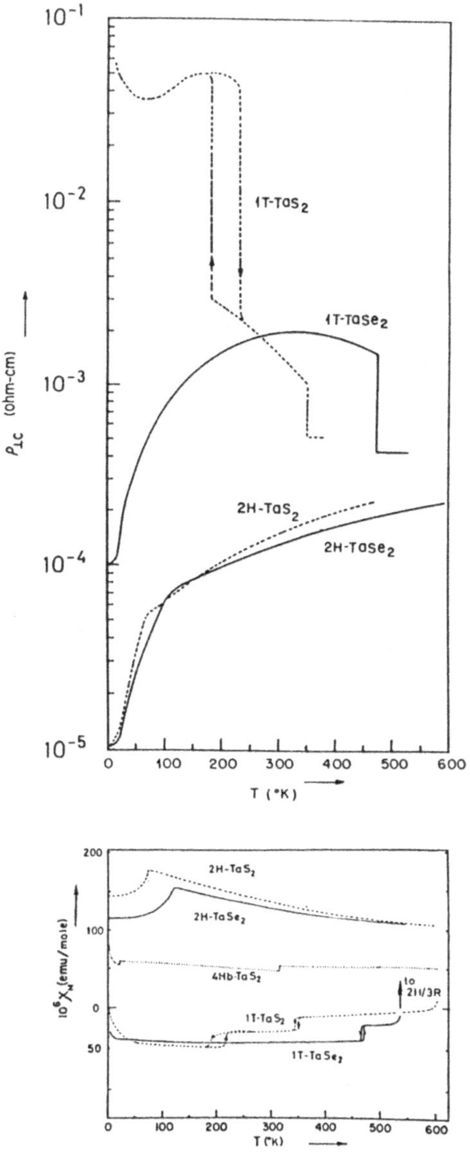

Fig. 19 Resistivity and susceptibility of 1T-$TaSe_2$ and 1T-TaS_2 [Wilson *et al.* [2]].

ment (e.g. Ta 5*d* conduction electron density, electric field gradient etc.) through their effects on the probe atoms. The nuclear spectroscopy is one of such means. Section 3.3 describes the results of the nuclear spectroscopic studies on the charge density waves of 1T-$TaSe_2$ and 1T-TaS_2, stressing the following two points. The first is to corroborate the formation of 13-atom star-of-David clusters conjectured by diffraction studies. The second is to find such differences in the charge density wave structure between 1T-$TaSe_2$ and 1T-TaS_2 as to give a reasonable explanation of the differences in their electrical properties.

The first subsection deals with the commensurate CDW state of 1T-$TaSe_2$. The next subsection starts with a detailed description of the commensurate CDW state of 1T-TaS_2 with a comparison between the sulfide and the selenide. Subsequently, further information regarding the nearly-commensurate, the incommensurate CDW, and the T-phase will be presented. Finally, first results on CDW phenomena in intercalation compounds of 1T-TaS_2 will be described.

3.3.1. *1T-TaSe*$_2$

ME: The first ME measurements on 1T-$TaSe_2$ were made on the 1T-$Fe_xTa_{1-x}Se_2$ system using the ^{57}Fe nucleus by Eibschütz and DiSalvo [41]. The obtained spectra definitely showed that the isomer shift and the quadrupole splitting of ^{57}Fe impurity sites are modulated by the CDW. But the spectra were rather smeared and the lines were not well-split. Eibschütz and DiSalvo suggested that the spectra can be fit with the "1:6:6" model, although their conclusion was that the electron density is minimum at the central site, which is opposite to the star-of-David cluster model.

More extensive ME measurements have lately been made by Pfeiffer *et al.* using the ^{181}Ta nucleus [42]. Figure 20 shows the obtained spectrum of 1T-$TaSe_2$ in the commensurate CDW phase, compared with the spectrum of 2H-$TaSe_2$ at 300K which is in the normal phase. The ME resonance of ^{181}Ta makes use of the 6.2keV γ transition ($T_{1/2}$ = 6.8μsec) between the I = 7/2 ground and I = 9/2 excited nuclear levels. Therefore the Mössbauer spectrum of a Ta site shows a series of quadrupole split lines. As seen in Figure 20, the spectrum in the normal phase of 2H-$TaSe_2$ has one series of quadrupole split lines, indicating that there exists only one Ta site. On the other hand, the spectrum in the CCDW phase of 1T-$TaSe_2$ shows two series of quadrupole split lines, namely the existence of two inequivalent Ta sites. A theoretical fit to the spectrum locates the isomer shifts of the sites at the arrows and gives the magnitudes of the quadrupole interaction which are accidentally very close to each other for the two sites. The asymmetry parameters were assumed to be zero in the data analysis. The resonance areas of the two sites are equal within the experimental accuracy. No third Ta site was found in this experiment, but its presence might be obscured at this level of statistical noise as suggested by Pfeiffer *et al.*. They also estimated the valence of Ta for the observed two sites from the empirical relation (Table 3.1) between the formal Ta valence and the isomer shift. The two sites have an isomer shift between Ta^{4+} and Ta^{5+}, and are separated by 0.34 electrons (see Figure 20). Furthermore, they tried a search for the missing third Ta site in the neighbourhood of Ta^{5+} with improved statistical accuracy, but failed. This implies that the unobserved site must be located on the Ta^{3+} side of the main site pair. This result on the Ta valence indicates that the electron density maximum is located at the central site, consistent with the 13-atom star-of-David cluster model.

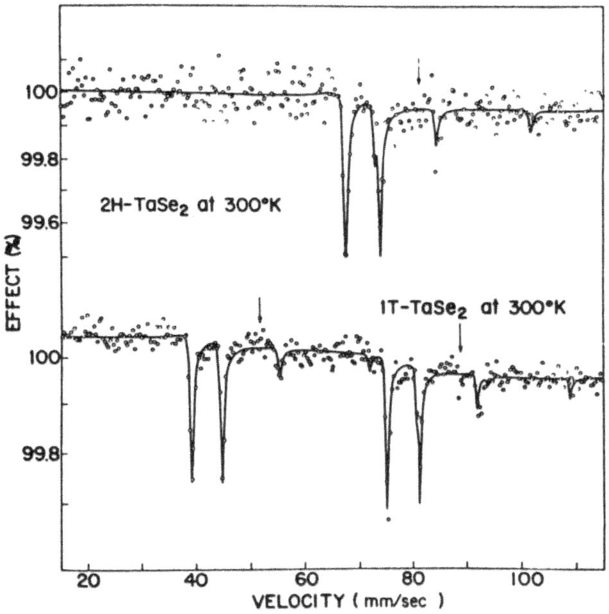

Fig. 20 Mössbauer spectrum of ^{181}Ta at 300K in 1T-$TaSe_2$ and 2H-$TaSe_2$ [Pfeiffer *et al.* [42]]. (Parameters in milimeters per second). 1T-$TaSe_2$ *inner*: linewidth=0.6(0.1), shift=+52.7(1), e^2qQ = 230(1); *outer* : width=0.7(0.1), shift=+89.1(1), e^2qQ = 239(1). 2H-$TaSe_2$: width=0.7(0.1), shift= +81.9(0.1), e^2qQ = 243(1).

TDPAC: TDPAC investigations of 1T-$TaSe_2$ have been performed by Saibene [43]. A room temperature Fourier transformed time spectrum of 1T-$TaSe_2$ is shown in Figure 21. These data were obtained with NaI(Tl)-detectors with a rather limited time resolution ($\sim$ 2.5ns full width at half maximum). Therefore, the first harmonics are strongly attenuated and the second harmonics are no longer visible. It is clear from this figure that the NQI at the two peripheral sites is very similar, as previously noted in the ME experiments, and the two sites are unresolved (the lineshape of the first harmonic around 1550 Mrad/s suggests a double line with equal

		Isomer shift [mm/sec]	Ta coordination	Bond length [Å]
Ta^{4+}	$TaSe_2$(2H)	82.7	6-Se	2.60
	TaS_2(2H)	70.3	6-S	2.46
	TaC	70.8	6-C	2.23
Ta^{5+}	Cu_3TaSe_4	32.6	4-Se	2.31
	Cu_3TaS_4	15.2	4-S	2.26
	$KTaO_3$	-8.1	6-O	1.99
	$NaTaO_3$	-15.4	6-O	1.94
	$LiTaO_3$	-17.9	6-O	2.11

Table 3.1 Relation between the valence and the isomer shift of Ta (ME).

intensity). The asymmetry parameters are around 0.25 for both sites provided the left/right part of the unresolved fundamental is associated to the left/right part of the first harmonic (this is correct, as shown by NQR studies). Apart from the prominent peak there is another small peak at 570 Mrad/s with 10(2)% relative intensity. Undoubtedly, this has to be assigned to the central site of the 13-atom star-of-David cluster. Due to insufficient accuracy, the asymmetry parameter of this site could not be determined (it was assumed to be zero for the data analysis in order to derive the population of this site; a small asymmetry parameter, as derived from NQR measurements, would have very little effect on this population).

Data have also been obtained at 77K and at temperatures above room temperature up to the shear transformation to 2H-$TaSe_2$ around 440K (there are also data for 1T-$TaSe_2$ in the temperature range from about 850K up to 1273K where no CDW exists). The derived V_{zz},η-values for the CDW phase agree with those derived form ME and NQR studies within the experimental accuracy. The resolution, of course, is rather limited.

NQR and NMR: NQR and NMR experiments on 1T-modifications for Ta dichalgonides were systematically performed by Naito *et al.* [44]. As mentioned before, the NQR and NMR spectroscopies have a resolution high enough to observe the fine stucture due to the stacking ordering of the CDW. Therefore, the results of NMR and NQR are described with emphasis on this point. Figure 22 shows the CDW stacking schemes possible in the 1T modifications of Ta dichalcogenides. Taking into account the stacking order, the peripheral two six-atom sites β and γ of the 13-atom star-of-David cluster may further split as shown in Figure 23. X-ray and electron diffraction studies show that 1T-$TaSe_2$ has the $c_0 + 2a_0$ stacking. Namely, the $(l+1)$-th layer CDW is translated by $2a_0$ relative to the l-th layer CDW. In this case, both the β and γ sites are split into three, and as a total, 7 inequivalent

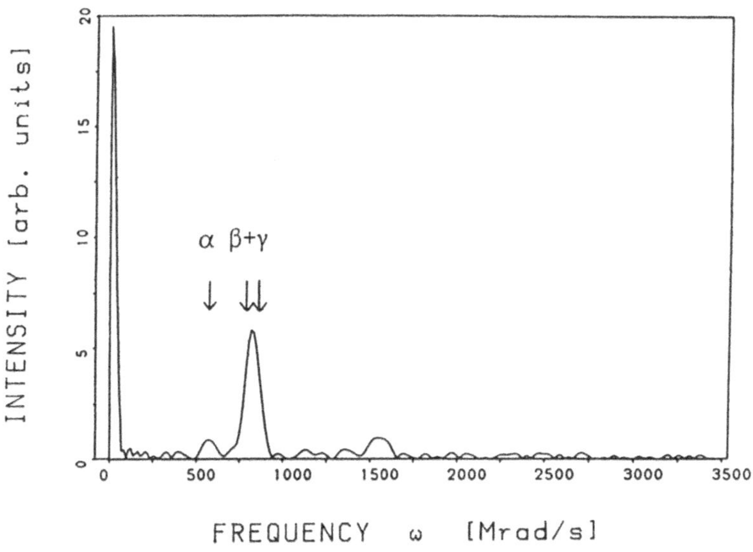

Fig. 21 Fourier transformed TDPAC spectrum for 1T-TaSe$_2$ at room temperature [43].

Ta sites of $\alpha : \beta_1 : \beta_2 : \beta_3 : \gamma_1 : \gamma_2 : \gamma_3 = 1 : 2 : 2 : 2 : 2 : 2 : 2$ result (Figure 23(b)).

Figure 24 shows the NQR spectrum of ^{181}Ta ($I = 7/2$) of 1T-TaSe$_2$ powders at 4.2K. Six fairy strong and sharp lines are observed in the frequency region between 90MHz and 115MHz, denoted by A'-F', and another six strong lines are observed in the frequency region between 160MHz and 175MHz, denoted by A"-F". Furthermore, weak lines are observed at 63.6MHz and 113.7MHz, denoted by Z' and Z". This spectrum is understood consistently with the $c_0 + 2a_0$ stacked star-of-David cluster model. Namely, the Z' and Z" are assigned to the $1/2 \leftrightarrow 3/2$ and $3/2 \leftrightarrow 5/2$ transitions of the central α site. The A'-F' and A"-F" lines are assigned to the $1/2 \leftrightarrow 3/2$ and $3/2 \leftrightarrow 5/2$ transitions of the β and γ sites, each split into three by the stacking ordering as mentioned above. Accidentally, the quadrupole frequencies and asymmetry parameters for the β and γ sites of 1T-TaSe$_2$ are very close to each other, which was already noted previously in the ME and TDPAC measurements, so the β and γ species are not well-separated even in the NQR spectrum.

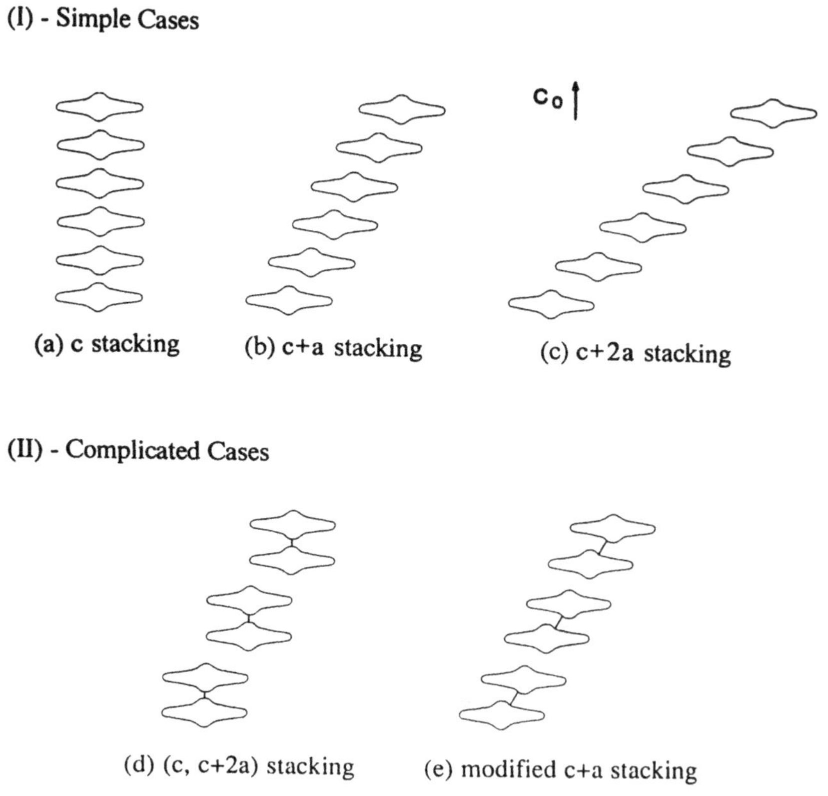

Fig. 22 Various stacking schemes for the CDW leading to an undistorted 1:6:6 pattern (star-of-David cluster) in 1T-TaX_2. *Upper*: simple stacking sequences. *Lower*: complicated stacking sequences which have some kind of "bonding" character (indicated by bars) between two adjacent CDW layers.

The NMR spectrum is quite consistent with the NQR result. Figure 25 shows the ^{181}Ta NMR spectrum of a 1T-$TaSe_2$ single crystal taken at 4.2K for the magnetic field perpendicular to the layer at 22MHz. The spectrum consists of six lines, A_s-F_s, which are assigned to the $\pm3/2$ transition of six inequivalent Ta sites corresponding to the A-F lines observed in the above mentioned NQR spectrum. The six lines observed in the NMR spectrum appear to have different intensities. However, if the

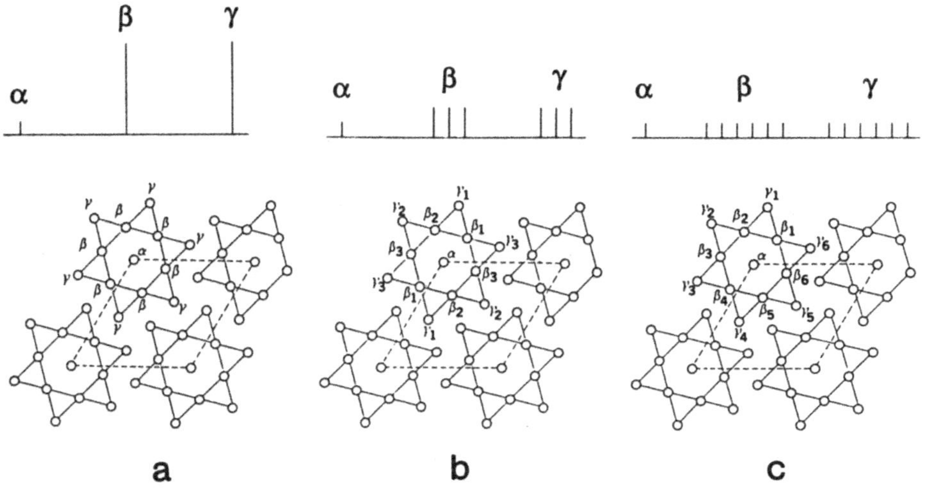

Fig. 23 Configuration of inequivalent Ta sites in a star-of-David cluster: (a) no interlayer coupling or c_0 stacking case ($\alpha : \beta : \gamma = 1 : 6 : 6$), (b) simpe $c_0 + a_0$ of $c_0 + 2a_0$ stacking case ($\alpha : \beta_1 : \beta_2 : \beta_3 : \gamma_1 : \gamma_2 : \gamma_3 = 1 : 2 : 2 : 2 : 2 : 2 : 2$), and (c) $(c_0, c_0 + 2a_0)$ or modified $c_0 + a_0$ stacking case ($\alpha : \beta_1 : \beta_2 : \beta_3 : \beta_4 : \beta_5 : \beta_6 : \gamma_1 : \gamma_2 : \gamma_3 : \gamma_4 : \gamma_5 : \gamma_6 = 1 : 1 : 1 : 1 : 1 : 1 : 1 : 1 : 1 : 1 : 1 : 1 : 1$).

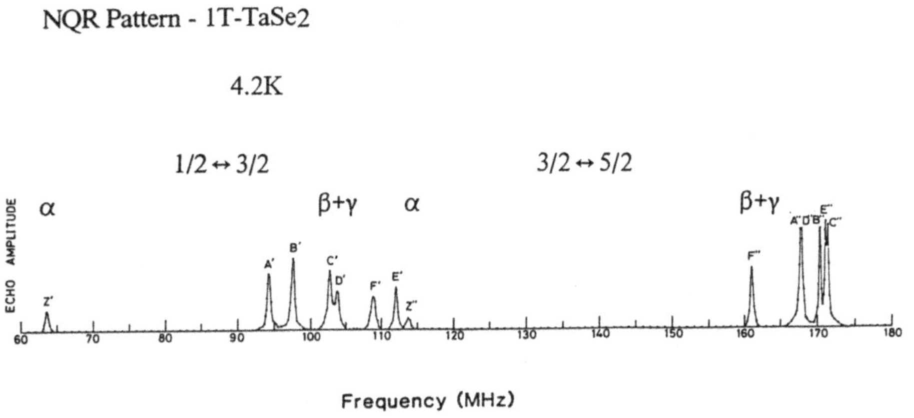

Fig. 24 NQR spectrum of ^{181}Ta in a 1T-TaSe$_2$ powder specimen taken at 4.2K [Naito *et al.* [44]].

apparent intensities are corrected by the η-dependence of the transition probability, then the populations of the six sites are equal within the experimental accuracy. The α site was not observed in this spectrum, which may be due to too low intensity.

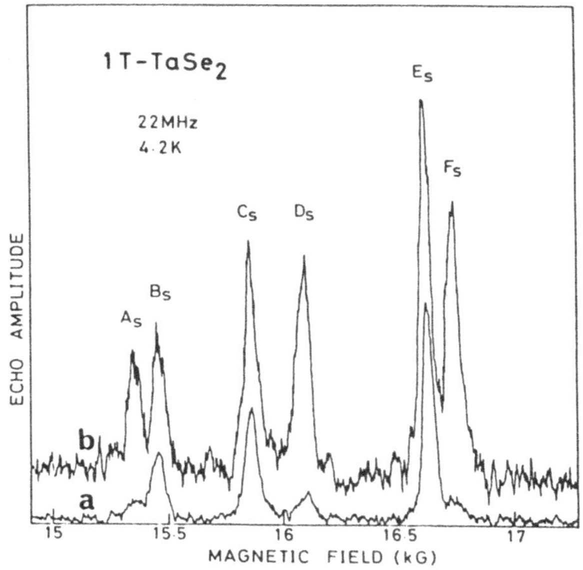

Fig. 25 NMR spectrum of ^{181}Ta in a 1T-$TaSe_2$ single crystal taken at 4.2K for the magnetic field perpendicular to the layer at 22MHz. The time interval between $\pi/2 - \pi$ pulse sequences is 20msec (a) and 1sec (b).

Figure 25 also compares two spectra which were taken under different experimental conditions: the time interval between the $\pi/2 - \pi$ pulse sequences is 20msec in (a) and 1 sec in (b). This comparison indicates that the spin-lattice relaxation time (T_1) of the lines A_s, D_s and F_s is significantly longer than that of the lines B_s, C_s and E_s. This significant difference in T_1 suggests one way to divide the six lines which are similar to one another both with respect to the quadrupole frequency and asymmetry parameter into two groups (β and γ).

In 1T-$TaSe_2$, there are a few conduction electrons left even in the commensurate CDW phase. In this case the dominant relaxation process of nuclear spins is due to the coupling to the conduction electrons, and the following Korringa relation holds for T_1,

$$\frac{1}{T_1 T} \propto \{N(E_F)\}^2 < |u_k(0)|^2 >_{E_F} \tag{3.4}$$

where $N(E_F)$ is the density of states of conduction electrons at the Fermi level and $< |u_k(0)|^2 >_{E_F}$ the Fermi-surface averaged density of the Bloch functions at the nucleus [45]. It can be supposed that the distribution of the conduction electrons left in the CCDW phase may not be uniform: high at the β and low at the γ sites or vice versa.

Summary: The above results of the four nuclear spectroscopies on 1T-$TaSe_2$ confirm that the star-of-David cluster model with the $c_0 + 2a_0$ stacking describes the commensurate CDW phase of this compound. Hyperfine parameters deduced from the ME, TDPAC, NQR and NMR spectroscopies are compared in Table 3.2. Here, the assignment to β- and γ-sites is based on the observed relaxation behaviour [44], and the β-sites are assumed to be subjected to the largest V_{zz}. The inequivalent Ta sites are labelled in ascending order of η. Note that the sign of V_{zz} is negative as derived from ME measurements [42]. The consistency among all four nuclear spectroscopies is very good.

Site	$V_{zz}[10^{18}V/cm^2]$				η			
	ME (300K)	TDPAC (77K)	NQR (4.2K)	NMR (4.2K)	ME (300K)	TDPAC (77K)	NQR (4.2K)	NMR (4.2K)
α	–	-1.07	-1.018	no	–	0*)	0.173	–
β_1			-1.532				0.191	0.192
β_2	-1.509	-1.47	-1.551	no	0*)	0.25(2)	0.225	0.2265
β_3			-1.567				0.287	0.290
γ_1			-1.503				0.176	0.189
γ_2	-1.542	-1.54	-1.551	no	0*)	0.27(2)	0.248	0.248
γ_3			-1.489				0.305	0.3045

Table 3.2 Comparison of the hyperfine parameters of 1T-$TaSe_2$ deduced form ME, TDPAC, NQR and NMR. *) assumed

3.3.2. 1T-TaS_2

ME: No successful ME measurement has been performed on 1T-TaS_2 thus far.

TDPAC: 1T-TaS_2 has been investigated by TDPAC in great detail [43, 46-49]. Apart form the early investigations, all experiments were performed with BaF_2-scintillators with a time resolution of about 500psec thus rendering the observation of all harmonics up to 2.5Grad/s possible. In this way, an unambiguous assignment of three sites with a population ratio of 1:6:6 is possible (the fine structure, to be discussed below, is not resolved). A typical TDPAC spectrum (left) together with the Fourier transform (right) of the CCDW at 13K is shown in Figure 26 for three different crystal orientations. A rather astonishing feature of these spectra is the fact that the unique site with low intensity (denoted α) is not located at a rather low frequency, contrary to 1T-$TaSe_2$, (in fact, the first harmonic of the unique site is located *between* the first harmonics of the two prominent sites). These data on the CCDW state recently led to a reassignment of the NQR data, as will be discussed below.

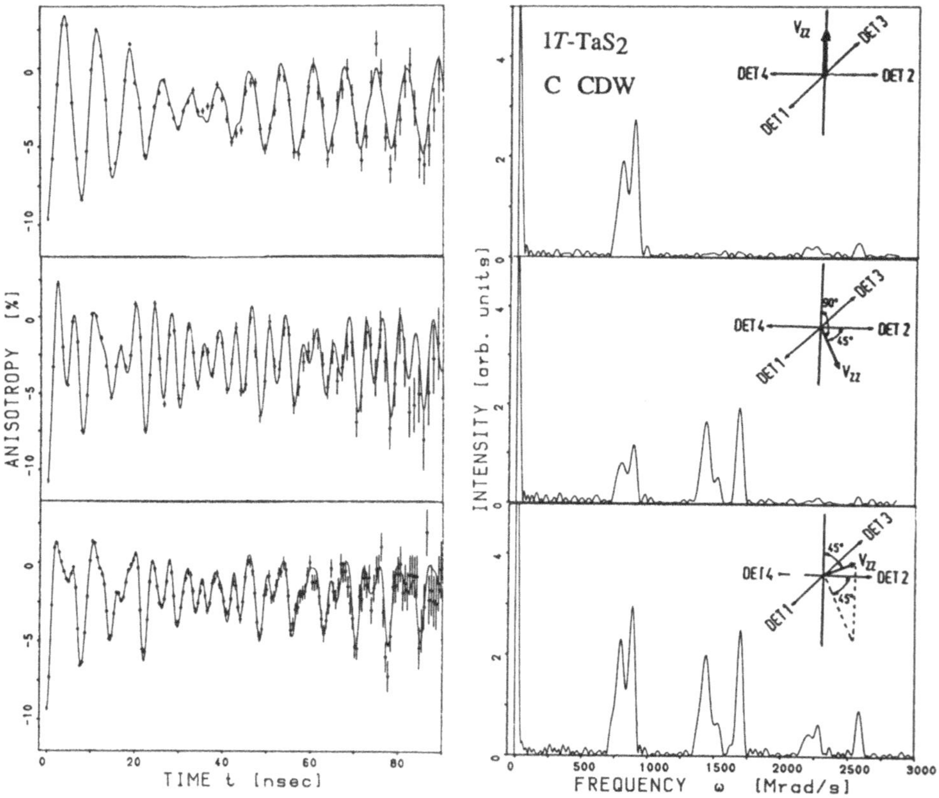

Fig. 26 TDPAC spectra (left) and Fourier transformations (right) for 1T-TaS_2 at 13K for various crystal orientations.

NQR and NMR: Whereas it is well established that the CDW of 1T-$TaSe_2$ has a simple $c_0 + 2a_0$ stacking, there is some controversy on the stacking order realized in the commensurate CDW phase in 1T-TaS_2. The early X-ray and electron diffraction experiments by Scruby *et al.* [50] suggested the simple $c_0 + a_0$ stacking (Figure 22 (b)). On the other hand, the later convergent beam electron diffraction studies by

Fung *et al.* [39] showed that the stacking of the CDW is disordered in the CCDW state of 1T-TaS_2. The more recent X-ray diffraction studies by Tanda *et al.* [51] proposed a complicated stacking sequence in which the c_0 and $c_0 + 2a_0$ stackings alternate with three equivalent "$2a_0$" vectors chosen randomly (Figure 22(d)). The NQR and NMR spectra below show that there is a significant difference in the stacking order between 1T-$TaSe_2$ and 1T-TaS_2, too.

Figure 27 shows the ^{181}Ta NQR spectrum of 1T-TaS_2 powders at 4.2K. The original assignment by Naito *et al.* [52] was incorrect in that they failed to notice the accidental overlap of the β and γ resonances for the $1/2 \leftrightarrow 3/2$ between 95MHz and 103MHz. Recently, Ganal *et al.* [48] proposed a revised assignment for the NQR spectrum as shown in Figure 27, on the basis of their TDPAC spectra in which the β- and γ-frequencies are well separated for the 482keV excited state of ^{181}Ta with $I = 5/2$. The erroneous assignment based on the NQR spectrum alone happened partly because of the lack of quantitative information on the site populations in the NQR experiments as mentioned in Table 1.1. Actually, if the structure located between 95MHz and 103MHz is examined closely, it consists of more than six lines, which strongly supports the revised assignment.

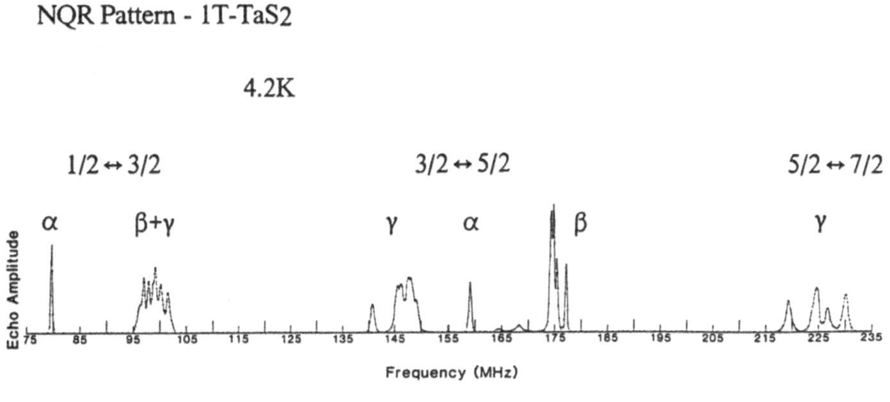

Fig. 27 NQR spectrum of ^{181}Ta in a 1T-TaS_2 powder specimen taken at 4.2K [Naito *et al.* [58]].

There is a remarkable difference noticeable in the NQR spectra between 1T-$TaSe_2$ and 1T-TaS_2. Namely, in 1T-TaS_2 the β and γ sites split into six respectively, whereas in 1T-$TaSe_2$ they split into three. Consequently, in 1T-TaS_2, the 13 atoms of the star-of-David cluster are all inequivalent (see Figure 23(c)). This observation is also supported by the ^{181}Ta NMR spectrum as shown in Figure 28. It is taken on a 1T-TaS_2 single crystal at 4.2K for the magnetic field perpendicular to the layer at 22MHz. The spectrum, correcting the apparent intensities by the η-dependent transition probability, shows the existence of more than 10 Ta sites, excluding the α site which is not picked up in the configuration of this experiment.

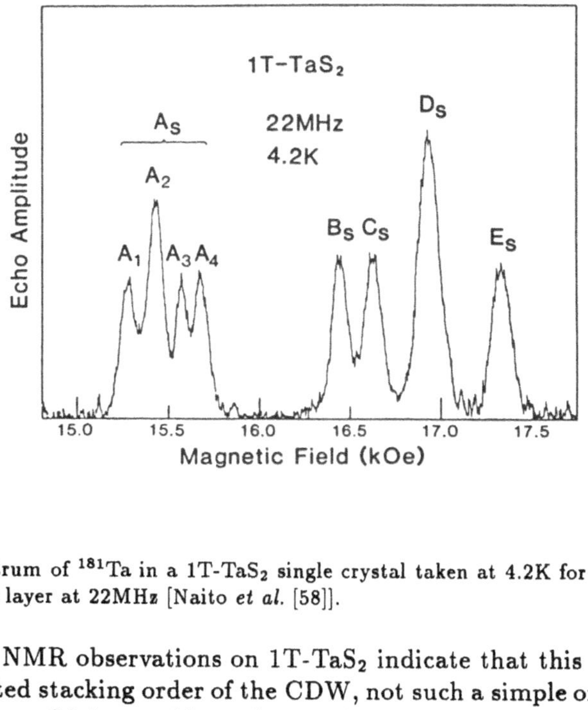

Fig. 28 NMR spectrum of ^{181}Ta in a 1T-TaS_2 single crystal taken at 4.2K for the magnetic field perpendicular to the layer at 22MHz [Naito *et al.* [58]].

The NQR and NMR observations on 1T-TaS_2 indicate that this compound possesses a complicated stacking order of the CDW, not such a simple one as the $c_0 + a_0$ or $c_0 + 2a_0$ stacking. Of the stacking schemes shown in Figure 22, Figure 22(d) and (e) may be appropriate for 1T-TaS_2. Figure 22(d) is the $(c_0, c_0 + 2a_0)$ stacking which has been suggested from the X-ray diffraction as mentioned above. Figure 22(e) shows the modified $c_0 + a_0$ stacking as another possibility in which the "bonding" and "non-bonding" $c_0 + a_0$ stacking alternate. The qualitative explanation for the occurence of these kinds of stackings will be made briefly below.

Nature of the stacking in 1T-TaS_2: When one considers the nature of CDW's in 1T-TaX_2 (X=S, Se), the large amplitude of charge density waves ($\sim$ 1 electron), and large lattice distortions (0.2 $\sim$ 0.3Å) lead one to think of a star-of-David cluster as a quasi-molecule ($Ta_{13}X_{26}$) with chemical bonding characteristics, rather than to think of conventional CDW's. This local quasi-molecule picture may also be applicable when considering the stacking of CDW's in 1T-TaX_2. One cluster (=molecule) contains an odd number (13) of Ta atoms and an equal number of

5d (d_{z^2} band) electrons. This cluster has some unstable nature, if it exists alone. Namely, although chemical bonding conceivably pairs off 12 out of 13 electrons from the d_{z^2} band, it should leave an electron unpaired in a cluster. This unpaired electron will have a tendency to make a covalent bond with another unpaired electron in a near cluster since electrons usually prefer to pair up in bonds. Such a covalent bond must be formed between the nearest-neighbor cluster *across* the Van der Waals gap, because the distance between the nearest-neighbor clusters within a layer is too far (12.1Å). Therefore, two adjacent CDW layers will prefer to make some kind of "bonding".

Obviously, the usual Coulomb repulsion has to be taken into account which keeps two clusters apart. Hence, two competing interactions between clusters are involved in the stacking problem of $1T\text{-}TaX_2$. When the Coulomb repulsion is stronger than the covalent-bonding attraction, the $c_0 + 2a_0$ stacking (Figure 22) will be realized as in the case of $1T\text{-}TaSe_2$. In the opposite situation, the stacking as schematically illustrated in Figure 22 (d) and (e) will be formed, which may be the case for $1T\text{-}TaS_2$ [52].

Furthermore, the electrical properties should also be very different for the above two situations. In the case of the $c_0 + 2a_0$ stacking, an electron unpaired in a cluster remains unpaired, leading to the metallic character. On the other hand, in the case where "bonding" is formed between two adjacent layers, the unpaired electrons will pair up, leading to the non-metallic character. This may explain the quite different behaviour of the low-temeperature electronic conduction between $1T\text{-}TaSe_2$ and $1T\text{-}TaS_2$.

Summary: The results of the nuclear spectroscopies on $1T\text{-}TaS_2$ are summarized as follows.

- The star-of-David clusters are formed in this compound as well as in the isostructural $1T\text{-}TaSe_2$.
- The stacking of the CDW along the c-direction is of a complicated type, in contrast to the $c_0 + 2a_0$ stacking in $1T\text{-}TaSe_2$.
- The stacking realized in $1T\text{-}TaS_2$ may have some kind of "bonding" character between two adjacent CDW layers, which may explain the non-metallic conduction at low temperatures in this compound.

Hyperfine parameters for $1T\text{-}TaS_2$ deduced from the combination of the TDPAC and NQR spectroscopies are summarized in Table 3.3, and compared with those for $1T\text{-}TaSe_2$, which are copied from the NQR results in Table 3.2. From a comparison, the following important differences should be noted:

- Whereas the values for V_{zz} between the β and γ sites of $1T\text{-}TaS_2$ differ markedly, they are almost identical for $1T\text{-}TaSe_2$. Whereas the value for V_{zz} at the α site is intermediate between the β and γ sites in $1T\text{-}TaS_2$, it is about 30% lower than those for β and γ sites in $1T\text{-}TaSe_2$ (see Figure 29). A reduction of the electric field gradient upon charge transfer is generally observed for the intercalation

Site	1T-TaS_2		1T-$TaSe_2$	
	V_{zz} [10^{18}V/cm^2]	η_{NQR}	V_{zz} [10^{18}V/cm^2]	η_{NQR}
α	-1.399	0.021	-1.018	0.173
β_1	-1.579	0.161	-1.532	0.191
β_2	-1.564	0.162	-1.551	0.225
β_3	-1.558	0.183	-1.567	0.289
β_4	-1.560	0.186	–	–
β_5	-1.566	0.193	–	–
β_6	-1.568	0.203	–	–
γ_1	-1.345	0.283	-1.503	0.176
γ_2	-1.361	0.289	-1.524	0.248
γ_3	-1.329	0.298	-1.489	0.305
γ_4	-1.342	0.311	–	–
γ_5	-1.363	0.319	–	–
γ_6	-1.301	0.345	–	–

Table 3.3 Hyperfine parameters of 1T-TaS_2 as compared with those of 1T-$TaSe_2$.

compounds of 2H-TaS_2 [53] and 1T-TaS_2 [47,49,54-56]. Hence, the result for 1T-$TaSe_2$ is quite expected and considered "normal". On the contrary, the high value for the α site for V_{zz} in 1T-TaS_2 is unexpected from the charge maximum at this site.

- The α site does not possess axial symmetry for both of the componds, which is a consequence of the distortion of the star-of-David cluster due to the stacking order. Whereas the asymmetry parameter η of the α site is fairly large in 1T-$TaSe_2$ it is very small in 1T-TaS_2.

These observations show that the charge distribution inside the star-of-David cluster (intra-cluster architecture) is dissimilar for both systems, although the overall shape of the cluster seems to be very similar, judging from diffraction experiments. Such a difference in the intra-cluster architecture may also account for the different stacking order in the two compounds. It should also be pointed out that the electric field gradient results from the $l = 2$ multipole component of the charge density distribution only and certainly does not simply scale with the charge density itself.

The following studies of the other CDW-phases of 1T-TaS_2 have all been carried out by TDPAC. A very important piece of information will be available from a very careful TDPAC study of 1T-TaS_2 over the entire temperature range from 13K up to the shear transformation to the 2H-modification around 450K [49].

The transition from the incommensurate CDW to the nearly commensurate CDW and finally to the commensurate CDW (upon cooling) as well as for the commensurate CDW via the T-phase [51] to the nearly commensurate CDW and finally

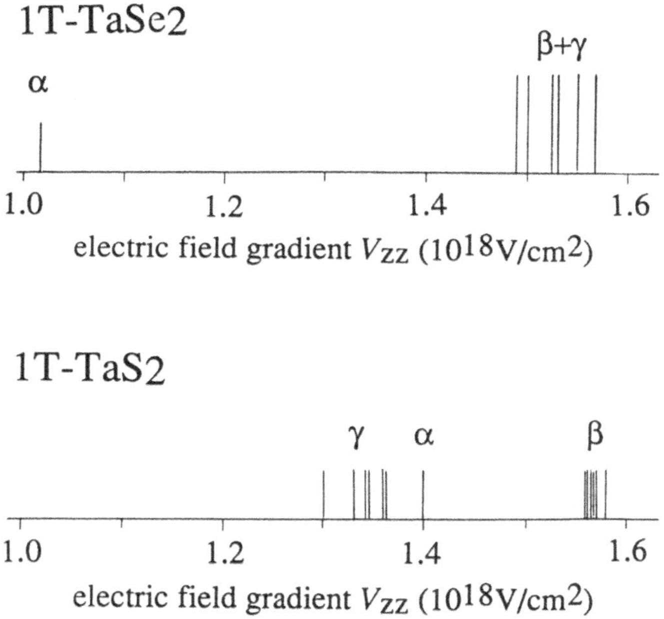

Fig. 29 Comparison of the distribution of V_{zz} in the star-of-David cluster between 1T-TaSe$_2$ and 1T-TaS$_2$.

to the incommensurate state are easily visible. Typical spectra for the different phases are shown in Figure 30. The temperature dependence of the derived NQI frequencies is shown in Figure 31. A striking feature is the linear variation of the NQI frequency within a specific CDW phase, which seems to be charactersitic for layered compounds with anisotropic lattice vibrations [57, 43]. A weak temperature dependence would indicate a rigid lattice whereas a larger temperature dependence corresponds to larger lattice vibration amplitudes. The lock-in to the commensurate state is obviously accompanied with a massive reduction of the mean squared displacement amplitudes perpendicular to the layers. A more detailed report on these data will be given in [49].

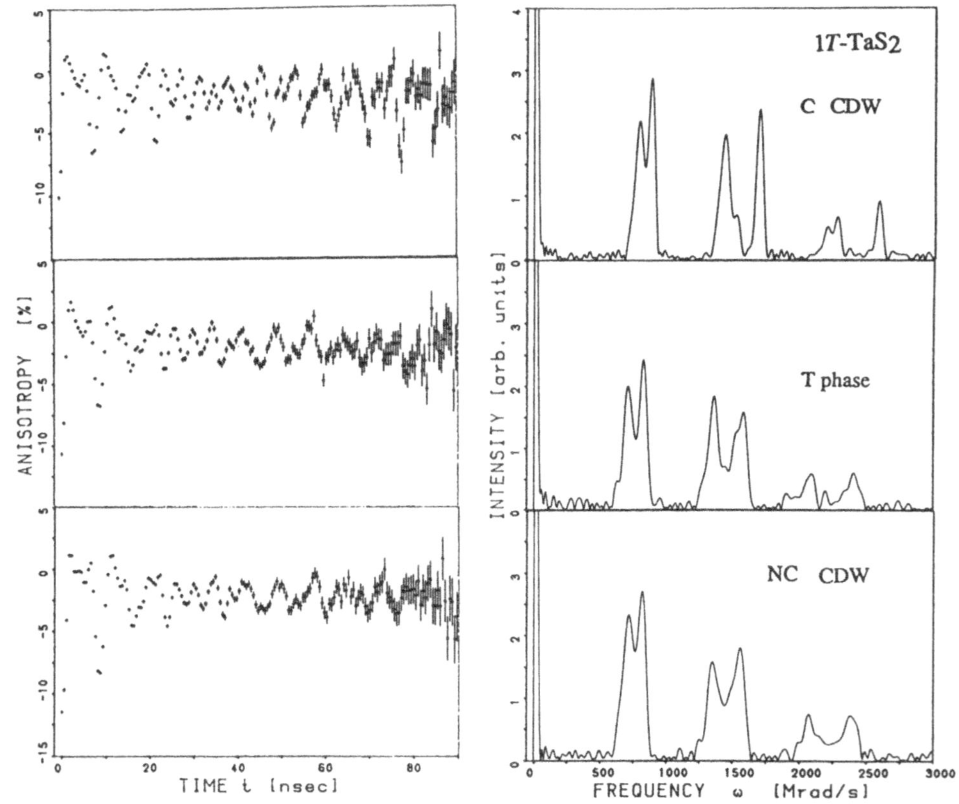

Fig. 30 Typical ^{181}Ta TDPAC spectra (left) and Fourier transforms (right) for the CCDW, the T-phase, and the NCCDW in 1T-TaS_2.

The incommensurate CDW phase has been studied by TDPAC in great detail, too [49]. A typical spectrum just above the NC-IC transition is shown in Figure 32. The characteristic ICDW line-profile is clearly visible, especially for the first and second harmonic. Unfortunately, the overall width is rather small compared to the line splitting in the NC phase, as has been noted previously [46], suggesting a massive decrease of the CDW amplitude at the NC-IC transition. A closer inspection of Figure 32 shows that the line positions and profiles are not consistent with a local picture in which the V_{zz}-distribution is proportional to the charge density distribution and the asymmetry parameter is zero. A simulation of a two-dimensional $V_{xx} - V_{zz}$ distribution function (this is equivalent to a $V_{zz} - \eta$-distribution) within

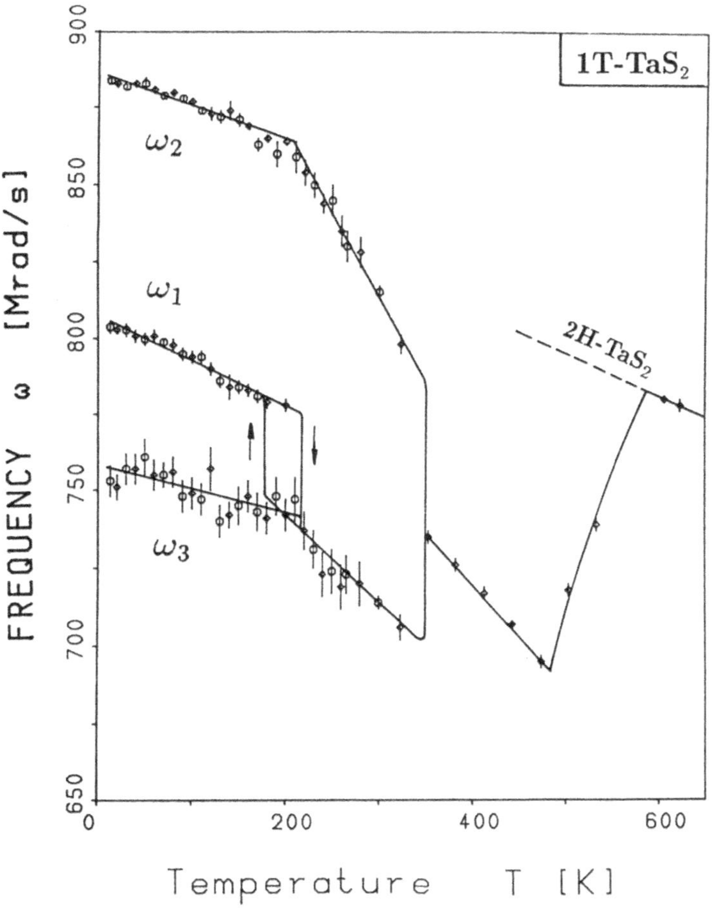

Fig. 31 ^{181}Ta precession frequency in 1T-TaS$_2$ versus temperature.

a simple point charge model with the CDW amplitude being the only non-trivial adjustable parameter yields a fairly good agreement with the observed data [49] as far as line positions and profiles are concerned. This clearly demonstrates that a non-local description is imperative.

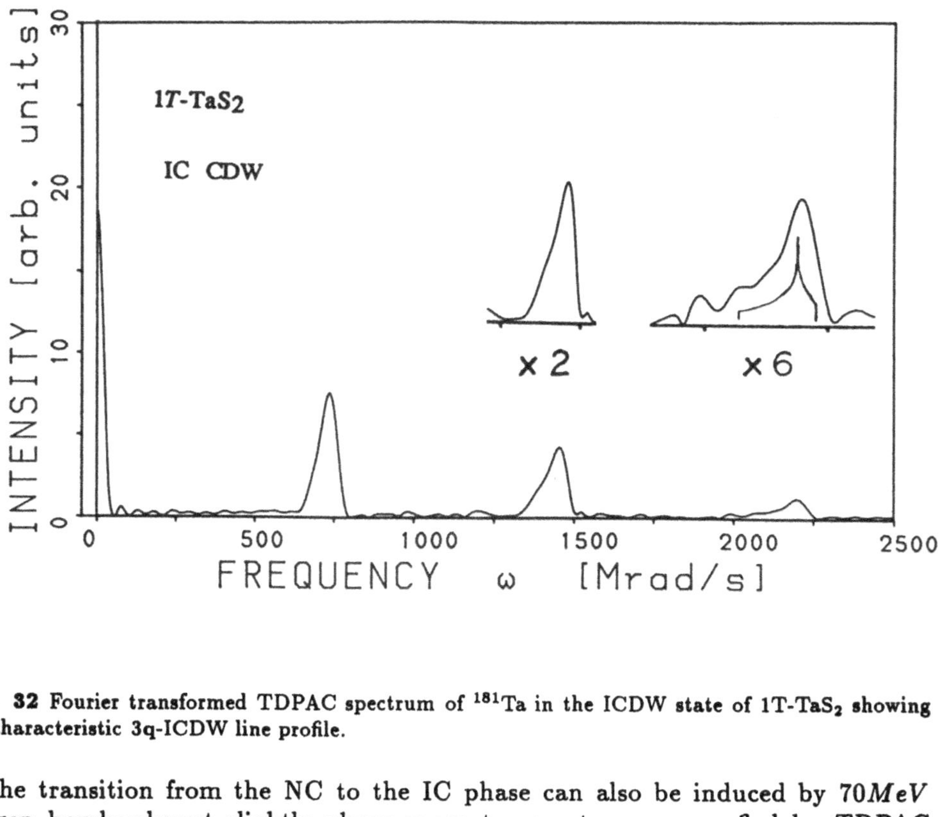

Fig. 32 Fourier transformed TDPAC spectrum of ^{181}Ta in the ICDW state of 1T-TaS_2 showing the characteristic 3q-ICDW line profile.

The transition from the NC to the IC phase can also be induced by 70*MeV* oxygen bombardment slightly above room temperature, as veryfied by TDPAC experiments [58]. The defects introduced by the oxygen bombardment stabilize the IC phase down to 12K.

Finally, very exciting CDW phenomena have been observed by TDPAC for intercalation compounds of 1T-TaS_2. Whereas the CDW is suppressed for Li-intercalation [34] and Hg-intercalation, massive CDW effects – even larger than in the empty 1T-TaS_2 – were observed for $Na_x(H_2O)_yTaS_2$ (1T) [56] and the intercalation of ethylenediamine [47]. Figure 33 shows some characteristic TDPAC spectra for $Na_x(H_2O)_yTaS_2$ (1T) at 300K. At low uptakes ($x \sim 0.2$) a commensurate $\sqrt{13} \times \sqrt{13}$ superstructure with orthorhombic phasing (i.e. the charge maximum between three Ta atoms) is proposed, whereas at higher uptakes ($x \sim 0.25$) a non-hexagonal $\sqrt{7} \times \sqrt{7}$ superstructure is found. At even higher uptakes the TDPAC spectra become unresolved, possibly due to the formation of an incommensurate phase or the slowing down of the intercalate diffusion or both. It should be mentioned that the stability of these various phases depends strongly on pH, the high uptake phase being stable only at $pH \simeq 14$. Thus the true composition of the compounds is not entirely clear. Apart from Na and H_2O an appreciable amount of $(OH)^-$ could cointercalate at high pH.

Although a detailed description of all observations is still missing it is clear that the simple view of tayloring the Fermi surface driven CDW by charge transfer upon intercalation is certainly inadequate.

4. Compounds with trigonal prismatic metal coordination

Among the transition metal dichalcogenides with trigonal prismatic metal coordination, 2H-$TaSe_2$ is the prototype which exhibits by far the largest variety of CDW-phases and has been most extensively studied. Consequently, we will start with the presentation of the results on 2H-$TaSe_2$, followed by the structurally very similar 2H-TaS_2, and conclude this paragraph with 2H-$NbSe_2$. These latter two componds are examples where no commensurate phase exists even at low temperatures. Nevertheless, the nuclear spectroscopy strongly supports the picture of locally commensurate regions separated by discommensurations.

4.1. Tantalum Dichalcogenides

4.1.1. 2H-$TaSe_2$ Apart from 1T-TaS_2, 2H-$TaSe_2$ shows the most complex variety of CDW-phases, as summarized in Figure 34. During cooling, two structural phase transitions take place successively from the normal (N) phase to the incommensurate (I) phase at 122K and to the $3a_0 * 3a_0$ commensurate (C) phase (triple q-state) at 90K [59]. At the onset of the I-phase its wave vector was reported to be $(1-\delta)\mathbf{a}_0^*/3$ with $\delta \simeq 0.02$; δ decreased with decreasing temperature. On heating from the C-phase, a new phase appears which exists in the temperature range from 90K to 113K[2] A detailed analysis of X-ray data [61] showed that this is a "striped" incommensurate phase in which one of the three components remains commensurate, while the other two are incommensurate. At 113K, this phase transforms into the normal I-phase (triple $\mathbf{q}$), which is the same as that seen on cooling below 122K. The periodicity along the c-axis is not changed by the transition within the CDW-states.

---

[2] The first evidence for a further phase transition came from thermal expansion measurements of Steinitz and Grunzweig-Genossar [60], who found an anomaly in the expansion at 113K.

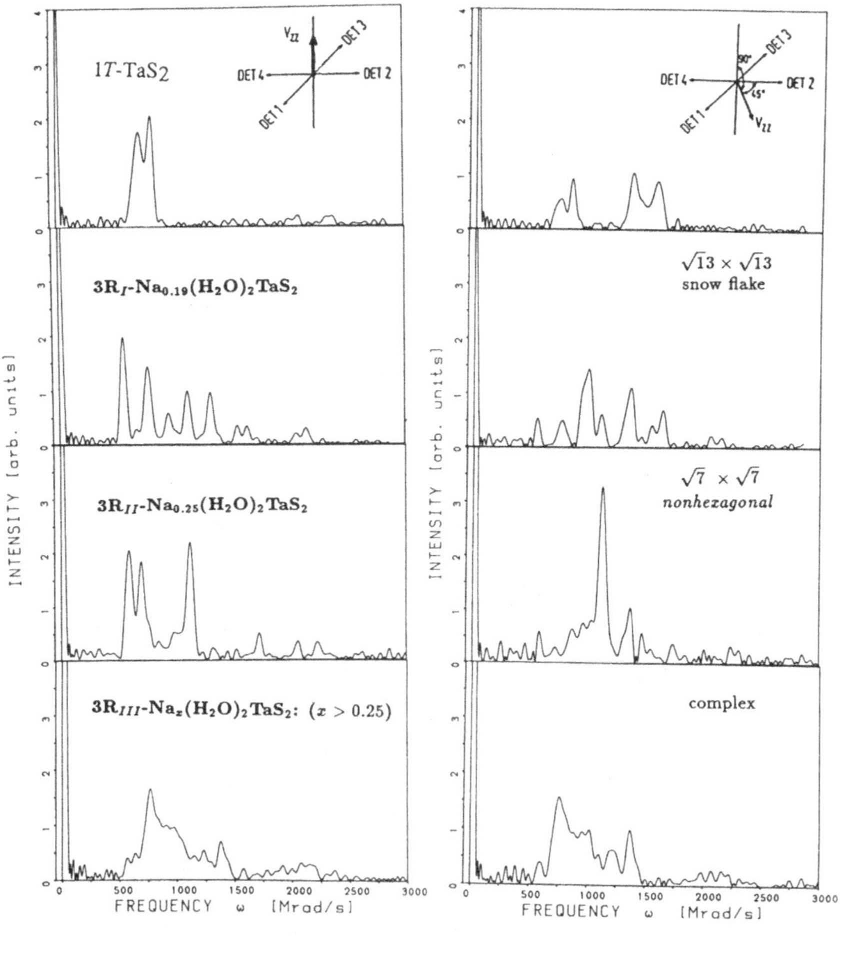

Fig. 33 Fourier transformed TDPAC spectra for ^{181}Ta in empty 1T-TaS_2 and various phases of hydrated Na intercalation compounds.
Left: transverse crystal orientation; fundamental peaks only.
Right: crystal *c*-axis in the detector plane; enhances first harmonic peaks.

The early neutron diffraction results of Moncton *et al.* [62] assumed that the CCDW state has hexagonal symmetry. This would yield 3 inequivalent Ta sites with a population of 1:6:2 (7-atom cluster + 2 isolated atoms). However, according to recent electron diffraction studies [63,64] and Raman-scattering results [65,66] in the C-phase the hexagonal symmetry is broken and the supercell is better described in terms of orthorhombic symmetry (group Cmcm (D_{2h}^{17})). This symmetry rules out the location of the charge density maxima on Ta sites. The alternative is a phasing of the CDW on interstitial sites (denoted β according to MacMillan; for this nomenclature see [67]) or on Se sites (γ). For both of these phasings three inequivalent Ta atoms with populations 3:3:3 are expected. Considering the stacking of the CDWs ($\beta\beta$, or $\gamma\gamma$) a further splitting of inequivalent sites into 2:1:2:1:2:1 results [68]. For the Se atoms the population of inequivalent sites differs for the $\beta\beta$ and the $\gamma\gamma$ stacking; they are 2:1:2:1:2:1 and 2:2:2:1:1:1, respectively. Bird and coworkers [69] determined the displacements of the Ta and Se atoms accurately using the information from higher order Laue zones of convergent beam electron diffraction patterns. Their results show that the $\beta\beta$ stacking is favoured over the $\gamma\gamma$ stacking. Bird *et al.* [69] find in addition a strong orthorhombic distortion which implies that the interlayer coupling is larger than the intralayer coupling. As a consequence, they developed a Landau theory for the CDW-state by treating the interlayer problem first and then taking into account the intralayer terms perturbationally. This leads to a new phasing of the CDW with the charge density maxima between *two* Ta atoms. This yields yet other numbers and populations of inequivalent Ta and Se sites. Wilson [70] preferred in a recent publication such a phasing with Ta populations of 5:1:1:2 and Se populations of 1:3:2:2:1 (δ-phasing).

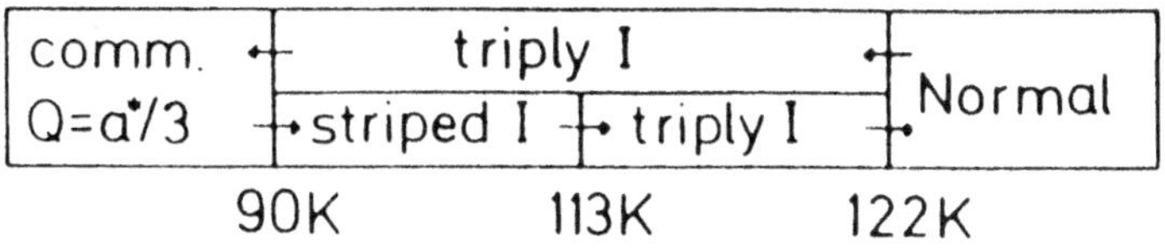

Fig. 34 Summary of the CDW-phases of 2H-$TaSe_2$[59].

ME: Mössbauer effect measurements of ^{181}Ta in 2H-$TaSe_2$ at 300K, 130K, and 52K were reported by Pfeiffer *et al.* [42] (see Figure 35). The authors analyzed their low temperature data in terms of three inequivalent Ta sites with equal populations. This strongly supports the concept of an orthorhombic CDW. From the observed isomer shifts the CDW amplitude was derived locating -0.049(20), +0.033(16) and +0.023(11) electrons relative to Ta^{4+} onto the three sites. The nuclear quadrupole interaction parameters were also given from which the electric field gradients can be derived as -1.598, -1.724 and -1.880*10^{18}V/cm^2. However, since in the data analysis $\eta = 0$ was tacitly assumed, no direct comparison with TDPAC, NMR and NQR data is possible, especially for the site with a large asymmetry parameter. Nevertheless, the site with the lowest electric field gradient exhibits the largest isomer shift and this should correspond to the site with the largest charge density due to the CDW. In this respect, there is agreement between the ME, TDPAC, NMR, and NQR results, as will be shown below.

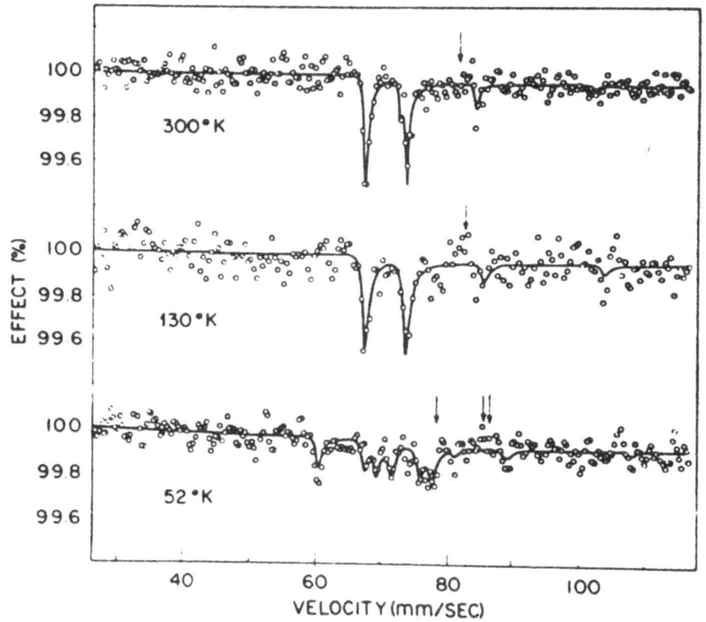

Fig. 35 Mössbauer ^{181}Ta spectra of 2H-$TaSe_2$ at 300K, 130K, and 52K. The low temperature data are analyzed in terms of three inequivalent Ta sites with equal populations [Pfeiffer *et al.* [42]].

A more detailed analysis of these ME data is not possible at present because of the limited statistical accuracy and the unusually large linewidth of 0.8(1)mm/s. Room temperature spectra of 2H-$TaSe_2$ have been obtained with a linewidth of

0.145(6)mm/s [71]. New ME measurements of 2H-$TaSe_2$ with this improved linewidth have been performed [72] but have never been published. The large asymmetry parameter of $\eta \sim 0.6$ as observed by TDPAC, NMR, and NQR (see below) has been questioned by Salomon [72]; however, the matter seems to be unsettled at present and has to await a detailed analysis of the ME data (which possibly do not yet cover the entire velocity range where resonances should occur).

TDPAC: First results by TDPAC with ^{181}Ta in 2H-$TaSe_2$ were reported by Butz *et al.* [73]. These measurements were performed with NaI(Tl)-scintillators with a limited time resolution. Consequently, only the fundamental frequencies and their first harmonics could be detected. The results were analyzed in terms of three inequivalent Ta sites with equal population within experimental accuracy. Because the second harmonic could not be observed, there was a choice of four different combinations of unresolved fundamental frequencies with their harmonics. Common to all four choices was a site with an asymmetry parameter η ranging from 0.53(2) to 0.68(2). These experiments were later repeated with BaF_2 scintillators with a much improved time resolution.

Fourier transformed TDPAC spectra verus temperature obtained with oriented crystals are shown in Figure 36. In order to enhance the statistical accuracy, 2-3 spectra at neighboring temperatures were added assuming that temperature variations in this temperature interval are negligible. The intervals were chosen in agreement with the known phase transition temperatures, of course. Typical spectra for the different CDW-phases are shown in Figure 37. In addition, an accurate spectrum was taken at 12K with the crystals oriented in such a way as to maximize the intensity of the second harmonics. In this way an unambiguous assignment of fundamentals to harmonics is possible. At 12K we obtained three sites with equal populations; one of them exhibits a large asymmetry parameter. One of the sites seems to be further split into two [43].

From these observations a nine-atom "snow-flake" cluster as shown in Figure 38 was proposed with the 3-atom cluster (A) being associated with the $\eta = 0.66$ site and the peripheral atoms (B) and (C). A possible choice to decide which frequencies should be associated to the nearest (B) and outermost (C) sites could be based on the following argument: one of the two sites is apparently split into two sites; this could be a result of the orthorhombic distortion which possibly affects the outermost sites strongest. However, a population ratio of 2:1:2:1:2:1 would be expected for an orthorhombic, distorted "snow-flake" cluster, which is not supported by the data. It could be that the orthorhombic distortion is too small to be resolved in the present experiments and the splitting of the C-sites has a different origin: according to Walker [74] there should be two different CCDW-phases in 2H-$TaSe_2$ which coexist.

Inspection of Figure 37 reveals further interesting details:

- There appears to be an unexpected additional change in the spectrum around 50K, right in the middle of the CCDW range [73] which is unexplained at present.

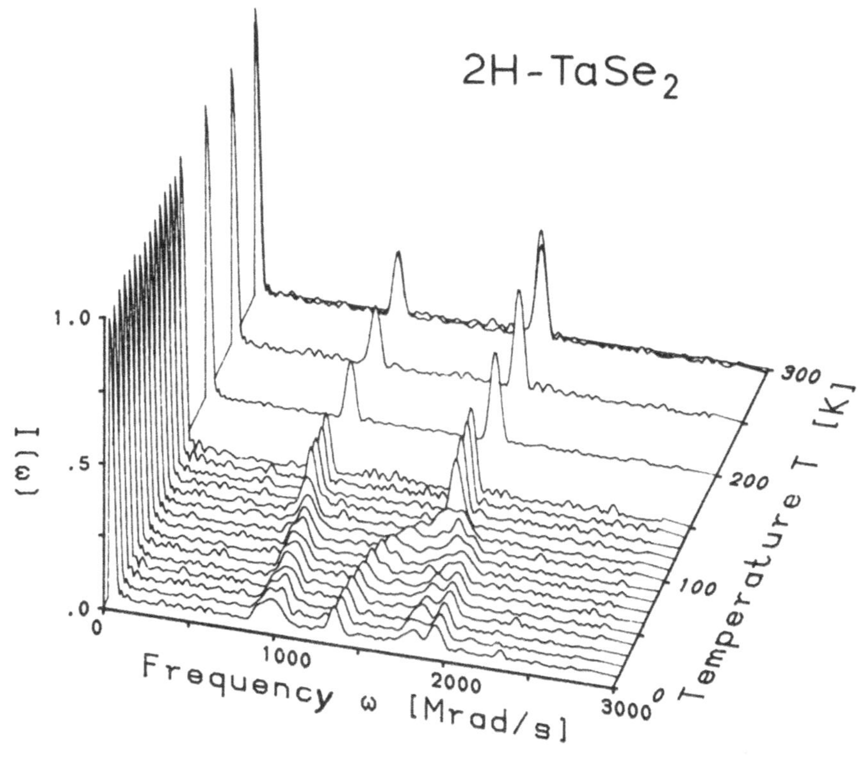

Fig. 36 Stacked plot of Fourier transformed TDPAC spectra for ^{181}Ta in 2H-$TaSe_2$ versus temperature showing clearly the CDW onset at 122K.

- The ICDW-phase is characterized by spectral intensity ranging from the first harmonic of the large-η-site around 1.4Grad/s up to the unresolved pair of harmonics of the B- and C-sites around 1.8Grad/s thus forming a "plateau". At the same time, the fundamental peak remains only little broadened. This is not the lineshape expected for a plane wave ICDW in the local picture. However, a distribution of charge density amplitudes would explain the observed line profiles, as discussed in [73]. A description in terms of commensurate regions with discommensurations is also possible, although the CDW amplitude -judging from the overall linesplitting of the first harmonic - has to be considerably smaller than at 12K.

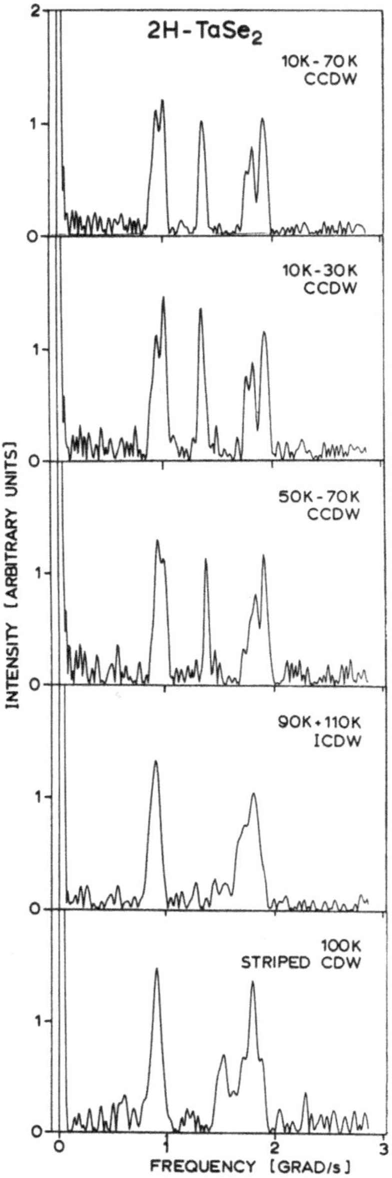

Fig. 37 Typical Fourier transformed TDPAC spectra for ^{181}Ta in various CDW-phases of 2H-TaSe$_2$. 2-3 spectra were added in the temperature ranges quoted in the upper corners of the spectra.

2H-$TaSe_2$			
Site	Population	ν_Q[MHz]	η
A	1/3	729(10)	0.66(2)
B	1/3	1022(8)	0.18(3)
C_1	1/6	804(15)	0.34(3)
C_2	1/6	882(9)	0.29(3)

2H-TaS_2			
Site	Population	ν_Q[MHz]	η
A	1/3	762(81)	0.37(2)
B	1/3	950(12)	0.10(8)
C	1/3	999(12)	0.11(8)

Table 4.1 NQI parameters derived from ^{181}Ta TDPAC of 2H-$TaSe_2$ at 12K and 2H-TaS_2 at 10K.

- The striped CDW spectrum is somewhat intermediate between the CCDW spectrum and the ICDW spectrum. The A-site first harmonic is still rather well separated from the "plateau" and thus fullfills the requirement for commensurate regions separated by discommensurations much better.

A comparison between the observed precession frequencies and asymmetry parameters for 2H-$TaSe_2$ and 2H-TaS_2 is given in Figure 39 showing striking similarities. Note that in the data analysis of the 2H-$TaSe_2$ data the site C was not treated as a split site due to insufficient accuracy. Hence, the average ν_Q is still reliable whereas the average η is less reliable.

Finally, experiments on 2H-$TaSe_2$ crystals with applied magnetic fields up to 0.9T parallel to the crystal *c*-axis were performed at 1.5K and 4.2K in order to look for the possible coexistence of charge and spin density waves, as proposed for 2H-TaS_2 [75]. The Fourier transformed TDPAC spectra are shown in Figure 40 where the data at 1.5K and 4.2K were added since they did not show significant differences. At 0.9T, the Zeeman split two groups of three lines each are well separated. For the data analysis a simultaneous fit of all spectra was performed with constant NQI parameters and adjustable slopes due to the applied magnetic field. In this way the paramagnetic enhancement factors listed in Table 4.2 were derived. It is clear that the A-sites enhance the external field whereas the B- and C-sites shield it. This variation of the local suspectibility - which happens to yield unity when averaged over all sites - can be taken as evidence for the coexistence of a charge and spin density wave. A similar analysis of the previously published data on 2H-TaS_2 [75] along similar lines gave fits with the same quality as those discussed in [75] without envoking a phase transition (which appears to be related to the choice of orthorhombic splitting). The paramagnetic enhancement factors are also listed in Table 4.2.

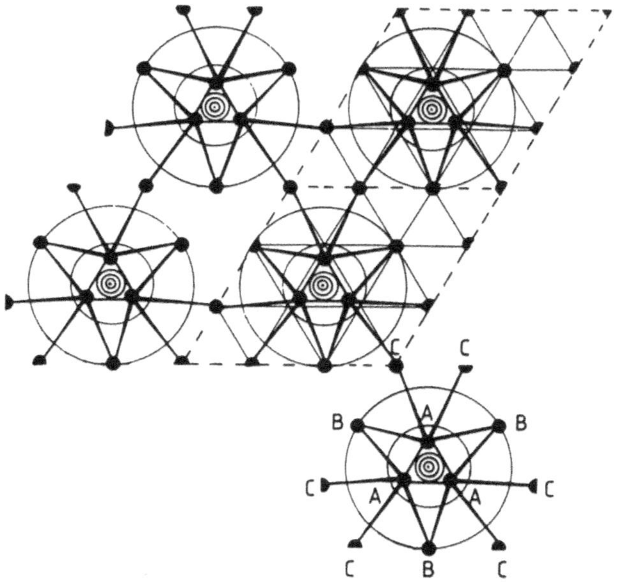

Fig. 38 Nine-atom "snow-flake" cluster proposed for the CCDW state of $2H\text{-}TaSe_2$ [73].

NMR and NQR: NMR investigations of the CDW state of $2H\text{-}TaSe_2$ by NMR of ^{77}Se were reported by several groups [76-78]. Borsa *et al.* reported the temperature dependence of the line width of ^{77}Se NMR by a steady state method for oriented crystals with the field parallel to the *c*-axis [76]. Below the critical temperature of 120K, an asymmetrical broadening of the ^{77}Se NMR was observed. A fractional CDW amplitude of 0.4 at the selenium site was estimated from the fractional Knight shift broadening at 25K by assuming that the Knight shift broadening is proportional to the charge redistribution. This value is larger than the value of 0.2 at the Se site in $2H\text{-}NbSe_2$ which was estimated in the same publication. Detailed

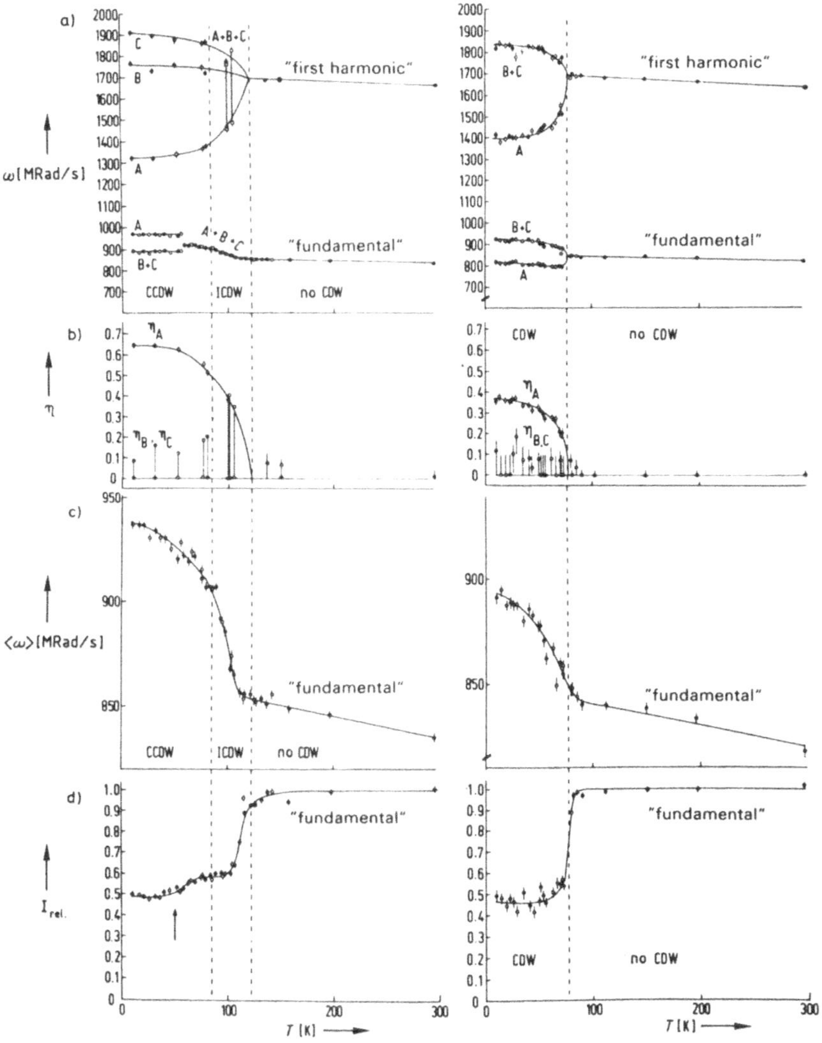

Fig. 39 Nuclear quadrupole interaction parameters for ^{181}Ta in 2H-$TaSe_2$ (left) and 2H-TaS_2 (right) versus temperature. I_{rel} denotes the intensity of the unresolved fundamental line normalized to the no-CDW state.

measurements of the temperature dependence of the line shape of the absorption derivative of the ^{77}Se NMR for a single crystal sample were reported by Suits *et al.* [77]. The spectrum observed for the incommensurate phase in the temperature

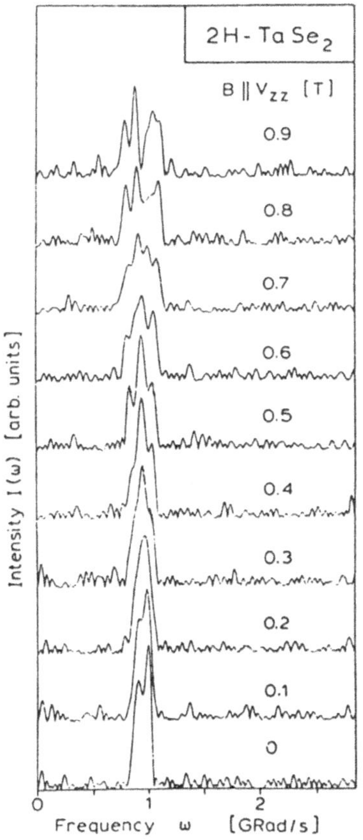

Fig. 40 Fourier transformed TDPAC spectra for oriented crystals of 2H-$TaSe_2$ in applied magnetic fields parallel to the *c*-axis. Data taken at 1.5K and 4.2K were added.

range 120-90K was not a smeared line but a gradual change was observed from a single line at high temperatures above 120K to three lines with roughly equal areas in the commensurate phase below 90K. Thus they confirmed McMillan's concept of

	$2H\text{-}TaSe_2$		$2H\text{-}TaS_2$
Site	1.5 K	4.2 K	1.5 K
A	1.14(3)	1.09(3)	1.01(2)
B	0.93(3)	0.91(2)	0.96(2)
C	0.98(2)	0.96(2)	0.95(2)

Table 4.2 Paramagnetic enhancement factors for the three inequivalent Ta sites in the CCDW state of $2H\text{-}TaSe_2$ and $2H\text{-}TaS_2$.

"discommensurations" where the so-called incommensurate CDW is best described by commensurate regions separated by regions of phase slip [79,80]. These two experiments were interpreted in terms of a hexagonal symmetry of the CDW, but later investigations showed that the CDW has orthorhombic symmetry. The consequences of this symmetry breaking were thoroughly discussed by Walker and Jacobs [81,82].

Pfeiffer *et al.* [78] reinvestigated the temperature dependence of the line shape of ^{77}Se NMR by a pulsed method using a sample of 70 aligned single crystal platelets. Typical results are shown in Figure 41. The line shape should be more reliable than the integrated pattern of absoption derivatives in continuous wave experiments [77]. The commensurate spectra are found not to be consistent with the 3:3:3 distribution expected for hexagonal symmetry. They fitted the spectra to the orthorhombic model by Walker and Jacobs [83] using six fitting parameters and concluded that there is a strong interlayer coupling and a significant interaction between discommensurations and the bulk CDW.

NMR investigations on the CDW states of $2H\text{-}TaSe_2$ by NMR of ^{181}Ta were reported by Nishihara *et al.* [84]. An example of a powder spectrum of low-field NMR of ^{181}Ta in $2H\text{-}TaSe_2$ at 4.2K and 17MHz is shown in Figure 42. It is qualitatively similar to the spectrum of ^{181}Ta in $2H\text{-}TaS_2$ shown in Figure 44. The $\pm 1/2$ transition for η nearly zero, and the $\pm 3/2$ transition for $\eta = 0.60$ and $\eta = 0.35$ are seen in Figure 42. The signal with $\eta = 0.35$ which is the magic value of the asymmetry parameter is very weak and it is not quite clear whether it is intrinsic or not. But there exist at least two distinct Ta sites with η nearly 0 and 0.60. The number of inequivalent Ta sites can be more than two, because Ta sites with nearly the same values of η give nearly the same NMR spectrum even if the values of ν_Q's are quite different, as was mentioned before. This experiment shows that there are at least two kinds of Ta sites; one has an EFG strongly deviating from axial symmetry whereas the other still keeps a nearly axially symmetric local charge density. The spectrum in Figure 42 is not as sharp compared to that of $2H\text{-}TaS_2$ shown in Figure 44. This seems to indicate that there is a larger distribution in η-values in $2H\text{-}TaSe_2$ than in $2H\text{-}TaS_2$. Preliminary NQR experiments on ^{181}Ta in $2H\text{-}TaSe_2$ [85] also suggest that the line width in $2H\text{-}TaSe_2$ is broader than that in $2H\text{-}TaS_2$, although the low-temperature CDW is commensurate in $2H\text{-}TaSe_2$ and is incommensurate

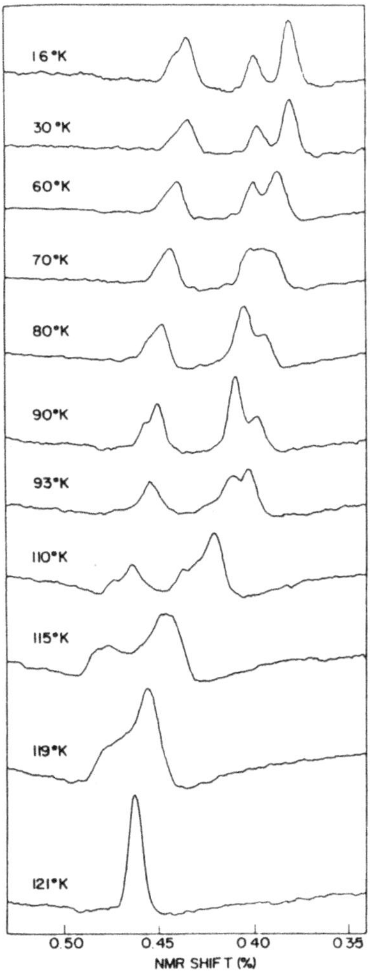

Fig. 41 ^{77}Se pulsed NMR spectra for 2H-$TaSe_2$ [78].

in 2H-TaS_2, and the sample quality of 2H-$TaSe_2$ is generally believed to be better than that of 2H-TaS_2.

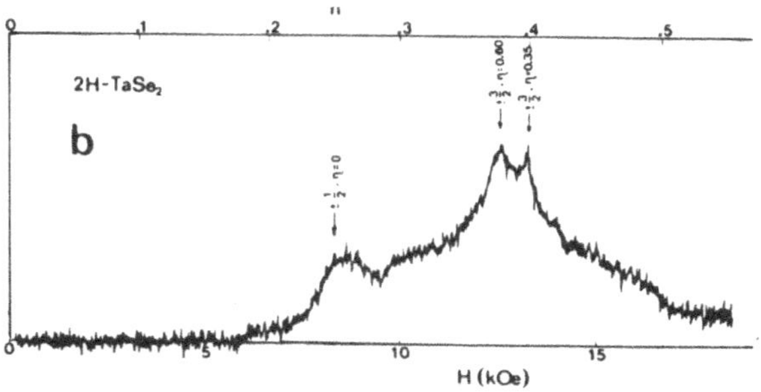

Fig. 42 Powder spectrum of low-field NMR of ^{181}Ta in 2H-$TaSe_2$ at 4.2K and 17MHz.

Summary: All four nuclear spectroscopies agree in one point: the CDW does not have hexagonal symmetry but orthorhombic symmetry. This follows from the fact that at Ta sites a 1:6:2 population was never observed. Furthermore, the Se sites exhibit a population of 1:2:2:2:1:1, inconsistent with the hexagonal phasing. There is less agreement concerning the orthorhombic distortion: judging from TDPAC results, the distortion is small; the ^{77}Se NMR data, on the contrary, suggest a substantial distortion. At present, it is not clear to what extent the discrepancy could be removed by allowing for a rather unharmonic CDW, as suggested by ^{181}Ta TDPAC; this solution with a small orthorhombic distortion was rejected in [78].

A striking feature based on the TDPAC and NMR/NQR results is the observation of a site with $\eta = 0.6$ (The ME data are inconclusive in this respect). This seems to be intimately connected to the orthorhombic CDW phasing with the charge density maximum located between 3 Ta atoms. This phasing which leads to the 9-atom snow-flake cluster is strongly supported by all nuclear spectroscopies applied thus far.

4.1.2. *2H-TaS₂* 2H-TaS_2 undergoes a structural phase transition at about 75K [86]. The superstructure in the low-temperature CDW state of 2H-TaS_2 is incommensurate with the basic lattice and the corresponding wave vector is 0.338 [87].

ME: No ME experiments on the CDW-phase of 2H-TaS_2 were reported thus far although high resolution spectra at 300K without CDW were reported [71].

TDPAC: The first results of ^{181}Ta TDPAC in 2H-TaS_2 were reported in 1981 [88]. These experiments were performed with NaI(Tl)-scintillators and a rather limited time resolution. A splitting into two sites was observed below 75K. The minority site with the lowest interaction strength exhibited $\eta = 0.35$. The pattern observed resembled very much the CCDW-pattern of 2H-$TaSe_2$. These experiments were later repeated with BaF_2-scintillators with a much improved time resolution [43]. A stacked plot of Fourier transformed TDPAC spectra versus temperature is shown in Figure 43. A spectrum taken at 10K with a crystal orientation such that all harmonics could be observed yields an unambiguous assignment of all observed peaks. The data yield three inequivalent Ta sites with roughly equal populations. The obtained NQI parameters are listed in Table 4.1 and show the striking similarity to 2H-$TaSe_2$. The temperature dependence of the NQI parameters is compared to those of 2H-$TaSe_2$ in Figure 39. These data suggest a *commensurate* CDW state at temperatures below 75K, in contrast to the diffraction results.

It is clear that the 9-atom "snow-flake" cluster proposed for 2H-$TaSe_2$ also applies to 2H-TaS_2; the CDW amplitude, however, is reduced compared to 2H-$TaSe_2$, judging from the reduced asymmetry parameter at the central A-sites.

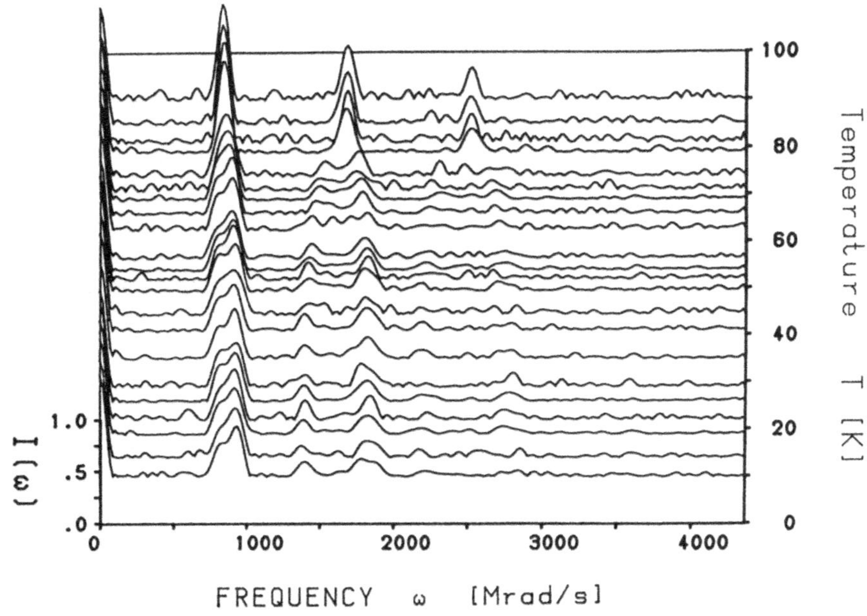

Fig. 43 Stacked plot of Fourier transformed TDPAC spectra for ^{181}Ta in 2H-TaS_2 vs. temperature.

Experiments performed on oriented crystals of 2H-TaS_2 at 1.5K in an applied magnetic field up to 0.7T were also performed [75] and reanalyzed, as discussed in paragraph 4.1.1., in terms of paramagnetic enhancement factors (see Table 4.2).

NMR and NQR: The NMR is not possible for S and the NMR for ^{181}Ta had also been expected to be difficult because of its large quadrupole moment. Nevertheless, 2H-TaS_2 was the first material for which NMR experiments on the ^{181}Ta nucleus had been started to investigate the local electronic structure in the low-temperature CDW state of tantalum dichalcogenides [89]. From the powder spectra of ^{181}Ta NMR in a swept magnetic field for the transition between degenerate $\pm m$ NQR states one can get values of the asymmetry parameter (η) of the EFG free from the magnitude of the EFG as far as the NMR frequency is much smaller than ν_Q as described in paragraph 1.2.2.. The shape of a powder spectrum depends strongly on η and the value of η can be determined easily if the Ta site has a definite value of η. The spectrum of the low-field NMR of ^{181}Ta in 2H-TaS_2 at 4.2K and 4MHz is shown in Figure 44. (It is different from the data in the first report [89], as was suggested in [90]. The sample in the first report was probably 4H_b-TaS_2 [91]). The $\pm 1/2$ transition for $\eta = 0$ and $\eta = 0.35$, and the $\pm 3/2$ transition for $\eta = 0.35$ are clearly seen in Figure 44. This indicates that there exist at least two distinct Ta sites with η of nearly 0 and 0.35 (just equal to the magic value). The number of inequivalent Ta sites can be more than two, because Ta sites with nearly the same values of η give nearly the same NMR spectrum even if the values of ν_Q's are quite different. No appreciable frequency (or field) dependence was observed. Direct nuclear quadrupole transitions for ^{181}Ta were also observed at 4.2K with a somewhat complex NQR spectrum shown in Figure 45. Clear peaks at frequencies of 95.5, 140, 190.2, 194.0MHz and weak ones at 105.3 and 284.6MHz are seen. The peak positions of the spectrum are reliable but the relative intensities should be considered to be only qualitative. A single Ta site creates three NQR transitions of $\pm 1/2 \leftrightarrow \pm 3/2$, $\pm 3/2 \leftrightarrow \pm 5/2$ and $\pm 5/2 \leftrightarrow \pm 7/2$ and the ratios of the three frequencies are determined by the asymmetry parameter η in the principal-axis system. If the assignment for the three transitions is done, the quadrupole frequency ν_Q can be determined. The peak frequencies of 95.5, 190.2, and 284.6 are in the ratio of nearly 1:2:3 indicating the existence of a Ta site with nearly axial symmetry ($\eta = 0.03$). But the assignment of the other peaks in Figure 45 to the NQR transitions is not unique due to the possibilities that some transitions overlap and that some transitions (probably $\pm 5/2 \leftrightarrow \pm 7/2$ transitions) are missing. Therefore, the data must be analyzed with the help of low-field NMR and TDPAC experiments, as described below. The Fourier-transformed TDPAC spectrum, as discussed above, shows three sites with equal populations with η's of 0.00, 0.10 and 0.36 [75] (or refined values of 0.10, 0.11 and 0.37 [43]). The present NQR and NMR data are consistent with the TDPAC data and unambiguous peak assignments are possible. The signals from the Ta site with $\eta = 0.03$ are assigned using the TDPAC data. The obtained NQI parameters are listed in Table 4.3. In particular, the possibility where the peaks at 140.0 and 194.0MHz are due to $\pm 1/2 \leftrightarrow \pm 3/2$ and $\pm 3/2 \leftrightarrow \pm 5/2$ transitions from a Ta site with $\eta = 0.35$, respectively, is excluded from the TDPAC experiments although the frequency ratio is equal to 1.39 which is expected for $\eta = 0.35$. It is to be noted that fine structures as observed in the NQR spectra of 1T-polytypes and interpreted as arising from the stacking order of CDW's [92], are not seen in the case of 2H-TaS$_2$.

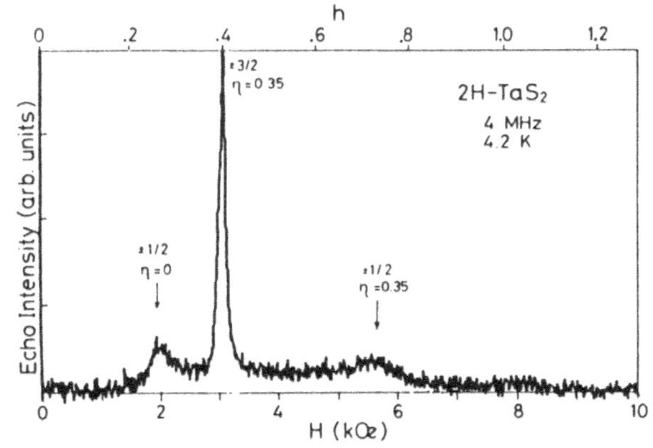

Fig. 44 A powder spectrum of low-field NMR of ^{181}Ta in 2H-TaS$_2$ at 4.2K at a frequency of 4MHz. $h = H/H_0$, where H_0 is the resonance field for a free ^{181}Ta. The assignments of the transitions are shown.

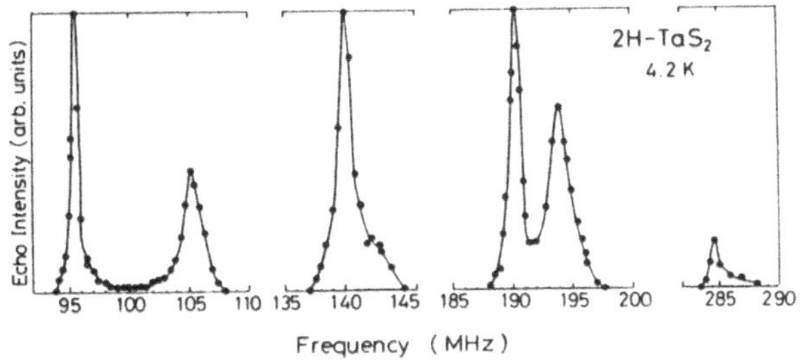

Fig. 45 NQR spectrum of ^{181}Ta in 2H-TaS$_2$ at 4.2K.

Ta Site	A	B	C
$f(\pm1/2 \leftrightarrow \pm3/2)$	105.3	105.3	95.5
$f(\pm3/2 \leftrightarrow \pm5/2)$	140.0	194.0	190.2
$f(\pm5/2 \leftrightarrow \pm7/2)$	(221)*	(293)*	284.6
η	0.37 ± 0.02	0.15 ± 0.02	0.03 ± 0.03
$\nu_Q(MHz)$	75.2 ± 0.5	98.4 ± 0.5	95.5 ± 0.2

Table 4.3 Assignment of the peak frequencies (MHz) in Figure 45 to the NQR transitions and derived values of η and ν_Q for three Ta sites in 2H-TaS_2 at 4.2 K. (* Signals at frequencies with parentheses were not observed.)

Summary: The overal features of NQR and low-field NMR spectra in 2H-TaS_2 are consistent with the 9-atom snowflake cluster model proposed from the TDPAC experiments. The three inequivalent Ta sites have definite values of quadrupole parameters in spite of the incommensurate CDW state observed by the diffraction experiments. The signals must arise from locally commensurate regions, separated by regions of phase slip, confirming again the concept of discommensurations. It seems to be a prominent feature of 2H-TaX_2 that there exists a tantalum site which has a local charge density distribution deviating strongly from axial symmetry. This is certainly a consequence of the fact that the maximum of the CDW is *not* located at a Ta site but rather in between three Ta atoms. Therefore, the expansion of the density around a Ta site contains substantial terms of the form $V_{2\pm1}$ and $V_{2\pm2}$, leading to a large asymmetry parameter. A calculation of the EFG taking into account the CDW is strongly needed.

4.2. 2H-$NbSe_2$

2H-$NbSe_2$ exhibits a second order phase transition from the normal to the incommensurate CDW phase at 35K [93,94]. The phase transition is acommpanied by a weak resistance anomaly. The incommensurateness is maintained down to the lowest temperature investigated. Superconductivity also appears below 7.3K in this compound in spite of the presence of the CDW. Below T_C, the CDW and the superconductivity coexist.

The ^{93}Nb nucleus ($I = 9/2$, $\gamma_N/2\pi = 1.0407$MHz/kOe, and $Q = -0.16 \times 10^{-24}$cm^2) has a good NMR sensitivity. Hence, NMR investigations of the CDW state in 2H-$NbSe_2$ have been reported by many groups [95-100]. The first report was made by Ehrenfreund *et al.* [95] in 1971, when the phase transition of this compound was not recognized as a CDW transition. At that time the phase transition of this compound was assumed to be of antiferromagnetic ordering, because the anomalies in the temperature dependence of the resistivity and the Hall coefficient were similar to those of antiferromagnetic NiO and α-Fe_2O_3. Ehrenfreund *et al.*

performed NMR measurements on powders of 2H-NbSe$_2$ at various temperatures. The ^{93}Nb NMR spectrum at 77K showed a typical quadrupole powder pattern ($I = 9/2$) which indicates the existence of a single Nb site with an axially symmetric EFG. On the other hand, the spectrum at 4.2K indicates the existence of two inequivalent Nb sites with asymmetric EFG's. Ehrenfreund *et al.* concluded from their NMR patterns that antiferromagnetic ordering does not occur in this compound, and that the low-temperature anomalies of 2H-NbSe$_2$ probably result from a structural transformation and associated conduction electron redistribution, which yields two inequivalent Nb sites.

Later, since the phase transition in 2H-NbSe$_2$ was identified as a CDW transition, several groups, for example, Berthier *et al.* [96], Stiles and Williams [97], and Wada [98], performed extensive investigations on this CDW system. The investigations include the determination of the charge distribution in the CDW state, and the temperature dependences of the Knight shifts, the electric field gradients and the relaxation times.

Figure 46 shows the evolution of the NMR line shapes of 2H-NbSe$_2$ single crystals for a magnetic field perpendicular to the layers taken by Berthier *et al.* [96]. Figure 46 (a) shows the central $1/2 \leftrightarrow -1/2$ transition and Figure 46 (b) shows the first quadrupole satellite $-1/2 \leftrightarrow -3/2$ transition. The former corresponds to a Knight shift distribution as deduced from measurements at different frequencies which show that the broadening is proportional to the frequency, and the latter corresponds to an EFG distribution. The solid lines in the figures are the calculated lineshapes based on the assumption of a triple incommensurate CDW. The calculations are performed in the same manner as eq. (3.3). The spatial modulation of the Knight shift and the quadrupole interaction can be expressed as

$$K(\mathbf{r}) = K^0 + K^1 \sum_{i=1}^{3} \cos(\mathbf{q}_i \cdot \mathbf{r} + \phi_i) \tag{4.1}$$

$$\nu_Q(\mathbf{r}) = \nu_Q^0 + \nu_Q^1 \sum_{i=1}^{3} \cos(\mathbf{q}_i \cdot \mathbf{r} + \phi_i) \tag{4.2}$$

where K^1 and ν_Q^1 are the modulation parts of K and ν_Q. Although the CDW phase period for 2H-NbSe$_2$ in the direction of the c-axis is $2c_0$, the existence of a center of inversion enables one to consider only a single layer when analyzing the line shape. This is in contrast with the situation in 1T-VSe$_2$.

In the case of a triple incommensurate CDW, the calculated lineshape exhibits two step singularities at $\nu = \nu_0 + 3\nu_1$ and $\nu_0 - 1.3\nu_1$ and an infinite singularity at $\nu = \nu_0 - \nu_1$, where ν_0 is the resonance frequency in the absence of the charge density wave, ν_1 is the modulation amplitude of the resonance frequency proprotional to K^1 or ν_Q^1. Berthier *et al.* claimed that this calculated linshape explains the observed spectrum well. However, as seen in both Figure 46 (a) and (b), the observed spectra

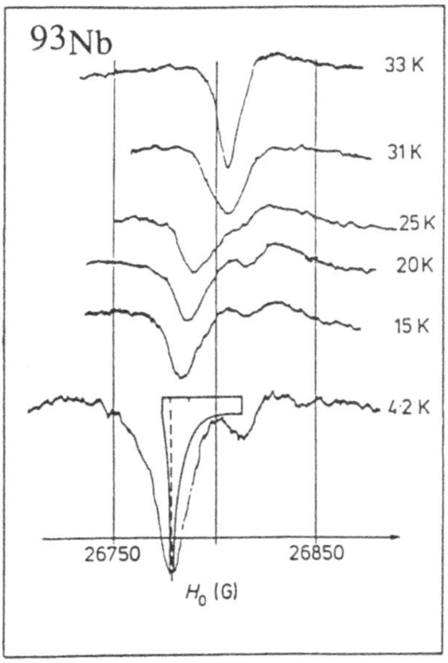

(a)

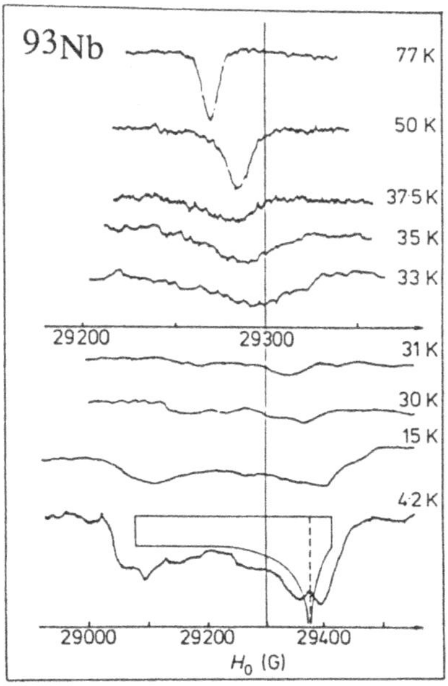

(b)

Fig. 46 Evolution of the lineshapes of (a): the central line ($1/2 \leftrightarrow -1/2$) and (b): the first quadrupole satellite ($-1/2 \leftrightarrow -3/2$) of ^{93}Nb NMR in 2H-$NbSe_2$ [Berthier *et al.* [96]].

have a hump instead of a step singularity at the position of $\nu = \nu_0 + 3\nu_1$. This discrepancy will be discussed below on the basis of the recent results by Skripov *et al.* [99].

Berthier *et al.* also discussed the amplitude of the Knight shift and the EFG distribution, and estimated the amplitude of the charge density wave as $\Delta n/n \sim 0.1$ in 2H-$NbSe_2$. Furthermore, they measured the temperature dependence of the nuclear spin-lattice relaxation time T_1, but did not observe such a sharp drop of $(T_1T)^{-1}$ below the transition temperature due to the opening of partial gaps associated with the CDW formation, as observed in 1T-VSe_2 and 1T-$TiSe_2$. This may suggest that the energy gap occurs over only a small part of the Fermi surface. Finally, Berthier *et al.* report on a pretransitional broadening of the $-1/2 \leftrightarrow -3/2$ transition (Figure 46(b)) below 70K, which is well above T_0. This phenomenon had also been pointed out by Stiles and Williams. The pretransitional broadening can be ascribed either to dynamical fluctuations of the CDW or to static fluctuations induced by impurities. Although the observed linewidth somehow follows such a $(T - T_0)^{-1}$ temperature dependence, as predicted by dynamical fluctuations, Berthier *et al.* concluded from nuclear spin-spin relaxation time (T_2) measurements that the broadening is of inhomogeneous character, namely it is due to static impurity-induced fluctuations.

The ^{77}Se NMR measurements of 2H-NbSe$_2$ were also made by Borsa *et al.* [100]. Their spectrum is shown in Figure 47, which was taken on oriented 2H-NbSe$_2$ powders with the magnetic field perpendicular to the layer at 18MHz. Borsa *et al.* claimed that the observed linshape is consistent with a triple incommensurate CDW (dashed curve) as calculated above for the ^{93}Nb NMR spectrum by Berthier *et al.*.

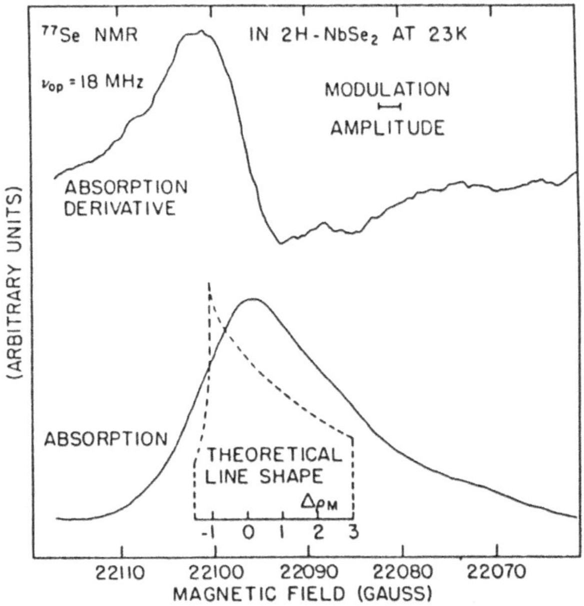

Fig. 47 ^{77}Se NMR line shape in the CDW state of 2H-NbSe$_2$ [Borsa *et al.* [100]].

Although both of the Nb and Se NMR studies in the early stage, as described above, seemed to agree with the simple model of a triple incommensurate CDW, the recent precise reexamination of ^{93}Nb NMR by Skipov *et al.* [99] led to a completely different picture. Their spectra were taken on single crystals of 2H-$NbSe_2$ at 21.4MHz for a magnetic field perpendicular to the layer. Although these spectra are essentially in satisfactory agreement with the spectra by Berthier *et al.*, Skripov *et al.* pointed out that the observed two-peak structure can be described more favorably by a locally commensurate CDW with orthorhombic symmetry rather than by a triple incommensurate CDW. This comparison is shown in Figure 48. Whereas Figure 48(a) represents a triple incommensurate CDW, Figure 48(b) and (c) (points) represent a locally commensurate CDW with orthorhombic symmetry ($\beta\beta$ and $\gamma\gamma$ type by the notation used in [67] in which the best fit parameters are $\phi_2 = \phi_3 = 2^o$, $\Delta\phi = \phi_2 - \phi_1 = 140^o$ in eq.(4.1) or (4.2). Skripov *et al.* could not conclude from their NMR spectra alone which type of $\beta\beta$ or $\gamma\gamma$ is realized in 2H-$NbSe_2$. But they suggest that the $\beta\beta$ state may be regarded as the preferred type because the Se NMR spectrum by Borsa *et al.* [76] is inconsistent with the $\gamma\gamma$ state.

The experimental results of NMR on 2H-$NbSe_2$ are summarized as follows, although there is still some controversy:

- The CDW in 2H-$NbSe_2$ seems to be locally commensurate with orthorhombic symmetry ($\beta\beta$ state). Thus the CDW resembles closely that in 2H-TaS_2 and 2H-$TaSe_2$.
- The relaxation time T_1T does not change significantly below the transition temperature in contrast to the case of 1T-VSe_2 and 1T-$TiSe_2$. This may suggest that the energy gap due to the formation of the CDW occurs over only a small part of the Fermi surface.
- A noticeable pretransitional broadening of the quadrupole satellites is observed well above T_0 which seems to be due to static fluctuations induced by impurities.

5. Compounds with mixed metal coordination

Information on the CDW-phases of compounds with mixed coordination have thus far been obtained by ^{181}Ta TDPAC only [43] (the only exception is 4H_b-TaS_2 [89]) on 4H_b, 6R-TaS_2 and -$TaSe_2$. In these studies a suprising observation was made which is related to the doping with ^{181}Hf which is required for this type of spectroscopy. This doping is achieved during the crystal growth at temperatures above 1000K in the 1T-modification. By quenching the samples to room temperature the 1T-modification can be retained whereas by slow cooling the 2H-modification is obtained via the 4H_b-intermediate phase. Upon heating the 2H-modification, the 1T-modification is obtained via the 6R-intermediate phase [43, 101]. It turns out that Hf which prefers the octahedral coordination remains on a Ta site upon cooling, even at the 4H_b-2H-transition whereas it leaves the trigonal prismatic lattice site of 2H-TaX_2 upon heating when approaching the 2H-6R-transition. This

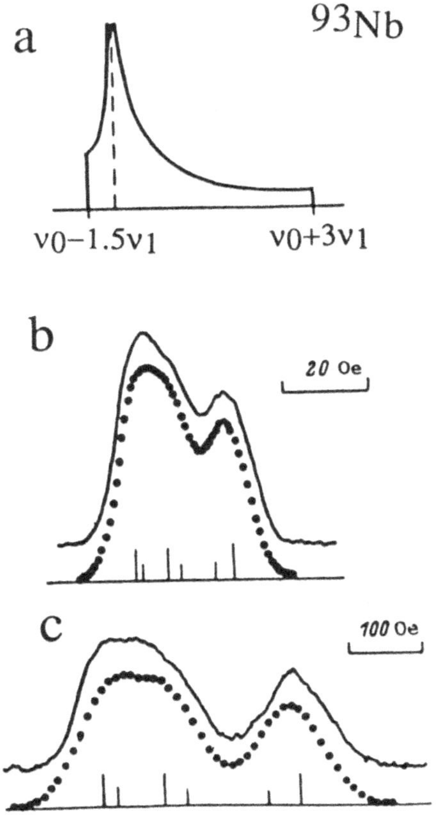

Fig. 48 (a): Calculated line profile for the model of three incommensurate charge density waves. (b,c): Solid lines represent the observed line profiles of the central line (b) and the first quadrupole satellite (c). Points represent the corresponding calculated line profiles for the model of a locally commensurate CDW with orthorhombic symmetry ($\phi_2 = \phi_3 = 2°, \Delta\phi(\equiv \phi_2 - \phi_1) = 140°$). In the calculations, the "background" broadening was taken into account using approximated Gaussian functions with width equal to the width of those lines in the normal phases for $T \approx T_0$ [Skripov *et al.* [99]].

latter transition is apparently high enough in temperature to allow this migration of the dopant whereas the $4H_b$-2H-transition occurs already at a low enough temperature to trap the dopant on the unwanted site. This surprising result is clearly demonstrated by a TDPAC spectrum for 6R-TaS_2, obtained by quenching from the high-temperature existence range, as shown in Figure 49 (bottom). This spectrum shows a well-resolved 1:6:6 pattern, characteristic for a commensurate $\sqrt{13} \times \sqrt{13}$

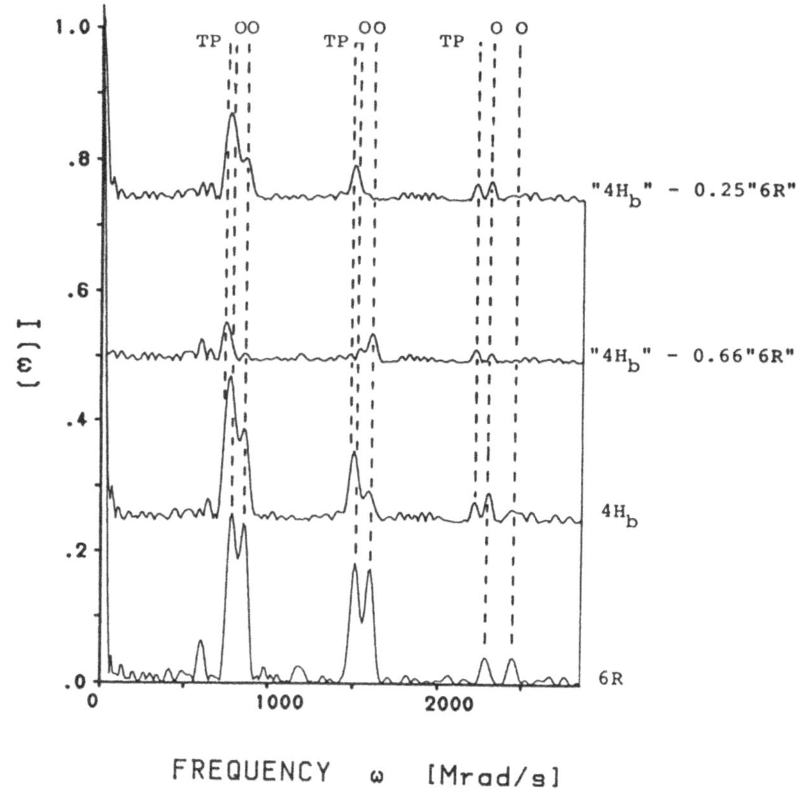

Fig. 49 Fourier transformed TDPAC spectra for ^{181}Ta in 6R-TaS_2 (bottom) and 4H_b-TaS_2 (second from bottom). The top section is the 4H_b-spectrum with 0.25 times the 6R-spectrum subtracted. In this way the first harmonic of the octahedral sites is cancelled. The second spectrum form top is the 4H_b-spectrum with 0.66 times the 6R-spectrum subtracted. In this way the fundamental of the octahedral sites is cancelled.

CDW in the octahedral layer. No signals from the trigonal prismatic layers are observable. There is only little temperature dependence in these NQI parameters down to 10K. A similar spectrum at 300K, though less well resolved, is obtained for 6R-$TaSe_2$.

On the contrary, the spectrum for $4H_b$-TaS_2, obtained by quenching from the high-temperature existence range, exhibits signals from both the octahedral as well as from the trigonal prismatic layers, as shown in Figure 49 (second from bottom). By subtracting appropriate amounts of the 6R-spectrum the contribution of the trigonal primatic layer (without CDW) can be separated from that of the octahedral layer (with CDW). The crystal orientation was not the same for both measurements; hence, different amounts of the 6R-spectrum have to be subtracted in order to remove the octahedral signal in the fundamental and the first harmonic, respectively. This TDPAC-spectrum changes only little upon cooling down to 10K with a possible small extra feature below 30K, where the CDW in the trigonal prismatic layers is expected to set on.

Similar results were obtained at 300K for $4H_b$-$TaSe_2$, although less accurate.

Summing up, the 6R-modifications – and with less accuracy also the $4H_b$-modifications – exhibit at 300K a simple 1:6:6 commensurate $\sqrt{13} \times \sqrt{13}$ pattern on the octahedral layers which corresponds to an undistorted star-of-David. The trigonal prismatic layers apparently decouple the layers sufficiently well that no stacking distortions occur. In this respect the $4H_b$- and 6R-modifications appear to be much better suited for ME-measurements than the 1T-modifications because they could be performed at 300K without CDW in the trigonal prismatic layers and a simple CDW-pattern in the octahedral layers. Interestingly enough, the 6R-spectra for both the selenide and the sulfide are "normal" in that the center site of the star-of-David cluster has by far the lowest NQI as expected from a single harmonic picture, contrary to the 1T-modification where 1T-TaS_2 does not behave "normal".

References

1. See references [2-4] for textbooks.
2. A. Abragam, *The Principles of Nuclear Magnetism*, Clarendon Press, Oxford, 1961.
3. C.P. Slichter, *Principles of Magnetic Resonance*, Harper & Row Publishers, New York, 1963 (expanded edition, Springer-Verlag, 1978).
4. A. Narath, *"Nuclear Magnetic Resonance in Magnetic and Metallic Solids" in Hyperfine Interactions*, ed. by A.J. Freeman and R.B. Frankel, Academic Press, 1967.
5. T. Moriya, *J. Phys. Soc. Japan* **18** (1963) 516.
6. A. Narath and T. Weaver, *Phys. Rev.* **175** (1968) 373.
7. H. Nishihara, I. Shirotani, and N. Inoue, *J. Phys. Soc. Japan* **48** (1980) 1957.
8. See, for a textbook, M.H. Cohen and F. Reif, in *Sol. Stat. Phys.* , Vol. 5, Academic Press, New York, 1957, ed. F. Seitz and D. Turnbull.
9. M.H. Cohen, *Phys. Rev.* **96** (1954) 1278.
10. H. Nishihara, G.A. Scholz, and R.F. Frindt, *Sol. Stat. Comm.* , **44** (1982) 507.
11. J. Szeftel and H. Alloul, *Phys. Rev. Lett.* **42** (1979) 1691.
12. A. Narath, *Phys. Rev.* **162** (1967) 320, and references therein.
13. H. Frauenfelder and R.M. Steffen in *Alpha-, Beta- and Gamma-Ray Spectroscopy*, K. Siegbahn ed., North-Holland, Amsterdam: 1965.

[14] T. Butz, *Hyperfine Interactions* **52** (1989) 189.

[15] T. Butz, A. Vasquez, and A. Lerf, *J. Phys. C* **12** (1979) 4509.

[16] Ph. Gütlich, R. Link, and A. Trautwein, *Mössbauer Spectroscopy and Transition Metal Chemistry*, Springer, Boston 1978.

[17] *Mössbauer Isomer shifts*, G.K. Shenoy and F.E. Wagner (eds.), North-Holland, Amsterdam: 1978.

[18] G. Kaindl, D. Salomon, and G. Wortmann in [17], p.561.

[19] G.T. Trammel and J.P. Hannon, *Phys. Rev.* , **180** (1969) 337.

[20] J.A. Wilson and A.D. Yoffe, *Adv. Phys.* **18** (1969) 193.

[21] D.W. Murphy, C. Cros, F.J. DiSalvo, and J.V. Waszczak. *Inorg. Chem.* **16** (1977) 3027.

[22] R.M.A. Lieth and J.C.J.M. Terhell in: *"Preparation and Crystal Growth of Materials with Layered Structures"*, R.M.A. Lieth (ed.), Vol.1 of *"Physics and Chemistry of Materials with Layered Structures"*, Reidel, Dortrecht 1977.

[23] F. Hullinger in *"Structural Chemistry of Layered-Type Phases"*, F. Lévy (ed.), Vol. 5 of *"Physics and Chemistry of Materials with Layered Structures"*, Reidel, Dortrecht 1976.

[24] J.A. Wilson, F.J. DiSalvo, and S. Mahajan, *Adv. Phys.* **24** (1975) 117.

[25] G.V. Subba Rao and M.W. Shafer in: *"Intercalated Layered Materials"*, F. Lévy (ed.), Vol. 6 of *"Physics and Chemistry of Materials with Layered Structures"*, Reidel, Dortrecht 1976, p. 9.

[26] F.J. DiSalvo, D.E. Moncton, and J.V. Waszczak, *Phys. Rev.* **14** (1976) 4321.

[27] J.A. Wilson, *Phys. Stat. Sol.* (b) **86** (1978) 11.

[28] R. Dupree, W.W. Warren,Jr., and F.J. DiSalvo, *Phys. Rev. B* **16** (1977) 1001.

[29] M. Bayard and M.J. Sienko, *J. Sol. Stat. Chem.* **19** (1976) 325.

[30] C.F. van Bruggen and C. Haas, *Sol. Stat. Comm.* **20** (1976) 251.

[31] F.J. DiSalvo and J.V. Waszczak, *Phys. Rev. B* **23** (1981) 457.

[32] K. Tsutsumi, *Phys. Rev. B* **26** (1982) 5756.

[33] A.H. Thompson and B.G. Silbernagel, *Phys. Rev. B* **19** (1979) 3420.

[34] T. Tsuda, Y. Kitaoka, and H. Yasuoka, *Physica* (Utrecht) **105B** (1981) 414.

[35] A.V. Skripov, A.P. Stepanov, A.D. Shevchenko, and Z.D. Kovalyuk, *Phys. Stat. Sol.* (b) **119** (1983) 401.

[36] R. Brouwer and F. Jellinek, *Physica* **99B** (1980) 151.

[37] D.M. Bird, D.J. Eaglesham, R.L. Withers, S. McKernan, and J.W. Steeds in: *Lecture Notes in Physics* **217** "Charge Density Waves in Solids", Gy. Hutiray and J. Sólyom (eds.), Springer 1985.

[38] P. Fazekas and E. Tosatti, *Philos. Mag. B* **39** (1979) 229.

[39] K.K. Fung, I.W. Steeds, and J.A. Eades, *Physica* **99B** (1980) 47.

[40] D.E. Moncton, F.J. DiSalvo, J.D. Axe, L.J. Sham, and B.R. Patton, *Phys. Rev. B* **14** (1976) 3432.

[41] M. Eibschütz and J.F. DiSalvo, *Phys. Rev. B* **52** (1977) 5281.

[42] L. Pfeiffer, T. Kovacs, and F.J. DiSalvo, *Phys. Rev. Lett.* **52** (1984) 687.

[43] S. Saibene, *PhD thesis*, TU München, Germany 1988, unpublished.

[44] M. Naito, H. Nishihara, and S. Tanaka, *J. Phys. Soc. Japan* **54** (1985) 3946.

[45] see 3, Chap.5, p.144.

[46] T. Butz, A.Vasquez, H. Ernst, and A. Lerf, *Phys. Lett.* **58A** (1976) 51 and ref. 15.

[47] P. Ganal, diploma thesis, TU München, Germany, 1989, unpublished.

[48] P. Ganal, T. Butz, A. Lerf, N. Naito, and H. Nishihara, *Z. Naturforsch.* **45a** (1990) 439.

[49] P. Ganal *et al.*, to be published.

[50] C.B. Scruby, P.M. Williams, and G.S. Parry, *Phil. Mag.* **31** (1975) 255.

[51] S. Tanda, T. Sambongi, T. Tani, and S. Tanaka, *J. Phys. Soc. Japan* **53** (1984) 476.

[52] M. Naito, H. Nishihara, and S. Tanaka, *J. Phys. Soc. Japan* **55** (1986) 2410.

[53] T. Butz and A. Lerf, *Rev. Chim. Min.* **19** (1982) 496.

[54] P. Ganal, T. Butz, and A. Lerf, *Synthetic Metals* **34** (1989) 641.

[55] P. Ganal *et al.*, to be published.

[56] P. Ganal, T.Butz, and A. Lerf, *Synthetic Metals* **34** (1989) 635.

[57] D.R. Torgeson and F. Borsa, *Phys. Rev. Lett.* **37** (1976) 956.

[58] P. Ganal, T. Butz, W. Karle, K.-H. Speidel, and A. Lerf, *Hyp. Int.* **60** (1990) 903.

[59] H. Shiba and K. Nakanishi in: *"Structural Phase Transitions in Layered Transition Metal Compounds"*, K. Motizuki (ed.), Reidel 1986.

[60] M.O. Steinitz and J. Grunzweig, *Sol. Stat. Comm.* **29** (1979) 519.

[61] R.M. Fleming, D.E. Moncton, D.B. McWhan, and F.J. DiSalvo, *Phys. Rev. Lett.* **45** (1980) 576.

[62] D.E. Moncton, J.P. Axe, and F.J. DiSalvo, *Phys. Rev. Lett.* **34** (1975) 734; *Phys. Rev. B* **16** (1977) 801.

[63] K.K. Fung, S. McKernan, and J.A. Wilson, *J. Phys. C* **14** (1981) 5417.

[64] C.H. Chen, J.M. Gibson, and R.M. Fleming, *Phys. Rev. B* **26** (1982) 184.

[65] J.A. Holy, M.V. Klein, and W.L. McMillan, S.F. Meyer, *Phys. Rev. Lett.* **37** (1976) 1145.

[66] S. Sugai, *Sol. Stat. Comm.* **44** (1982) 1141.

[67] J.A. Wilson and R. Vincent, *J. Phys. F* **14** (1984) 123.

[68] M.B. Walker and A.E. Jacobs, *Phys. Rev. B* **25** (1982) 4856.

[69] D.M. Bird, S. McKernan, J.W. Steeds, and R.L. Whithers, *J. Phys. C* **18** (1985) 499.

[70] J.A. Wilson, *J. Phys. F* **15** (1985) 591.

[71] M. Eibschütz, D. Salomon, and F.J. DiSalvo, *Phys. Lett.* **93A** (1983) 259.

[72] D. Salomon, private communication.

[73] T. Butz, S. Saibene, and A. Lerf, *J. Phys. C* **19** (1986) 2675.

[74] M.B. Walker in: *Lecture Notes in Physics* **217** "Charge Density Waves in Solids", Gy. Hutiray and J. Sólyom (eds.), Springer 1985.

[75] 2H-TaS_2: T. Butz, K.-H. Ebeling, E. Hagn, S. Saibene, E. Zech, and A. Lerf, *Phys. Rev. Lett.* **56** (1986) 639; 2H-$TaSe_2$: unpublished results.

[76] F. Borsa, D.R. Torgeson, and H.R. Shanks, *Phys. Rev. B* **15** (1977) 4576.

[77] B.H. Suits, S. Couturie, and C.P. Slichter, *Phys. Rev. Lett.* **45** (1980) 194 and *Phys. Rev. B* **23** (1981) 5142.

[78] L. Pfeiffer, R.E. Walstedt, R.F. Bell, and T. Kovacs, *Phys. Rev. Lett.* **49** (1982) 1162.

[79] W.L. MacMillan, *Phys. Rev. B* **12** (1975) 1187 and *B* **14** (1976) 1496.

[80] K. Nakanishi and H. Shiba, *J. Phys. Soc. Japan* **44** (1978) 1463.

[81] K.K. Fung, S. MacKernan, J.W.Steeds, and J.A. Wilson, *J. Phys. C* **14** (1981) 5417.

[82] M.B. Walker and A.E. Jacobs, *Phys. Rev. B* **24** (1981) 6770.

[83] M.B. Walker and A.E. Jacobs, *Phys. Rev. B* **25** (1982) 4856.

[84] H. Nishihara, G.A. Scholz, M. Naito, R.F. Frindt, and S. Tanaka: *J. Mag. Mag. Mater.* **31-34**, 717 (1983).

[85] H. Nishihara *et al.*, unpublished.

[86] J.P. Tidman, O. Singh, A.E. Curzon, and R.F. Frindt, *Phil. Mag.* **30** (1974) 1191.

[87] G.A. Scholz, O. Singh, R.F. Frindt, and E. Curzon, *Sol. Stat. Comm.* **44** (1982) 1455.

[88] T. Butz, A.Hübler, A. Lerf, and W. Biberacher, *Mat. Res. Bull.* **16** (1981) 541.

[89] H. Nishihara, G.A. Scholz, and R.F. Frindt, *Sol. Stat. Comm.* **44** (1982) 507.

[90] H. Nishihara, G.A. Scholz, M. Naito, R.F. Frindt, and S. Tanaka, *J. Mag. Mag. Mater.* **31-34**, 717 (1983).

[91] H. Nishihara, K. Hayashi, M. Naito, T. Butz, A. Lerf, and S. Saibene, *Z. Naturforsch.* **45A** (1990) 412.

[92] M. Naito, H. Nishihara, and S. Tanaka, *J. Phys. Soc. Japan* **55**, 2410 (1984) and references therein.

[93] D.J. Hundley and R.F. Frindt, *Can. J. Phys.* **52** (1974) 861.

[94] I.M.E. Harper, T.H. Geballe, and F.J. DiSalvo, *Phys. Lett. A* **54** (1975) 27.

[95] E. Ehrenfreund, A.C. Gossard, F.R. Gamble, and T.H. Geballe, *J. Appl. Phys.* **42** (1971) 1491.

[96] C. Berthier, D. Jérome, and P. Molinie, *J. Phys. C* **11** (1978) 797.

[97] J.A.R. Stiles and D.L.G. Williams, *J. Phys. C* **9** (1976) 3941.

[98] S. Wada, *J. Phys. Soc. Japan* **40** (1976) 1263.

[99] A.V. Skripov, A.P. Stepanov, A.D. Shevchenko, and Z.D. Kovalyuk, *Sov. Phys. Sol. Stat.* **28** (1986) 1107.

[100] F. Borsa, D.R. Torgeson, and H.R. Shanks, *Phys. Rev. B* **15** (1977) 4576.

[101] A. Lerf and T. Butz, *Angewandte Chemie*, Int. Ed. English **26** (1987) 110.

Note added in proof: M. Eibschütz performed high resolution ^{181}Ta ME experiments on 1T-$TaSe_2$ (*Phys. Rev. B, in press*). The central site hyperfine parameters are in complete disagreement with NQR and TDPAC data. The asymmetry parameters are in reasonable agreement as are the V_{zz}-values for the outermost sites. The isomer shift provides a very powerful way to assign resonances to β- and γ-sites.

NUCLEAR SPECTROSCOPY STUDIES OF QUASI-ONE-DIMENSIONAL TRANSITION METAL CHALCOGENIDES

JOSEPH H. ROSS, JR.
Texas A&M University, Department of Physics, College Station
CHARLES P. SLICHTER
Department of Physics, University of Illinois, Urbana-Champaign
USA

1. Introduction

In this chapter will be described quasi-one-dimensional Nb and Ta chalcogenides. These include materials of the MX_3 and MX_4 formulation, where M denotes Nb or Ta, and X denotes S, Se, or Te. Also included are related structures such as the iron-containing $FeNb_3Se_{10}$, and several iodine-containing structures, such as $(NbSe_4)_3I$.

The one-dimensional Nb and Ta chalcogenides exhibit a variety of behavior, but charge density waves (CDW's) are the most common low-temperature phase. At least 12 CDW-bearing materials in this class have been identified. A large number of interesting related structures have been prepared, including Ti, Zr, or Hf chalcogenides, and other stoichiometries such as the MX_5 materials. The abundance of crystal variations in these materials makes this class quite interesting from a chemical standpoint, and also from the point of view of identifying new and important electrical properties. For a review of the structures and properties of these materials, consult Meerschaut and Rouxel [1].

The feature that has generated the most interest in these materials is the CDW conductivity phenomenon. Particularly well-studied for this aspect has been $NbSe_3$, although similar features are found in a variety of materials (including the blue bronze materials, described elsewhere in this volume). CDW conductivity is a response in which CDW motion contributes an electric current, accompanied by a variety of electronic properties.

1.1. Nuclear spectroscopy techniques

We assume that the reader already has some familiarity with NMR, NQR, and Mössbauer spectroscopies, which have been described in previous chapters. A brief description of general aspects and notation is given below.

T. Butz (ed.), Nuclear Spectroscopy on Charge Density Wave Systems, 113–175.

In ^{93}Nb NMR [2], the two primary spectral shifts are the Knight shift and the quadrupole shift. The Knight shift K is a magnetic shielding caused by electrons in the vicinity of each site. In this category we include the contact shift due to s-electron magnetization, and also the core polarization due to mixing of s-states with magnetized d-states, the dipolar shift, and the orbital shift. The latter will often be termed a chemical shift. The orbital and dipolar shifts in Nb will be dominated by the d-electrons. The 3×3 tensor K contains the ratio of the frequency shift and the bare NMR frequency, referenced to the crystal orientation. K can be diagonalized in a principal axis system (PAS) for each site, and the average of the three principal values is the isotropic shift, K_{iso}.

The quadrupole shift in NMR is an interaction due to electric field gradients (EFG's) at each site. For NMR of ^{93}Nb, spin $I = 9/2$, 9 lines are observed for each crystal orientation. These consist of a narrow central line near the bare NMR frequency, flanked by eight satellites, equally-spaced to first order in the quadrupole interaction. The central line is shifted only to second and higher orders in the quadrupole interaction. Quadrupole and Knight shifts can be separated through their characteristic dependences on the magnetic field.

NQR, or pure quadrupole resonance, has spectral lines determined only by the EFG principal values. The absence of orientation signatures means that even a powder will have narrow, single-crystal-like spectra. For ^{93}Nb, four lines are observed per site.

The EFG's are derivatives of the form, $V_{xy} = \partial^2 V/\partial x \partial y$, etc., where V is the electric potential. In these spectroscopies, the sign of V_{ij} is not determined. The EFG tensor is traceless, and may be specified by two principal values, plus the orientation of the PAS. EFG principal values are observed as a product, eQV_{ij}, in which Q is the nuclear quadrupole moment. The notation for the largest principal value is $eq \equiv V_{zz}$, and the other principal values are defined through the asymmetry factor, $\eta \equiv (V_{xx} - V_{yy})/V_{zz}$ (where V_{xx} is a principal value which is intermediate between V_{zz} and V_{yy}, so that $0 \leq \eta \leq 1$). Also used for the largest principal value is ν_Q, defined by $h\nu_Q = \left(\frac{3}{2I(2I-1)}\right)\frac{e^2qQ}{h}$, which is the maximum NMR satellite separation. For ^{93}Nb, $\left(\frac{3}{2I(2I-1)}\right) = \frac{1}{24}$.

Mössbauer spectroscopy [3] exhibits line shapes determined by the electric quadrupole interaction and by the isomer shift, as well as magnetic hyperfine interactions. For ^{57}Fe, the nucleus observed for all studies reported here, the quadrupole interaction induces a symmetric doublet splitting linear in eq, whereas magnetic effects induce further splitting. The quadrupole doublet energy splitting is given by $\Delta E = \frac{1}{2}e^2qQ\sqrt{1+\eta^2/3}$, referenced typically in mm/s according to the Doppler shift of the photon. The isomer shift is a shift of the center of mass of the signal, related to the $r = 0$ charge density at each site. This shift, denoted δ, is referenced to the signal from iron metal.

1.2. MATERIALS

A more detailed description of each material under investigation is given in later sections. The most basic feature of these materials is the structural chains. The one-dimensional transition metal chalcogenides form well-defined crystals consisting of cross-linked chains of MX_3 or MX_4 units. Furthermore, the one dimensional electronic properties are related to strong valence-electron contributions from d_{z^2} orbitals threading the chains.

The $NbSe_3$ structure [4] is shown in Figure 1. The trigonal prismatic chain structure is common to the quasi-one-dimensional MX_3 materials. In the case of $NbSe_3$, three inequivalent chains make up the unit cell. These chains are similar in that the Nb ions are surrounded by triangular Se cages. There is an important distinction, though, in the spacing of Se ions on the three chains. As noted by Wilson [5], closely-spaced Se pairs on chains I and III can be considered $(Se_2)^{2-}$ charged pairs, whereas the other ions are more closely described as Se^{2-}. The electron transfer implied in this argument requires chain II to contain Nb^{5+} ions, with no valence d-electrons, while the other two chains have roughly quarter-filled d-bands. This analysis agrees well with NMR measurements, described below.

Wilson [5] has classified the three chains as "red," "orange," and "yellow" (for sites II, I, and III, respectively in $NbSe_3$) according the the relative Se pair bonding. This classification also extends to the other MX_3 materials, and gives a first-order measure of the on-chain electron count, which determines the CDW structure via the Peierls mechanism.

The Peierls mechanism [6] has been described previously in this volume. For a one-dimensional metal, this mechanism produces a modulated distortion with wavevector $2k_F$. Peierls' original example was the dimerization of the linear chain of valence one, giving a CDW of period two lattice constants. To form a one-dimensional CDW that is incommensurate, non-integer on-chain electron count is required. In the trichalcogenides, this is accomplished through electron transfer between inequivalent chains, and for the most part the CDW wavelengths agree with electron-counting arguments. For instance, $NbSe_3$ exhibits two separate incommensurate CDW's of close to four lattice constants, in good agreement with the two chains having quarter-filled bands, predicted by Wilson [5].

In addition to the trichalcogenides, the tetrachalcogenide-iodide materials will be discussed below. These materials have the form $(MX_4)_nI$, and consist of chains separated by regularly-spaced iodine ions. The electron count in these materials is determined by the spacing of I^- ions between the chains. While the chains are structurally quite different from the trigonal MX_3 chains, the d-electron bands at the Fermi level are similar to those in the trichalcogenides [7]. The electronic properties are hence quite similar in these two types of structure, and the CDW conductivity is exhibited in materials of both types.

The transition metal chalcogenides discussed here are generally less one-dimensional in character than, for instance, the organic or KCP materials. Indeed, ideal

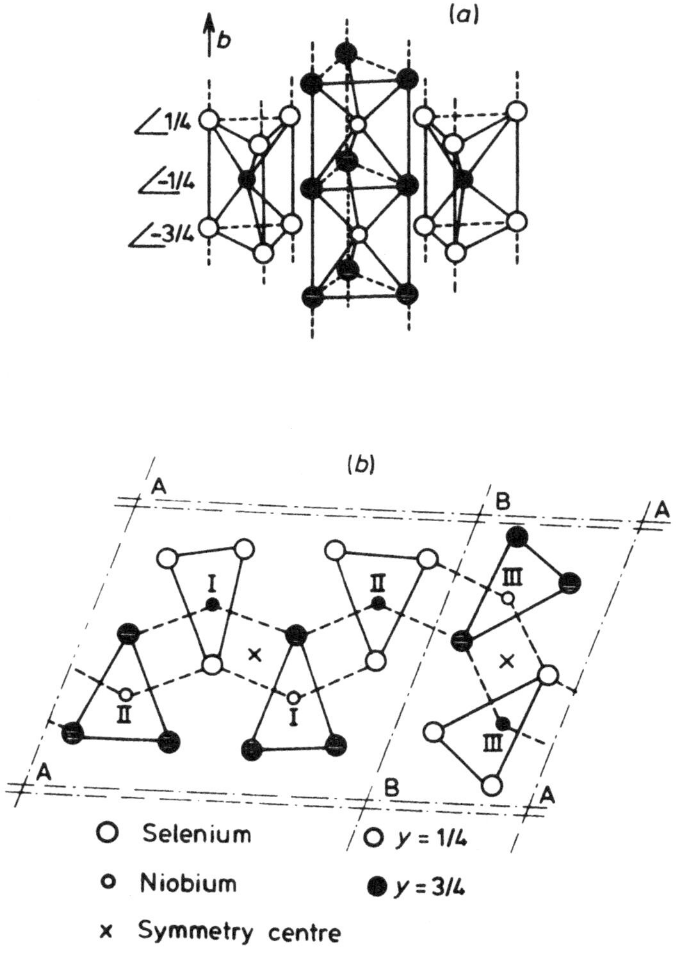

Fig. 1 Structure of $NbSe_3$. View nearly perpendicular to the chains, and parallel to the chain axis (*b*-axis). Se-Se bond lengths are indicated (in Å). Chains according to Wilson's conventions [5]: yellow = III, orange = I, red = II. [from Hodeau *et al.* [4]]

one-dimensional materials can never exhibit long-range order, as is well known [8]. On the other hand, most of the transition metal chalcogenides have well-defined CDW transitions, which are stabilized by three-dimensional coherence between chains. One measure of the one-dimensionality is the anisotropic electrical response; for $NbSe_3$ the resistance anisotropy [9] is approximately 30. Such measurements, though, depend strongly on macroscopic defects. An excellent local probe of one-dimensional electron motion [10] is provided by the NMR spin lattice relaxation time T_1, and such measurements will be described below for $NbSe_3$.

1.3. CDW CONDUCTIVITY EFFECTS

Only the main features of CDW conductivity are described here; for more details the reader is referred to recent reviews [11, 12] of the transport properties of these materials. CDW conductivity has been a focus for much of the interest in the one-dimensional transition metal chalcogenides. The effects described here were first observed in $NbSe_3$. Other transition metal chalcogenides exhibiting such phenomena include monoclinic and orthorhombic TaS_3, one form of NbS_3, $(NbSe_4)_{10/3}I$, $(NbSe_4)_2I$, $(TaSe_4)_2I$, and $NbTe_4$. Additionally, the blue bronze materials (described in chapter 5) exhibit quite similar properties, and certain organic CDW materials, such as TTF-TCNQ [13], have exhibited nonlinear conductivity characteristic of CDW motion.

The CDW consists of electrons arranged into a periodic array, plus a periodic lattice distortion. Simple translational motion of the CDW with respect to the lattice thus would be described by oscillatory motions of the ions about their equilibrium positions, as well as a net translation of the electrons. This type of motion can therefore be detected as a DC electrical current. In 1954, Fröhlich [14] associated such a mechanism with superconductivity. While measurements have confirmed electrical currents associated with CDW motion, superconductivity has not been observed in these materials.

Among the hallmarks of CDW conductivity are non-Ohmic current-voltage characteristics, "narrow-band noise" signals, resonant rf response in the microwave region, and huge dielectric constants. Non-Ohmic conductivity is related to "pinning" of the CDW due to impurities, defects, or commensurability. These provide barriers to CDW motion, allowing no DC CDW current in small applied fields.

In good crystals the electrical conductivity is non-Ohmic only above a well-defined threshold field, E_T. E_T typically varies from crystal to crystal, since the CDW pinning is generally due to random impurities or defects, which depend on sample morphology. Good crystals of $NbSe_3$ exhibit the lowest thresholds, a few mV/cm at temperatures near 50 K. This is determined mainly by impurities and defects, as described below.

Narrow-band noise is observed in $NbSe_3$, and some of the other materials. This response consists of an AC voltage induced by DC current above the threshold. The AC voltage, in a spectrum analyzer, appears as a fundamental with several harmonics; up to 23 harmonics have been observed in some samples of $NbSe_3$ [15]. The fundamental frequency is clearly related to the CDW velocity, as it scales with the CDW current as the voltage is increased. (The CDW current is defined as the excess over the ohmic current established at low voltages.) The origin of the narrow-band noise remains undecided; some models are described below. Narrow-band noise is observed only in better crystals, and upon sample damage the fundamental has been observed to split into two or more peaks, each with associated harmonics. This can be attributed to a broken coherence within the crystal. Transverse coherence is difficult to achieve in these quasi-one-dimensional materials, and indeed $NbSe_3$, exhibiting the sharpest narrow-band noise, is one of the least one-dimensional, according to its resistivity anisotropy [9].

Other electrical features of note include large dielectric constants (measured at voltages below threshold), and frequency-dependent conductivity in the rf to microwave range. These effects are in keeping with the response of a very weakly pinned CDW condensate. In addition, hysteresis and pulse-memory effects [16] have been observed indicating that the CDW retains a memory of a previous conducting state. This most likely indicates that local deformations of the CDW are involved.

1.3.1. *CDW Pinning* One of the most fundamental aspects of CDW motion is the CDW pinning mechanism. Pinning can be due to commensurability, strong or weak impurities, crystal defects, or surfaces. Commensurate regions of the CDW are strongly fixed to the lattice, and generally cannot be moved. The discommensurate structure [17], as described elsewhere in this volume, contains commensurate-pinned regions divided by more mobile phase solitons. Commensurate pinning is probably important to CDW motion in orthorhombic TaS_3, at lower temperatures [18], and in polyacetylene [19]. In most of the materials considered here, though, impurities or defects are the dominant pinning mechanism.

The model of Fukuyama, Lee, and Rice (FLR) [20, 21] represents a basis for many of the theories relevant to the quasi-one-dimensional transition metal chalcogenides. These ideas were developed independently by Gor'kov [22]. FLR recognized that impurities could pin a CDW, independent of the size of the impurity potential, and introduced two categories of pinning: *strong* and *weak* pinning. These concepts generally apply to other crystal defects in addition to impurities.

Impurities placed at random will not pin a rigid CDW (in macroscopic crystals), but a deformable CDW can be pinned, as established by FLR. Pinning comes about by local adjustment of the phase so that impurities tend to fall at energy minima. The CDW pinned in this manner is characterized by a FLR domain length, L. Within a domain the phase is, on average, rigid, but these domains are bounded not by sharp domain walls but by a smooth change in the phase. The situation is quite different from the discommensurate structure exhibiting sharp solitons. The strong and weak pinning categories are based on the relative strength of the impurities, elastic strength of the CDW, and impurity density. For weak pinning, the domain length contains many impurities; for strong pinning, the coherence length contains one impurity, which fixes the CDW phase.

Initial estimates indicated that $NbSe_3$ was weakly-pinned, predominantly by isoelectronic Ta impurities [20, 21], although it is possible that strong pinning centers play a role, as well. FLR predicted that the threshold field dependence on impurity concentration (n_i) is $E_T \sim n_i$, for strong pinning, and $E_T \sim n_i^2$, for weak pinning. Experiments [23, 24] in doped $NbSe_3$ establish the weak pinning exponent for Ta impurities, but also show a size effect indicative of surface pinning. In addition, a small number of strong-pinning centers are important to some models, and it is difficult to exclude the possibility of such sites. Dark-field electron microscopy [25] shows a chain-axis domain length of order 10 μm, consistent with weak pinning due to known Ta impurity concentrations.

Nuclear spectroscopy is quite sensitive to discommensurations, whereas impurity-induced static CDW distortions are more difficult to detect. NMR studies of discommensurations yield a spectrum consisting of sharp lines above a broad background, as has been well established [26, 27]. The sharp lines characterize the commensurate regions. For FLR-type CDW structures, the CDW remains incommensurate, and spectroscopy of the host nuclei generally will not detect small changes in wavelength. Spectroscopy of the impurities, on the other hand, has the potential to distinguish strong and weak pinning, on the basis of the CDW phase distribution at the impurities.

1.3.2. *Models for CDW motion* Several models have been developed to treat CDW motion. Only those theories most applicable to the transition metal chalcogenides are described here. For more details, the reader is directed to recent reviews [11, 12] and references therein.

One model applied to CDW conductivity phenomena, the overdamped oscillator model [28], approximates the CDW as a rigid semiclassical body, with one degree of freedom. A sinusoidal potential due to pinning, with the addition of a phenomenological damping term, gives a physical situation identical to that of an overdamped classical pendulum. While this model does exhibit a threshold field, as found experimentally, the differential conductance becomes singular at the threshold voltage, in strong disagreement with observations. Better agreement is obtained for the ac response, and the pinning frequency (natural frequency of the CDW in its pinning barriers) can be estimated by fitting the observed ac conductivity to the overdamped oscillator model.

A model due to Barnes and Zawadowski [29], gives a CDW pinning energy proportional to $\sin^2(\varphi_{CDW})$, rather than first-order in the sinusoidal charge density, from an analysis of the CDW-impurity interaction. Combined with a rigid overdamped CDW equation of motion, this model yields results quite similar to the overdamped oscillator model. The narrow-band noise frequencies, though, are a factor of two larger than predicted with other pinning models.

It is generally accepted that a model which includes CDW deformations is necessary for a complete description of CDW conduction phenomena in these materials. Two models have been developed in which CDW deformations due to weak pinning are an essential feature. Weak pinning, and the FLR description of pinning-induced deformations, have been described in section 1.3.1.. These two models, the quantum tunneling model and the deformable overdamped oscillator model, take different views of the microscopics of CDW motion within a domain.

The quantum tunneling model has been developed by Bardeen [30, 31]. In this model, CDW motion is driven by macroscopic quantum tunneling of electrons across CDW domains. These domains are weak-pinning FLR domains, with somewhat loosely-defined domain walls, but since for weak pinning one domain contains perhaps 10^6 impurities, the pinning characteristics averaged over the coherence volume can be considered quite uniform. Calculations based largely on the properties within one domain yield a number of results in good agreement with experiment.

In the quantum tunneling model, the domain length represents a cutoff distance for electron tunneling. The cutoff length implies a threshold field, E_T, which is proportional to $1/L$ (where L is the FLR domain length). The calculated CDW dc conductivity, similar to Zener tunneling behavior, has the form,

$$\sigma_{CDW} = \sigma_\infty (1 - \frac{E_{pol}}{E}) e^{-E_o/(E+E_{pol})}. \tag{1.1}$$

This expression gives excellent agreement with measured curves in the one-dimensional transition metal chalcogenides, over most of the temperature range. The parameters are E_{pol}, the local polarization field (a function of position), E_o, which is closely related to the Zener activation energy, and σ_∞, the limiting CDW contribution for large fields. Typically E_o is 1-10 times E_T, and σ_∞ produces a high-field conductivity comparable to that of the normal metal, extrapolated from above the CDW transition. The polarization field, E_{pol}, is due to the response of the charged condensate. This quantity is negligible above about $2E_T$, and exactly opposes the applied field at $E = E_T$.

From application of photon-assisted tunneling theory [32], a scaling law between the ac and dc conductivity was found in the quantum tunneling model. This result fits measurements [33] in orthorhombic TaS_3. Calculations of harmonic mixing due to combined ac and dc fields also have found agreement in experiment [34]. Hysteresis and pulse-memory behavior, which require nonuniformity between domains, correspond in this model to local variations in E_{pol} below and near E_T.

It should be emphasized that, in the quantum tunneling model, motion of an entire CDW domain involves many single-particle tunneling events, occurring at random. The phase advance is gradual, and does not correspond to a sudden tunneling of the entire domain. Bardeen proposes a specific form for this motion [31], which involves alternating motion of adjacent domains. The motion of one domain changes the CDW phase profile over a wide region, and modifies the phase configurations energetically allowed for neighboring domains. Bardeen's model is illustrated in Figure 2, where the two curves (Φ_A and Φ_B) are alternating phase profiles for the CDW. Resulting narrow-band noise frequencies are expected to be twice those for rigid CDW motion at the same average velocity. NMR in the blue bronze system, described elsewhere in this volume, gives results closer to the value from the classical model.

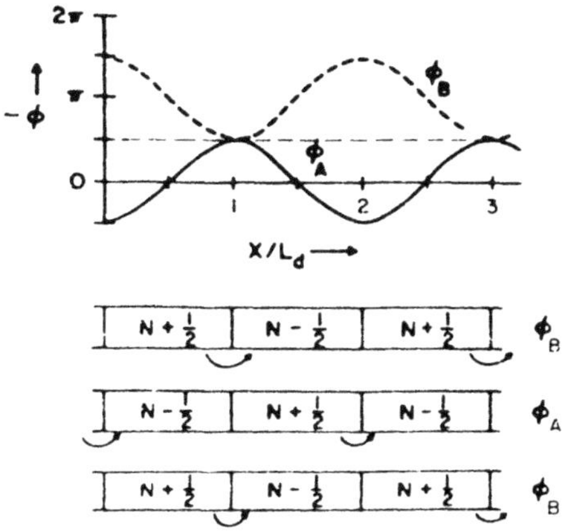

Fig. 2 CDW phase advance in the quantum tunneling model. Shows phase profile at successive times, and the two alternating configurations Φ_A and Φ_B. [from Bardeen [31]]

A classical model, the deformable overdamped oscillator model, has been developed, in which CDW motion is described as motion of weak-pinned domains. In this model, the CDW has a local semiclassical equation of motion similar to that of the overdamped oscillator model, described above. However, the CDW is regarded as an elastic continuum, allowed to deform in the manner of the FLR model, and the deformations significantly modify the results found for the rigid semiclassical CDW.

Retaining all of the degrees of freedom implied by the deformable CDW introduces a great deal of complexity, and the $I-V$ behavior of this model has been calculated only in certain limits. Sneddon, Cross, and Fisher [35] found through perturbation theory that at large electric fields, the first non-constant term in the conductivity is proportional to $E^{-1/2}$. This, however, has not generally agreed with experimental tests [36]. For fields near the threshold, Fisher [37], using critical phenomena methods in a mean-field theory, showed that the CDW current obeys a power law, $(E-E_T)^{3/2}$. Experiments in several materials have shown that a power law description works well near threshold [11, 38], with exponents between 1.2 and 2.2. Sibani and Littlewood [39] calculate numerically an exponent $1.16 \pm .04$ in three dimensions, in good agreement with results in clean $NbSe_3$ crystals.

Another consequence of the deformable classical model with weak pinning [40] is that narrow-band noise oscillations vanish in the large-sample limit, by cancellation of random signals from different domains. Conflicting experimental findings have been reported on this point [41, 42].

Numerical simulations of Littlewood [43] have shown the complicated time-dependence of the phase advance for a moving CDW. The simulations of Figure 3 utilized the deformable overdamped oscillator model, in the strong pinning regime. From these simulations, and from calculations of Fisher [37], the deformable CDW is found to possess many metastable phase configurations. These nearly-degenerate configurations may account for the observed hysteresis and "pulse memory" effects. Figure 3 shows a phase profile for the moving CDW, in which each curve represents a phase difference referenced to the starting condition. Note some similarity to Figure 2, but here the differences between individual domains are quite pronounced. Although each domain moves at a different time, the motion is periodic, with the same fundamental frequency as for rigid CDW motion (half the frequency of Bardeen's model).

A broad class of models has been proposed for which dislocations in the CDW are important. These involve motion of discommensurations, or of dislocations in the CDW related to strong pinning.

Discommensurations, or phase solitons in commensurate systems, have already been described. These have been well-established as static configurations for some systems. If, however, a discrete section of a one-dimensional commensurate CDW is advanced in phase, a soliton-antisoliton pair can be created at the endpoints. (The term "discommensuration" will be reserved for phase solitons representing equilibrium-state distortions.) In two dimensions, the solitons form dislocation lines. Solitons are electrically charged [44], and CDW current can be accounted for by soliton motion. Wilson [5, 45, 46] has advanced the idea that solitons in a discommensurate CDW are responsible for CDW motion in $NbSe_3$. Maki [19] has developed a theory based on tunneling of solitons induced by the field.

NMR results described in this chapter show that discommensurations are not present, at least for the high-temperature $NbSe_3$ CDW. However, soliton-like solutions can also be found for a purely incommensurate CDW, constrained by strong pinning centers. This type of soliton was recognized by Lee and Rice [21], and is a dislocation in the CDW, similar to dislocations in crystalline materials. In fact, there is a strong connection between CDW dislocation motion and plastic flow in crystals [47]. Models involving such motion are closely related to the behavior pictured in Figure 3, although isolated dislocations generally imply much stronger pinning. Stretched or field-generated CDW dislocations may account for observed electrical hysteresis phenomena in these materials.

Ong, Verma, and Maki [48] have examined generation of CDW dislocations at the electrical contacts. These authors note that the CDW must be destroyed (or created) as CDW current leaves (enters) the crystal at the contacts. They propose that this occurs through transverse dislocations, which may propagate down the crystal and provide the mechanism for CDW motion. This is illustrated in Figure 4, where the CDW dislocations are vertical solid curves, and current moves horizontally. The dislocations are nucleated by vortices beneath the contacts (circles

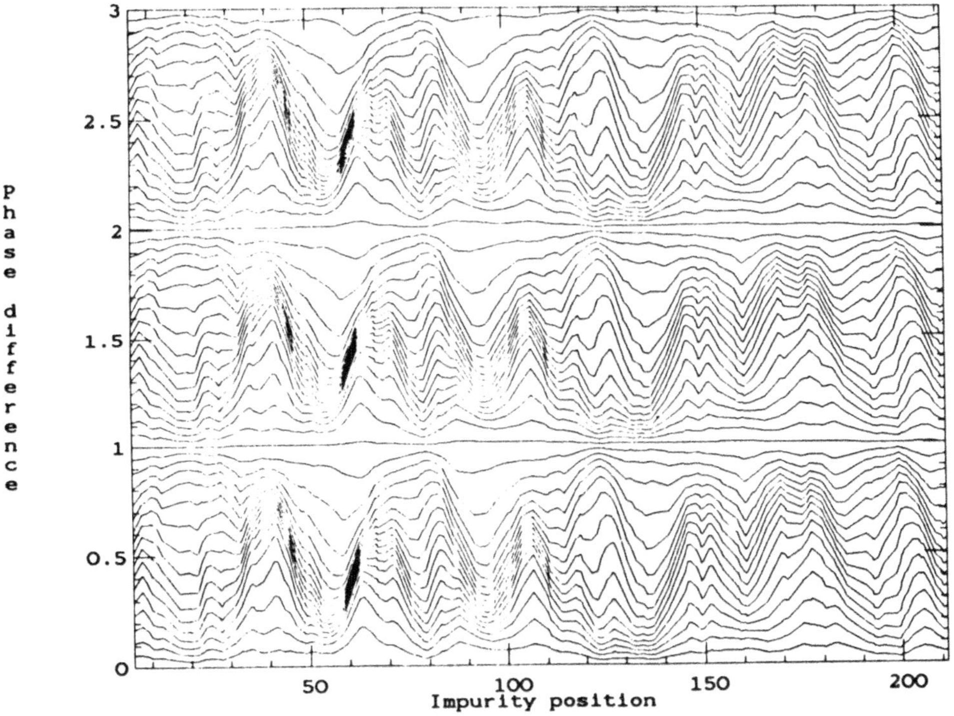

Fig. 3 CDW phase advance from a numerical simulation of the deformable overdamped oscillator model. Curves represent the phase difference from the $t = 0$ configuration. [from Littlewood [43]]

in the figure). Both surface pinning (responsible for the vortices) and bulk pinning control the CDW current; the sample length determines which is more important. The vortices, however, generate the narrow-band noise signals, which are predicted to be independent of the crystal length. The latter question remains unresolved experimentally [41, 42].

Plastic flow of CDW dislocations has been treated by Feinberg and Friedel [47]. They include several types of dislocation pinning. Their estimates of the threshold voltage based on contact pinning gives reasonable agreement with observations in several materials. The treatment includes surface pinning, as well as defect pinning.

Tucker, Lyons, and Gammie [49] treat not dislocation lines, but phase jumps at fixed strong impurity centers. These widely-spaced pinning centers must be

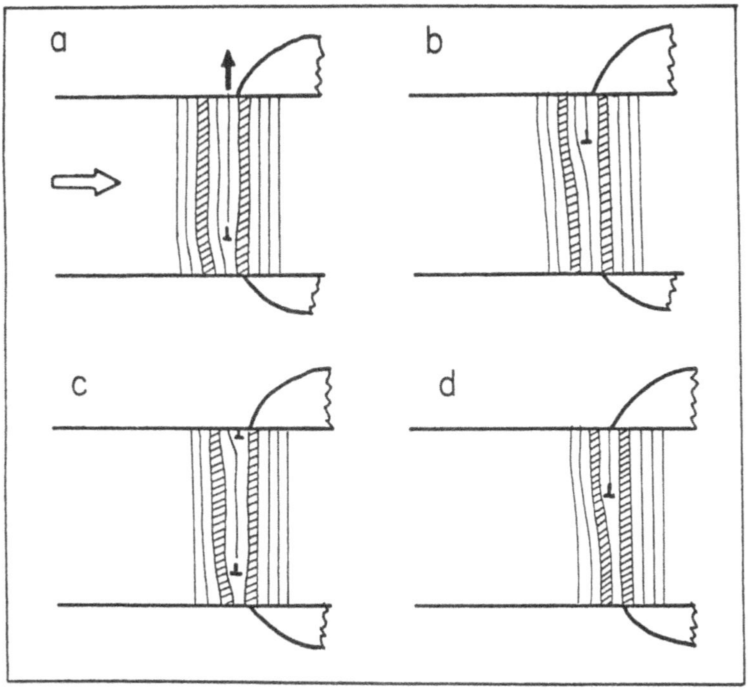

Fig. 4 CDW depinning in the vortex model. Parts a through d are sequential CDW profiles, as the CDW moves to the right. An electrical contact is shown at the right side of the sample [from Ong *et al.* [48]].

strong enough to break the CDW phase locally. Tütto and Zawadowski [50] have given a microscopic treatment of such pinning. Calculated results [49] for $NbSe_3$ agree with many of the observations, including dielectric displacements and the DC conductivity.

A time-resolved study of the narrow-band noise signals in blue bronze [51] has found voltage pulses which propagate in both directions, under current flow in one direction, just above threshold. Reversed velocities are expected for the motion of solitons and antisolitons, described above. This gives support to dislocation-motion theories. Furthermore, voltage signals generated in three-probe measurements [52] also support such motion.

2. $NbSe_3$

Nuclear spectroscopy in $NbSe_3$ has been studied by several groups [53-59]. These measurements utilized ^{93}Nb NMR and NQR, on both powders and aligned multicrystalline samples of the fibrous crystals. As a result of this work, the quadrupole and Knight shift tensors have been mapped out, and broadening due to incommensurate CDW's has been observed at both the 145 K and 59 K transitions. Relaxation measurements have been reported both in pure $NbSe_3$, and in Ta-doped crystals near the CDW-superconducting transition. Furthermore, from studies where voltages in the nonlinear regime are applied, NMR measurements have given details of the CDW dynamics associated with CDW electrical transport.

The $NbSe_3$ structure is shown in Figure 1, and the general features have already been described in section 1.2. Since its initial preparation [60] in 1975, $NbSe_3$ has been perhaps the most studied of the transition metal chalcogenides for its CDW conductivity behavior. There are three inequivalent chains, for a total of 24 atoms per unit cell. The selenium ions occupy isosceles triangles containing two paired and one lone Se. As first argued by Wilson [5], the lone selenium is in a (2^-) charge state, whereas two of the pairs are close enough to bond covalently, implying a $(Se_2)^{2-}$ charge state. In Wilson's terminology, the closely bound Se pairs lie on the "yellow" chain, the well-separated pairs are "red," and the "orange" is the intermediate case. In this section, the spectral sites will generally be identified using Wilson's scheme (red = site II, orange = site I, and yellow = site III).

$NbSe_3$ exhibits two separate CDW transitions, at temperatures of 144 K and 59 K. At each transition X-ray measurements show a new incommensurate lattice distortion [61]. The wave vectors are 0.24117, and 0.26038 (relative to $2\pi/b$) for the high and low temperature CDW, respectively. These reflect nearly quarter-filled bands on each of two structural chains. Although $NbSe_3$ is the prototype for CDW conductivity, it is unique in several respects. For one, there is no complete metal-insulator transition in $NbSe_3$. In addition, quasi-one-dimensional $NbSe_3$ also has strong two-dimensional characteristics, both structural and electronic.

Several electronic structure calculations have been performed [62-66] for the normal state of $NbSe_3$. The most significant electronic features are conduction electrons concentrated mainly on the yellow and orange chains, with a strong d_{z^2} component. These two chains contain the two CDW's, as demonstrated by the NMR and NQR results. The low-temperature metallic electrons may result from an isolated metallic band at the red site, or pockets due to partial destruction of the Fermi surface in CDW transitions on the other two sites.

Examination of the $NbSe_3$ structure (Figure 1) shows a strong cross-linking of chains. The chains are bound tightly along the c axis, while the coupling is van der Waals in the third direction. Electronically, one-dimensional metal-metal bonding orbitals along the chain axis play an important role, but in fact the small resistance anisotropy (approximately 30) indicates that electron motion in the second direction is quite important.

The space group of $NbSe_3$ is monoclinic, and site symmetry only partially determines the PAS orientations for the Knight shift and EFG tensors. The structure (Figure 1) has a reflection plane perpendicular to the b-axis (chain axis), cutting each site. This fixes the b-axis as one of the principal axes. The other two principal axes lie in the a-c plane, with orientations not determined by symmetry. Significantly, the CDW transitions break the b-axis reflection symmetry, allowing the principal axes to point in arbitrary directions.

2.1. $NbSe_3$ Normal-State Spectra

For $NbSe_3$ in the normal metallic state, the three Nb sites have been identified, and from measurements of the Knight shifts and quadrupole parameters, the electronic structure has been well-characterized. These measurements have been performed by Devreux [53], Suits and Slichter [54] and Ross, Wang, and Slichter [55-57]. Relaxation properties are described in a later section.

Devreux [53] measured the high-field NMR using a powder sample and also a partially-aligned sample. The partially-aligned sample consisted of fibers having the b-axis (chain axis) aligned, and random alignments of the other two directions; with the applied field along b, this will produce single-crystal-like spectra. These studies used a field-sweep technique, at fields near 6.5 and 14 T. Measurements were taken at 300 K, 77 K, and 4.2 K, temperatures which bracket the CDW transitions of 144 K and 59 K.

Measurements of Devreux covered all transitions of the quadrupolar ^{93}Nb spectrum. Spectra are shown in Figure 5. From the first-order quadrupole splitting of the satellites, and the dependence on the magnetic field, Devreux found the b-axis principal values of both the Knight shift and EFG tensors. In addition, using the powder sample, an isotropic Knight shift was measured, as an average over all sites.

An NQR study of $NbSe_3$ was completed by Suits and Slichter [54]. This study identified two of the three Nb sites; it is now realized that the third site NQR resonance falls at frequencies below the range of that study. The sample was a powder sample; crystals were grown at Illinois and were similar to those used in the later NMR study [55-57]. NQR spectra are plotted in Figure 6, scaled by the temperature divided by the frequency. All measurements were taken using spin-echo techniques. The measurements covered temperatures above and below the 145 K CDW transition. Changes induced by the CDW transition will be discussed in section 2.3. Room-temperature EFG principal values from these measurements are, for one site, $\nu_Q = 0.925 \pm .02$ MHz and $0.25 > \eta > 0$, and for the other site, $\nu_Q = 1.41 \pm .02$ MHz and $\eta = 0.95$.

Combined room-temperature results of Devreux and of Suits and Slichter are shown in Table 1. In the table, $V_{\|}$ refers to the b-axis EFG, whereas $V_{\alpha\alpha}$ and $V_{\beta\beta}$ refer to principal values for axes in the $a - c$ plane. The directions of these axes have been determined at lower temperatures, as described below, and generally the α and β axes are different axes for each site. The site identifications, listed in

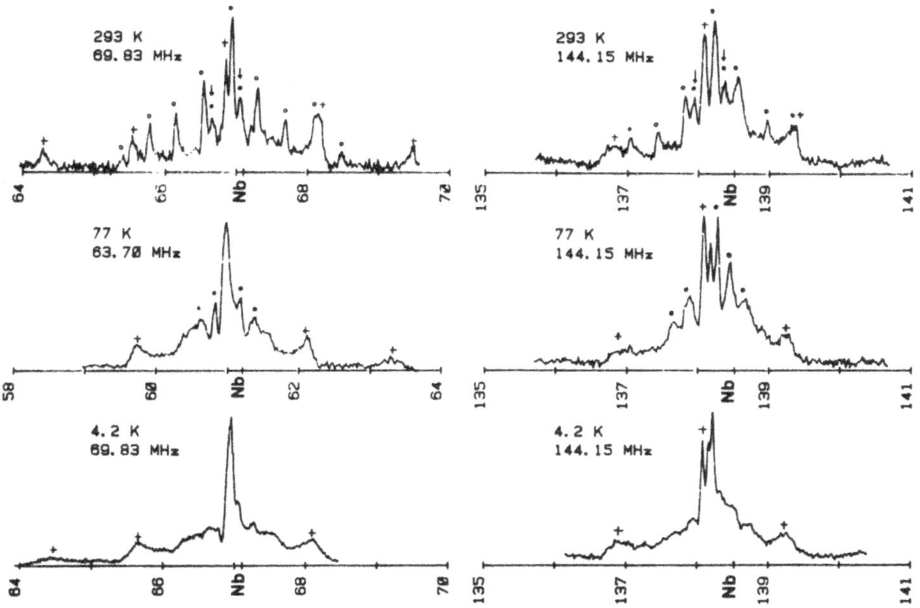

Fig. 5 ^{93}Nb NMR spectra of $NbSe_3$, from Devreux [53]. Crystal orientaton is $H||b$, and the horizontal axis is in kG. Site identity as described in the text, • = yellow, ○ = orange, + = red.

Table 1, are described in section 2.2.. Also given in Table 1 are data for $(NbSe_4)_3I$, described below [67], and for comparison, data from the layer compound 2H-$NbSe_2$, described in an earlier chapter [68].

NMR line shapes were obtained by Ross, Wang, and Slichter [55-57] using a multicrystalline sample in which all three crystalline axes were oriented. The oriented sample was made by aligning ribbon-shaped crystals using the needle (b)-axis, and the ribbon facets which contain the b and c axes. Two orientations of the a-axis make this sample effectively a twinned crystal. From NMR measurements it could be deduced that the a and c axes had approximately a 5° full-width distribution, whereas in the b direction, central lines were limited mainly by dipolar broadening (a few kHz width).

Using this sample, line shapes were obtained between 77 K and 175 K, bracketing the 145 K CDW transition, with limited measurements below the 59 K transition. Spectra were measured at several fields between 2.5 and 8.4 T, using pulsed NMR techniques. For most of the analysis, only the central transitions were measured.

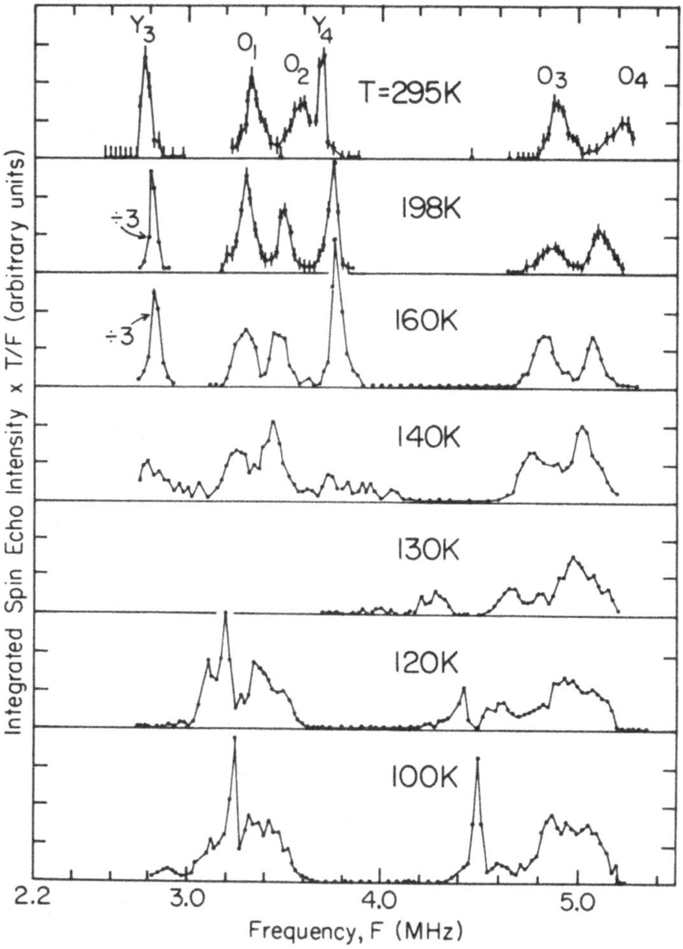

Fig. 6 ^{93}Nb NQR spectra of $NbSe_3$, from Suits and Slichter [54]. Signals labeled Y represent two highest-frequency transitions for the yellow site. Signals labeled O represent the red site. Orange site is out of range of the spectra. (Site identification described in text.)

The latter are shifted by the second-order quadrupole shift, as well as the Knight shift.

The 175 K results of this study are indicated in Table 1, along with 300 K results described above. Little temperature dependence is seen between the two sets of

	NbSe$_3$ @ 300 K (ref 53,54)			NbSe$_3$ @ 175 K (ref. 55-57)			(NbSe$_4$)$_3$I	2H-NbSe$_2$
	yellow (III)	orange (I)	red(II)	yellow (III)	orange (I)	red(II)	293 K (ref. 101)	273 K (ref 72)
$eQV_{\|\|}/h$ (MHz)	9.5±0.5	3.5±0.5	33.8±0.4	10.44±0.05	3.96±0.05	33.6±0.3		60.12±0.2
$eQV_{\alpha\alpha}/h$ (MHz)	12.7±0.9		-2 4±0 4	12.4±0.6	12.0±0 6	-1.2±0.6	12.5±0.1	-30.06±0.1
$eQV_{\beta\beta}/h$ (MHz)	-22.2±0.4		-31 4±0.4	-22 8±0.6	-16.0±0.6	-32.4±0.6		-30.06±0.1
η	0.14±0.06		0 93	0.09±0.03	0.50±0.06	0.93±0.03	0.56±0.01	0
c-α angle				±(43 ± 5)°	±(80 ± 5)°	±(70 ± 5)°		—
$K_{\|\|}$ (%)	0.20±0.04	0.20±0.02	0.32±0 02	0.215± 012	0 181±.012	0.32±.02		0.40 ± 0.02
K_{iso} (%)	–0.10±0 04 (average)–			0.067± 010	0 058±.010	0 19±.02		
K_γ (%)				0 061± 010	0.046± 010	0.12±.02		
c-γ angle				±70°	90°	—		

Table 1 ^{93}Nb NMR parameters for NbSe$_3$ and (NbSe$_4$)$_3$I, above the phase transitions. Data for the layer compound 2H-NbSe$_2$ are given for comparison. Absolute signs for quadrupole parameters are arbitrary. $V_{\|\|}$ and $K_{\|\|}$ are chain-axis principal values (c-axis for NbSe$_2$). The principal axes α, β, and γ, for NbSe$_3$, are perpendicular to the chains and different for each site.

data. Orientation information available from the aligned-crystal study at 175 K is the basis for the site identification of Ross, Wang, and Slichter [55-57]. In the notation of Table 1, α and β refer to the two EFG principal axes for each site in the a-c plane. (These are different axes for each site.) The $c - \alpha$ angle is given, uncertain to within a sign due to the twinning described above. Similarly, $K_{\|\|}$ is the b-axis Knight shift, while K_γ is a second Knight shift principal value, and K_{iso} is the isotropic value (average of the three principal values). For each site, γ is a different axis, and not related to the EFG principal axes.

For consistency in Table 1 and succeeding tables, Knight shift data have been scaled to reflect a ^{93}Nb gyromagnetic ratio of 10.407 MHz/T, which is the Varian table value. A different value, 10.405 MHz/T, suggested by Carter *et al.* [69], was used in the aligned-sample work [55-57]. Relative changes in the Knight shifts by a few percent result from this choice, and the former value is consistent with hyperfine field estimates from the literature, used below.

2.2. SITE IDENTIFICATION

The three Nb NMR sites can be identified by comparing the Knight shifts with electronic structure calculations, and also by utilizing the symmetry of the EFG tensors. Two of the sites have nearly identical Knight shifts, while the other has a much larger value in all orientations. Devreux [53] noted that the core polarization will give a negative Knight shift contribution, and may dominate the spin susceptibility terms. Thus the large positive shift on one site represents an orbital shift with no cancellation from core polarization, while the smaller shifts represent a difference of orbital and core polarization terms, the latter due to a more metallic character to the other two sites. This is in excellent agreement with the conjecture of Wilson [5], and band-structure calculations [63, 64].

The latter argument identifies site II (the red site of Wilson) as that with the largest shift. Wilson also suggested [5] that site III (the yellow site) has the largest electron count, and should correspond to the 145 K CDW transition. Devreux finds that at the 145 K transition one of the spectral sites disappears, as described below, and correspondingly identifies this spectral site with the yellow. Wada *et al.* [58] use a similar electronic structure argument based on their T_1 data. These are the site identifications indicated in Table 1.

Further confirmation of these site assignments comes from the aligned-sample measurements [55-57]. The EFG tensor is found to be oriented approximately 43° from the *c*-axis for one spectral site (Table 1). Comparison to the structure (Figure 1) shows that the yellow site has the same orientation for its near-neighbor seleniums, whereas for the other two sites the selenium triangle aligns perpendicular to *c*. The EFG tensor is due to all nearby charges, not valence electrons only, and so is strongly related to the site symmetry.

In an EFG calculation, Ross *et al.* [55-57] found numerical results that agree closely with the observed red site EFG, in a calculation based on purely ionic sources. In this calculation, all ions were approximated by point charges. The red Nb ions were given a charge of +5e, while the yellow and orange Nb were given charges of +4.5e, and seleniums were given -2e or -e according to Wilson [5]. A Sternheimer shielding factor $\gamma_\infty = -15$ was used [70]. The result for the red site was $\eta = 0.85$, with a symmetry such that the maximum gradient, $eQV_{zz}/h = +31.2$ MHz, was along the *b*-axis. This is in close agreement with observations (Table 1). Similar calculations for the other two sites fail to agree with the observations; for instance, the yellow site numerical result is a nearly axial EFG ($\eta = 0$), with the maximum gradient along *b*. The inference is that a significant valence-electron contribution determines the EFG contributions for the orange and yellow sites, whereas the red site has a much smaller density of such electrons, and so its EFG is given to a good approximation by the ion positions. This agrees well with band structure calculations [63, 64], and further confirms the site analysis based on the Knight shift data.

2.3. CDW Transitions

The NMR data show that the CDW transitions are localized on the yellow and orange sites, leaving the red site unchanged. This is found in the data of Devreux [53] (Figure 5), in which the first-order quadrupole-split satellites disappear for one site at each of the two transitions. The spectral lines corresponding to the yellow site disappear at the 145 K transition, and those of the orange site at the 59 K transition. In fact, further studies [55-57] show that the yellow satellites disappear within about 2 K of the transition. Thus it appears that the two transitions correspond to localized CDW structures on these two sites, in agreement with the conjecture of Wilson [5]. Table 2 shows Knight shift and EFG information resolved at 175 K, 77 K, and 4.2 K. As in Table 1, data for the layer material, 2H-$NbSe_2$ ($T_c = 33$ K), are included for comparison.

	Temperature	$K_{\parallel}$ (%)	ref.	e^2Qq/h (MHz)	$eQV_{\parallel}/h$ (MHz)	ref.
$NbSe_3$ Yellow (III) site	175 K	0.215±0.012	56	22.8±0.6	-10.44±0.05	56
	77 K	0.239/0.247†	57	28.0±0.3	-6.1*	54,56
	4.2 K	0.241±0.020	57			
$NbSe_3$ Orange (I) site	175 K	0.181±0.012	56	16.0±0.6	-3.96±0.05	56
	77 K	0.184±0.005	57	‡	‡	54,56
	4.2 K	0.218±0.020	57			
$NbSe_3$ Red (II) site	175 K	0.32±0.02	56	33.6±0.3	33.6±.3	56
	77 K	0.312±0.005	57		31±1	53
	4.2 K	0.305±0.020	57		30±1	53
2H-$NbSe_2$	300 K	0.40±0.02	68	60.12±0.2	60.12±0.2	68
	30 K	0.28±0.02	68	60.96±0.2	60.96±0.2	68

† Two edges of line shape.
* At CDW maxima and minima (high-frequency line edge).
‡ Values nearly identical to 175 K value.

Table 2 Changes in ^{93}Nb Knight shift and electric field gradients in $NbSe_3$ induced by the CDW transitions at 145 K and 59 K. Changes in 2H-$NbSe_2$ parameters are also given, for comparison.

While the NMR satellite structure disappears at the two transitions, the central lines (1/2,-1/2 transitions) remain, and a well-resolved incommensurate line has been observed for the high-temperature CDW [55-57]. The development of the incommensurate line shape at the yellow site below 145 K is seen in Figure 7, where central lines for the two other sites are also shown. The orange- and red-site central lines are unchanged by the transition in all orientations, whereas the yellow-site line exhibits the double-edged line shape characteristic of an incommensurate CDW [26, 27]. The field-dependence of the edges of this line indicated that the broadening mechanism is predominantly second-order quadrupole.

In contrast to the broadened yellow NMR lines, the NQR data (Figure 6) exhibit a large shift in the resonance for the yellow site, but no broadening. The yellow site is found to shift from a nearly axial EFG to an EFG of $\eta \approx 0.55$ and $e^2qQ/h \approx 27$ MHz. While this appears contradictory, comparison shows that both results can be explained by an EFG modulation in the principal axis *orientations* only.

NMR lines for the field orientation $H \| b$ were fit to such an orientation modulation, as shown in Figure 8. Detailed fits were obtained for 77 K and 95 K, at several applied magnetic fields. Defining the x, y, and z principal axes in the usual way [2], so that $|V_{zz}|$ is the largest gradient, and V_{xx} is intermediate between V_{zz} and V_{yy},

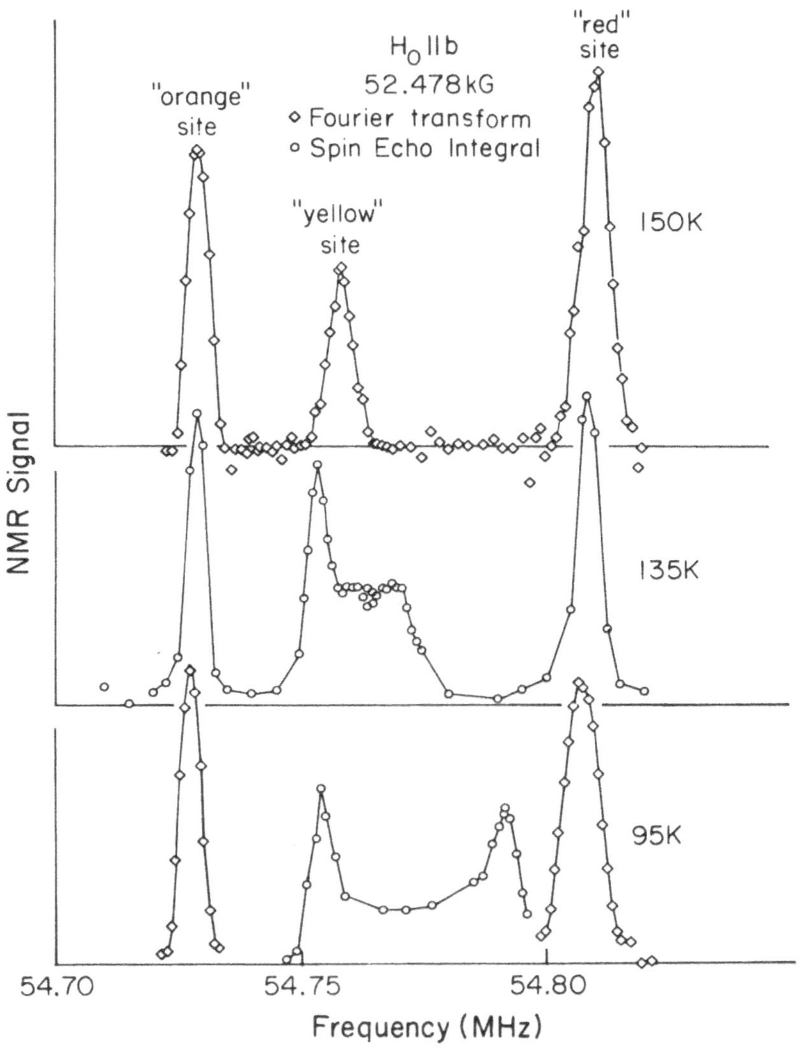

Fig. 7 ^{93}Nb NMR central transitions for $NbSe_3$. Development of an incommensurate CDW on the yellow chain is signaled by the characteristic line shape change. [from Ross, Wang and Slichter [55]]

the fitted modulation has the y principal axis fixed in orientation perpendicular to the b-axis. The orientation modulation used in Figure 8 consists of rotations of the x and z principal axes in a plane containing b. That is, b in the polar coordinates of the principal axes has the orientation $\Phi = 0$, and $\Theta = 90° + \Delta\Theta$, where the orientation modulation $\Delta\Theta$ has the form,

$$\Delta\Theta = \alpha \cos(\varphi_{CDW}). \tag{2.1}$$

In this equation, φ_{CDW} is the linearly-advancing CDW phase. The modulation amplitude α was fit to 27.8° at 95 K, and 30.0° at 77 K. Other parameters entering these fits are, $\eta = 0.60$ and $\nu_Q = 1.176$ MHz (for 95 K), and $\eta = 0.60$ and $\nu_Q = 1.18$ MHz (for 77 K). In addition, a small Knight shift modulation, described below, was added; the latter gives agreement with the field-dependence of the line width, but little effect on the shape. An improved fit was obtained by adding to the cosine in equation (2.1) a second harmonic of relative amplitude 4%; this is represented by the dashed curves in Figure 8.

The modulation described above is not a unique fit, for a y-axis modulation added to the above modulation up to approximately 45° also gives reasonable agreement, if constrained so that at some point x coincides with b. The latter condition is related to the observed high-frequency edge of the line shape, and is also a symmetry constraint, for while the CDW generally destroys the mirror plane perpendicular to b, the reflection symmetry still holds for sites at maximum and minimum positions of the CDW. This symmetry requires b to be a principal axis at some point, and thus the extremes of the CDW must correspond to the orientation x along b (high-frequency edge of the line shape).

The EFG rotation should not be taken to imply physical bending of the chains. For instance, in an ionic EFG calculation as described in section 2.2., the authors obtained large rotations of the yellow EFG from an axial displacement of the Nb site by 0.01 Å. There is no symmetry argument, however, that precludes an additional modulation in the EFG principal values.

Comparison of the high-temperature yellow EFG symmetry with the CDW-modulated EFG on the same site indicates that the sudden change in EFG symmetry indicated by the NQR may be a motional narrowing effect, as suggested by the NQR [54]. For the yellow site, below 145 K, the y principal axis remains fixed in the EFG orientation modulation, and its principal EFG is nearly identical to the largest EFG component for the yellow site above T_c. (Using the 77 K parameters, this principal value is $eQV_{yy}/h = 22.6$ MHz.) The direction of this axis is perpendicular to b, the same as the direction of the axial field above T_c. Rapid rotation of the 77 K EFG will give a motionally narrowed axial EFG tensor having the same symmetry and magnitude as the high-temperature yellow spectra. This requires, though, rotation through angles near 360°, rather than the 30° modulation found at 77 K. Thus the sudden shift in EFG for the yellow site is consistent with a valence electron contribution to the modulated EFG for this site which becomes time-dependent above T_c.

With the resolution available in spectra such as Figure 8, the discommensurate CDW structure can be ruled out, at least for temperatures above 77 K. The high-temperature $NbSe_3$ CDW differs from a four-atom commensurate wavelength by

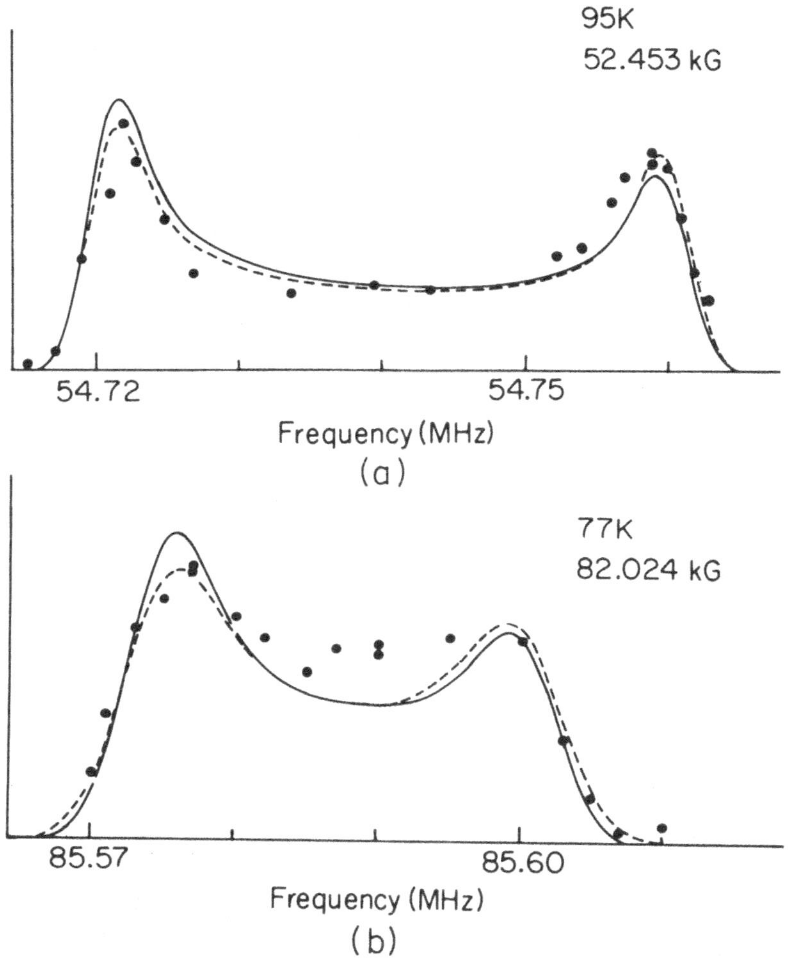

Fig. 8 Theoretical fits to the broadened $NbSe_3$ yellow-site central transitions, using rotated field gradients. Parts (a) and (b) are fits to the data at the indicated temperatures and fields. The dashed curves include a small second harmonic in the CDW phase, as described in the text [from Ross, Wang and Slichter [56]].

only a few percent, and commensurate locking might be expected, as predicted by Wilson [5, 45]. However, the line shape of Figure 8, particularly the lower-field spectrum, shows only two edge singularities, and no sign of the additional peaks expected for the discommensurate case [26, 27]. A quantitative measure is given by replacing the linear CDW phase in equation (2.1) with the solution of a sine-Gordon equation,

$$\frac{d^2\varphi_{CDW}}{dx^2} = \alpha^2 \sin\left(4\left(\varphi_{CDW} - \frac{2\pi}{4b}x\right)\right), \tag{2.2}$$

after McMillan's [17] formalism. In equation (2.2), $4b$ denotes four times the lattice constant b, the commensurate pinning wavelength. The phase φ_{CDW} is also specified by its average slope (set equal to the observed CDW wave vector), and an additive phase constant. Figure 9 shows numerical simulations of the line shape in which the additive phase constant was chosen to give two edge peaks, and α parameterizes the discommensuration width. The data are inconsistent with α larger than approximately 1.5 (in units where the discommensuration separation is 2π), and this corresponds to little or no commensurate distortion, as shown in the inset to Figure 9.

Clearly, then, NMR measurements have demonstrated that the high-temperature CDW transition involves an incommensurate modulation localized on the yellow structural chain. Furthermore, the low-temperature CDW is associated with changes at the orange structural chain, with no changes at the red site. The complete disappearance, at 4.2 K, of the quadrupole satellite lines associated with the orange site [53], provides strong evidence that the 59 K transition is associated with that site. Furthermore, a broadening of the orange central line, characteristic of an incommensurate CDW, has been observed at 4.2 K [57]. The latter is shown in Figure 10, for the orientation $H\|c$, although the broadening mechanism has not been characterized. No change in the red line shape is observed at these temperatures.

In measurements of the yellow line at 4.2 K, both the NQR lines [59] and Knight shift (Table 2) are little changed from the 77 K values. These confirm that the 59K transition does not change the electronic structure of the yellow site. A phase-locking of the two CDW's is more difficult to rule out through the NMR results. Evidence for and against such an interaction is discussed by Monceau [11].

2.4. KNIGHT SHIFTS AND SUSCEPTIBILITIES

Devreux has estimated the orbital and spin contributions to the susceptibility [53], providing a quantitative measure of the changes in density of states caused by the transitions. These estimates utilize bulk susceptibility results in combination with Knight shift measurements. Devreux assumes that d-electrons are the main contribution to the spin susceptibility, with a negligible selenium contribution. (Devreux quotes unpublished NMR work by Pannisod giving some estimate of the latter.) Susceptibility information was extracted using the hyperfine fields [71] $H^{orb} = 0.28 \times 10^6$ G, $H^{cp} = -0.21 \times 10^6$ G, and $H^{dip} = 0.04 \times 10^6$ G (the orbital, core polarization, and dipolar terms, respectively).

The isotropic and b-axis Knight shifts were measured in all temperature regimes (Tables 1 and 2), and compared to the $NbSe_3$ susceptibility measurements of Kulick

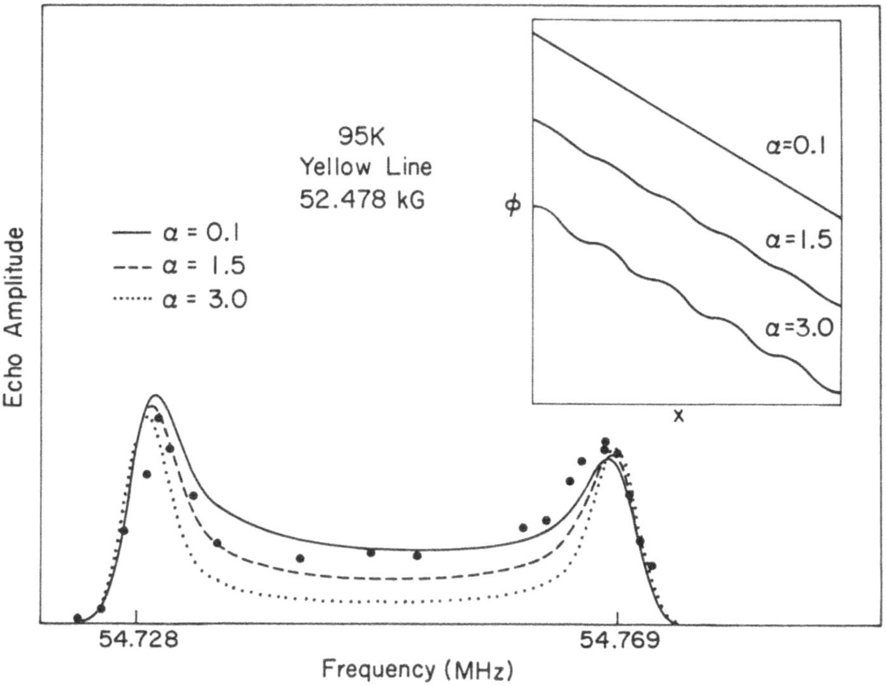

Fig. 9 **CDW line shape broadening in $NbSe_3$, fit to a model including discommensurations. Inset: CDW phase profile for three values of α corresponding to the three curves in the main graph. [from Ross, Wang and Slichter [55]]**

and Scott [72]. For the isotropic case, the Knight shift is given by:

$$\beta A\ \overline{K}_{iso} = H^{cp}\ \overline{\chi}^{s} + H^{orb}\ \overline{\chi}^{vv}_{iso}, \tag{2.3}$$

where $\overline{K}$ and $\overline{\chi}$ denote averages over the three Nb sites, β is the Bohr magneton, and A is Avogadro's number. In combination with the measured isotropic susceptibility, $\overline{\chi}_{iso} = \overline{\chi}^{s} + \overline{\chi}^{vv}_{iso}$, values given in Table 3 are obtained. The data in this table are taken directly from Devreux' analysis, except for the addition of data for 175 K, extracted in the same way. The average spin susceptibility decreases with temperature, consistent with the opening of Fermi-surface gaps due to the CDW transitions.

Further comparison of the b-axis Knight shift components gives a site-selective measure of these changes. The relationship for these components is,

$$\beta A\ K_{\|,i} = (H^{cp} + 2H^{dip})\chi^{s}_{i} + H^{orb}\ \chi^{vv}_{\|,i}, \tag{2.4}$$

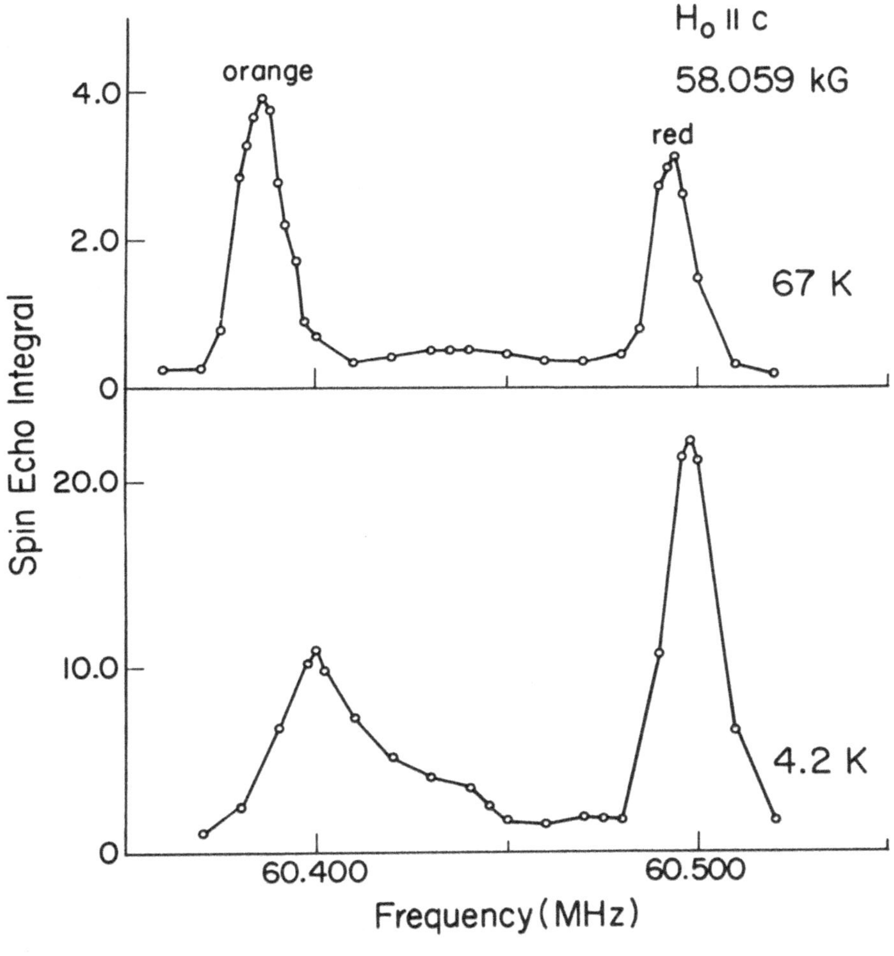

Fig. 10 ^{93}Nb NMR central lines in $NbSe_3$, showing broadening of the orange site below the 59K CDW transition. Crystal orientation, $H\|c$, rotated with respect to previous figures. Yellow-site resonance is out of range. [from Ross [57]]

where the subscript i refers to the three sites, $\chi_{\|,i}^{vv}$, denotes the b (or parallel) component of the van Vleck susceptibility, and the dipolar portion is assumed to be axial along b (appropriate for a d_{z^2} band). The mean spin susceptibility extracted

	295 K	175 K	77 K	4.2 K
$\bar{\chi}^{s} \times 10^{-6}$ emu/mole	20	18	16	9
$\bar{\chi}^{vv}_{iso}$ "	35	34	31	36
$\chi^{s}_{R,\parallel}$ "	0	0	0	0
$\chi^{s}_{O,\parallel}$ "	30	35	39	18
$\chi^{s}_{Y,\parallel}$ "	30	19	9	11
$\chi^{vv}_{R,\parallel}$ "	63	64	63	62
$\chi^{vv}_{O,\parallel} = \chi^{vv}_{Y,\parallel}$ "	54	52	54	52

Table 3 Susceptibility components for $NbSe_3$ extracted from measured Knight shifts, as given by Devreux [53]. These are supplemented by 175 K values extracted from data of Table 2. R, O and Y refer to the red, orange and yellow sites, respectively (site identification as in section 2.2).

above, $\overline{\chi}^{s} = (1/3)\sum \overline{\chi}_i^{s}$, provides a fourth relationship determining the susceptibility components. To completely determine the six components, Devreux assumed that the spin susceptibility was zero for the red site, and that the Van Vleck components were equal for the yellow and orange sites. The result of this analysis, given in Table 3, clearly shows a reduced spin susceptibility for the yellow and orange sites below the respective transitions.

For the *b*-axis components in Table 3, the numbers again are those of Devreux, with the addition of 175 K data from Table 1, analyzed in the same way. The most notable difference between the room temperature and 175 K spin susceptibilities is a reduction at the yellow site at 175 K. The reduction in yellow-site susceptibility is consistent with the temperature dependence of the bulk susceptibility [72]. This change may be due to sharp features in the Fermi-surface density of states, as observed in other transition metals [73], or due to the development of a "pseudogap" due to critical CDW fluctuations.

The susceptibility changes of Table 3 indicate that roughly half the Fermi surface at both the yellow and orange sites is destroyed at the corresponding transition. As Devreux notes, these figures must be regarded as rough indications because of uncertainties in extracting the core diamagnetism from the susceptibilities, and also in the hyperfine fields. For instance, choosing $H^{cp} = -0.15 \times 10^{6}$ G (consistent with the measurements of Berthier *et al.* [74] in $NbSe_2$) produces a yellow-site spin susceptibility of zero at 77 K using the above analysis.

It is important to note that magnetic fields of a few Tesla or more have been found to affect the non-Ohmic transport properties of $NbSe_3$ strongly, below the 59K transition [75, 76]. No such effects occur in the high-temperature CDW regime. Explanations that have been advanced include Fermi surface changes due to magnetic breakdown [77], and field-induced spin density waves. No field-dependence of the 4.2K Knight shift is observed, however, as might accompany these processes. The Knight shift measurements of the two groups are in close agreement, and include measurements in fields up to 14 T.

2.5. SPIN-LATTICE RELAXATION

Spin-lattice relaxation measurements in $NbSe_3$ include a ^{93}Nb NMR study by Wada, Aoki, and Fujita [58] and the NQR measurements of Suits and Slichter [54]. Key findings include a drop in relaxation rate at each CDW transition, frequency-dependent relaxation indicative of one-dimensional electron diffusion, and evidence of fluctuations on the yellow site below the 145 K transition.

NMR T_1 data are shown in Figure 11. For these data Wada *et al.* used a powder sample, and the spectra are superpositions of signals from the three sites. The NMR frequency was 10.5 MHz, and the measurements were taken near the center of the line shape, where the (1/2,-1/2) transitions are strongest. In this situation, the relaxation for a single site is multi-exponential [78], and for magnetic relaxation the signal recovery has the form,

$$M = M_o(1 - a_1 e^{-2Wt} - a_2 e^{-12Wt} - a_3 e^{-30Wt} - a_4 e^{-56Wt} - a_5 e^{-90Wt}). \quad (2.5)$$

The coefficients a_i depend on saturation conditions, and $W = (2T_1)^{-1}$, where T_1 is the conventionally-defined recovery time. Wada *et al.* find that for a comb of pulses saturating the central line, the a_1 term has a significant contribution, while for a single pulse applied to the central transition the a_1 term is almost zero, giving effectively a much faster relaxation. The measurements of Figure 11 utilized a saturating comb at 4.2K and below, and a single pulse for higher temperatures, and were fit in each case to the appropriate recovery curve. The result is a single T_1 representing a combination of the three sites.

Three temperature regions of Figure 11 are identified, in which the relaxation is fit to the Korringa law, (T_1T) = constant. These temperature regions are separated by sudden changes in T_1 at the CDW transitions. Below each transition, T_1T increases to a new value, consistent with a destruction of part of the Fermi surface, but with a peak in T_1T near each transition. This behavior may be due to a reduced density of states, and hence relaxation rate, for the chain associated with each transition. However, this is not consistent with NQR observations of the yellow-site relaxation at the 145 K transition, and this result awaits further interpretation.

Relaxation of the yellow-chain resonance in NQR was measured by Suits and Slichter [54]. In this study, the relaxation was measured above and below the 145 K transition, following the yellow-line frequency changes as in Figure 6. Results

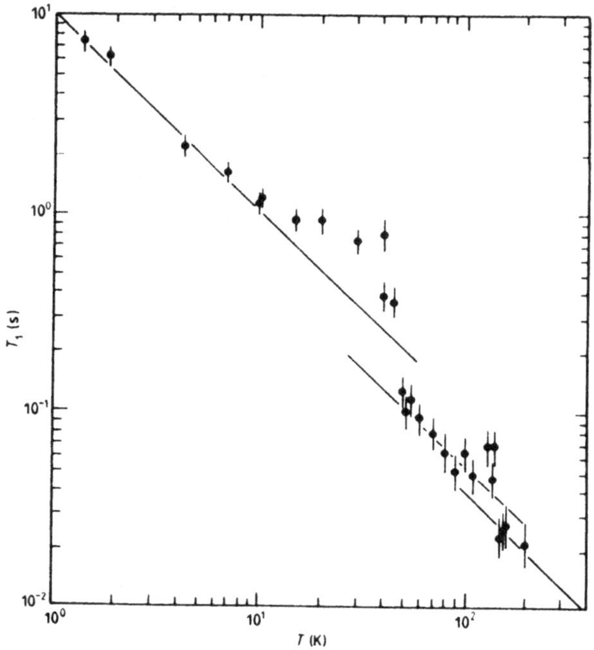

Fig. 11 ^{93}Nb spin-lattice relaxation times in $NbSe_3$, from NMR measurement. Signal is a superposition of three sites, and straight lines are Korringa-law fits. [from Wada, Aoki and Fujita [58]]

are shown in Figure 12, where the T_1 represents the time for the signal to recover by approximately 60%, following a one-pulse saturation of the highest-frequency transition. The measurements cover the range 100 K to 300 K, bracketing the 145 K transition. The peak in relaxation rate below T_c is interpreted as the result of EFG fluctuations, and an analysis based on fluctuations of the yellow-site EFG [54] yields rates consistent with those observed. The increase in relaxation rate is consistent with thermally-driven normal modes of the CDW [26], and the peak is quite similar to that observed in the blue bronze material [79].

Wada, Aoki, and Fujita have examined the frequency-dependence of the T_1 at 4.2K, and find that it fits a model of one-dimensional electron diffusion. Relaxation due to this mechanism has been treated by Devreux [10], and Lyo [80], in a Hubbard approximation. For three-dimensional diffusion, the T_1 is frequency-independent, whereas when the resonance frequency exceeds $1/\tau_\perp$, a frequency-dependent T_1

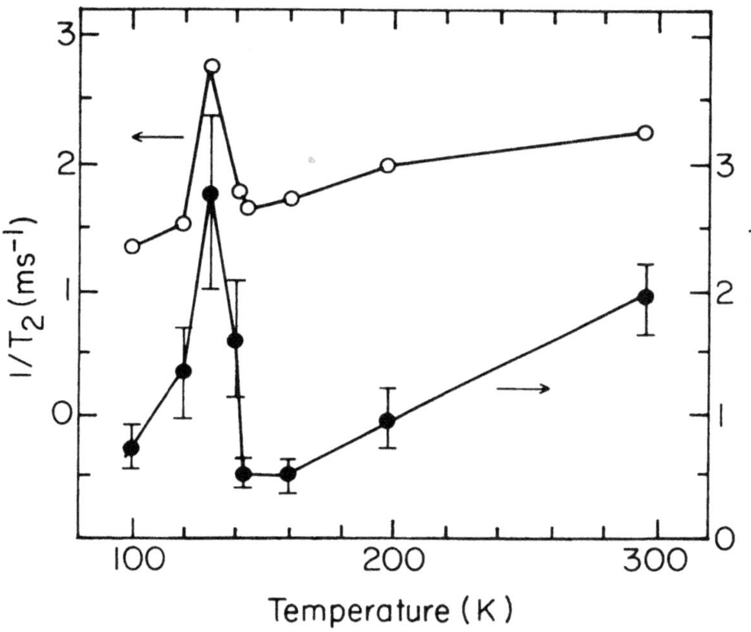

Fig. 12 ^{93}Nb relaxation rates in $NbSe_3$ for the yellow site, from NQR measurements. CDW transition occurs at 145 K, and the relaxation peak below T_c is described in the text. [from Suits and Slichter [54]]

can be observed, related to $\tau_{\|}$. (The times $\tau_{\|}$ and $\tau_{\perp}$ are on-chain and inter-chain relaxation times, not Hubbard transfer integrals). The experimental data have a crossover near an NMR frequency of 10 MHz, as shown in Figure 13.

Relaxation in this model contains spectral density terms at both the nuclear and electron resonance frequencies (ω_n and ω_e, respectively). The ω_e term comes from core polarization, and the ω_n term from orbital couplings. However, the high-frequency $(T_1T)^{-1}$ data are proportional to $\omega^{-1/2}$ with no additive constant, and from the magnitude of the crossover frequency Wada *et al.* deduce that the ω_e (core polarization) contributions dominate the relaxation. From the crossover frequency is then found $\tau_{\perp} \approx 5.7 \times 10^{-12}$ s, at 4.2K.

An estimate of $\tau_{\|}$ was obtained with the additional measurement of the anisotropy of the 4.2K T_1. For the anisotropy measurement [58], a partially-aligned sample was used having all b-axes parallel. From these data is deduced that the d-orbital admixture responsible for core-polarization contains approximately 80% d_{z^2}, uti-

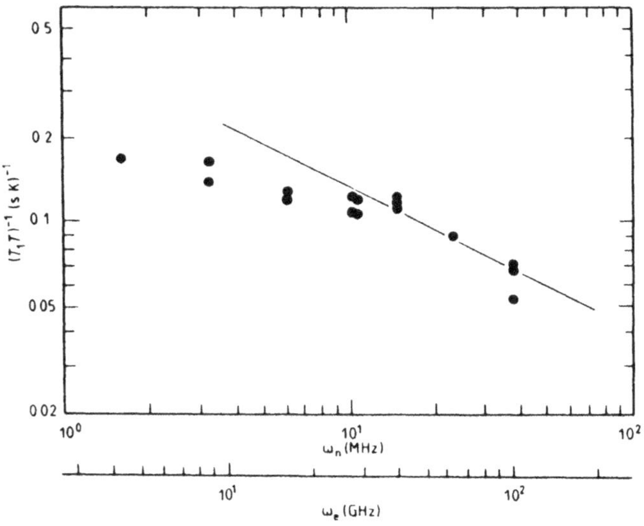

Fig. 13 Frequency-dependent ^{93}Nb relaxation rates in $NbSe_3$, from NMR at 4.2K. Straight line indicates a rate proportional to $\omega^{-1/2}$, and the crossover frequency is indicated. [from Wada, Aoki and Fujita [58]]

lizing the calculations of Obata [81]. They find $\tau_{\parallel} \approx 3 \times 10^{-14}$ s, and using an estimated density of states, the ratio of Hubbard transfer integrals is found to be, $\tau_{\perp}/\tau_{\parallel} \approx 0.07$.

2.6. NMR OF THE MOVING CDW

In a test of the properties of the moving CDW, Ross, Wang, and Slichter [55-57] have measured NMR in a $NbSe_3$ sample in which current could be applied simultaneously. These studies provide a local probe of the dynamics of CDW conductivity, which can give information unavailable with macroscopic probes such as transport measurements. These studies are closely related to measurements of the blue bronze material, described separately in this volume. The NMR measurements have provided a demonstration of bulk CDW motion through motional narrowing, and information about the CDW motion spectrum, as well as CDW response in the ohmic regime.

2.6.1. *NMR Sample* The sample for these measurements, aligned along the *b*-axis only, contained approximately 30 crystals from one growth batch. Electrical contact was made with low-melting-point solder, and electrical leads included small inductors acting as rf chokes to minimize damping of the NMR tank circuit. The crystals, 1 cm in length, were connected electrically in parallel, in a two-probe arrangement, and had minimum CDW conductivity thresholds between 40 and 100 mV/cm near 110 K, comparable to high-quality crystals reported by other groups. A differential resistivity measurement of this sample is shown in Figure 14(a) at 111 K; a distributed threshold characteristic of the multiple crystals is apparent. At 77 K, at which the NMR measurements were taken, the threshold is less sharp, and distributed between approximately 100 and 200 mV/cm. Both the rounding and increase in the threshold are typical for this material.

Narrow-band noise signals were also observed in the NMR sample. At 110 K the spectra consist of several closely-spaced peaks, with harmonics. The separate peaks can be attributed to different crystals in the sample. At 77 K the noise peaks become more rounded, and the spectrum is shown in Figure 14(b). The lower trace is the spectrum with no sample voltage, and the upper trace is the spectrum with a sample bias of 711 mV/cm. The structure common to both traces is external noise. A prominent fundamental is observed, and its harmonic, along with a broad enhanced noise background. A small third harmonic can also be observed, but is not apparent in this figure.

This sample, used for all of the CDW motion measurements, contains approximately 3 μg of $NbSe_3$, or 6×10^{16} spins per site. This sample required typically 10^6 spin echoes signal-averaged to produce an observable signal at 77 K and 8.4T, although 30 ms repetition rates were possible due to the ^{93}Nb relaxation rate. All measurements were taken at 77 K with the sample immersed in liquid nitrogen, and the pulse generator for sample bias was current-regulating due to a large series resistance.

2.6.2. *Motional Narrowing* Motional narrowing of the yellow-site central line, and limited spin-spin relaxation time T_2 measurements, have provided a demonstration that the CDW exhibits bulk translation in the non-Ohmic conduction regime [55-57]. For these measurements, crystalline *b*-axes were aligned along the magnetic field, and the field was fixed at 8.4533 T. The (1/2,-1/2) NMR line for the yellow site at this field and 77 K is essentially the same as shown in Figure 8(b). The 30 kHz linewidth consists of second-order quadrupole broadening due to the incommensurate CDW (as described in section 2.3), convoluted with an intrinsic (non-CDW) linewidth of 8 kHz. The intrinsic linewidth, mostly dipolar in origin, is equivalent to the linewidth of the red and orange lines in the same field. Differential resistance measurements in the NMR magnet and in ambient field confirm the absence of field-induced changes in non-Ohmic behavior at these temperatures.

Motional narrowing of the CDW line, measured via the spin-echo width, is shown in Figure 15. Two echoes are shown, with and without an applied current in the sample. The observation frequency was in the center of the broadened yellow line, and high frequency oscillations in the echoes are beats between the oscillation

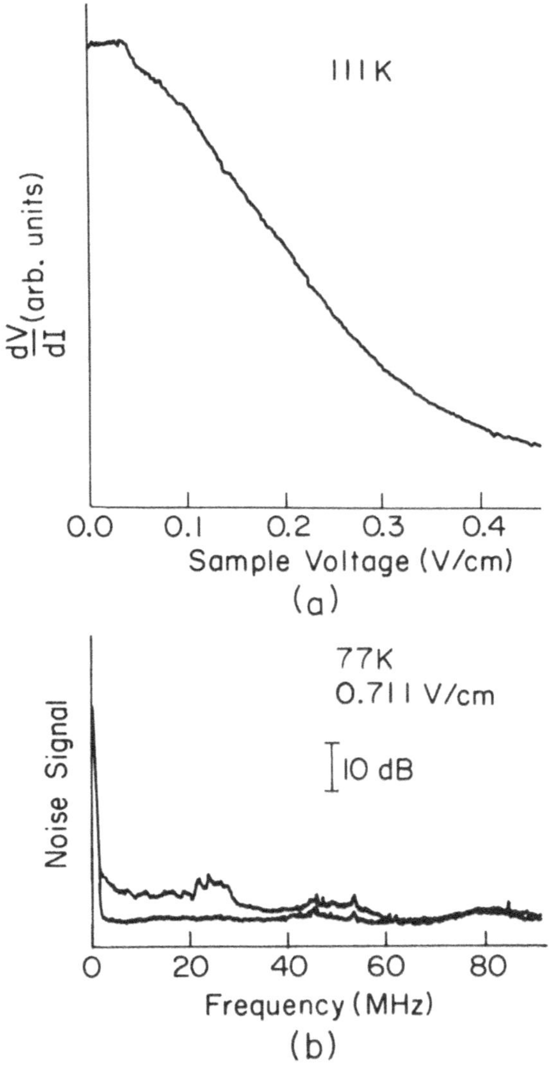

Fig. 14 Electrical characteristics of the $NbSe_3$ sample used for NMR studies of CDW motion. (a) Non-ohmic conductivity observed a 111 K. (b) Narrow-band noise, with broad backround, observed at 77 K. [from Ross, Wang and Slichter [55-57]]

frequency and echoes from other sites. With no applied voltage (lower curve), the narrower echo corresponds to a broader line shape, and the "wings" on this echo are due to edge peaks in the line shape. With 100 mA in the sample (or a 1.4 V/cm applied field, about 10 times the threshold), the echo becomes broad, and the wings disappear. The broadened echo corresponds to a narrowed line shape; limited studies of the spin-echo Fourier transform indicate a featureless narrowed line at large currents.

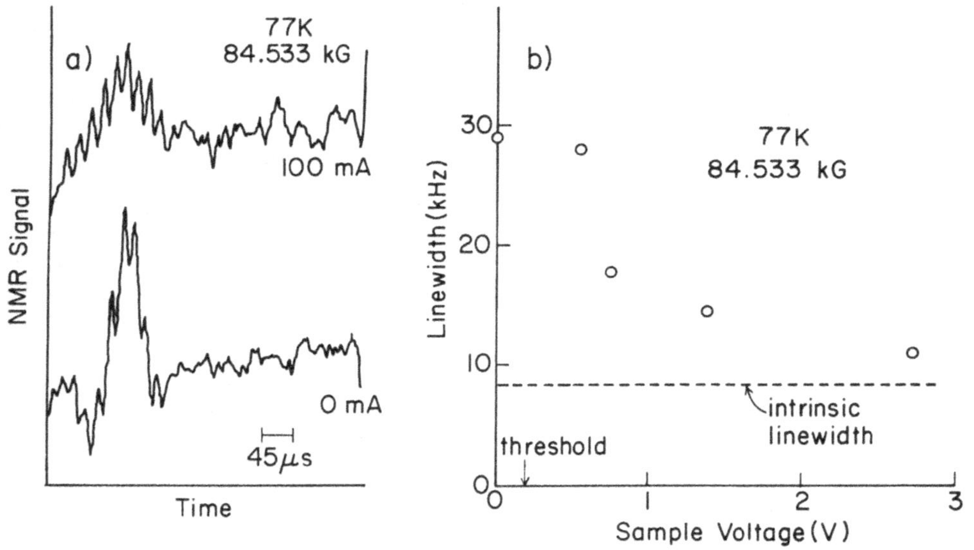

Fig. 15 Motional narrowing of NMR signal in $NbSe_3$ due to CDW motion. (a) Spin echoes observed with and without current in the sample; the broad echo corresponds to a narrowed line. Oscillations on top of the echo are due to red and orange sites. (b) Linewidth extracted from the echo, plotted *vs.* sample voltage. Intrinsic linewidth is that of non-CDW sites. [from Ross, Wang and Slichter [55]]

For Figure 15, the rf pulse separation was 150 μs. A simple two-pulse sequence was used, alternating the phase of the first pulse for an improved baseline. Echoes were acquired at 30 ms intervals, and current was applied only for the duration of the echo sequence. In this manner, saturation effects (described below) were avoided.

A measure of the motional narrowing dependence on sample voltage is shown in Figure 15(b). The quantity plotted is $\frac{0.882}{t_{1/2}}$, where $t_{1/2}$ is the full-width at half maximum (FWHM) of the echo. This corresponds to the spectral FWHM for a Gaussian line shape, and serves as a relative measure of the line narrowing for the more complicated actual line shape. The intrinsic linewidth, indicated in the figure, represents the width of the red and orange central lines at the same field, measured in the same way. The yellow-site linewidth approaches the intrinsic linewidth at large sample bias, indicating the disappearance of the CDW contribution to the linewidth.

Motional narrowing at large sample currents in $NbSe_3$ is consistent with translational motion of the CDW. Since the nuclei remain essentially stationary during such motion, the intrinsic (dipolar) linewidth is not averaged, contrary to the usual case of motional narrowing in liquids [2]. Ideal uniform translation of the CDW will produce a periodic EFG at each yellow site. This situation should produce sidebands in the NMR line [82], at intermediate CDW translation velocities. Such features have not been observed in $NbSe_3$, but a distribution in CDW velocity between crystals could easily mask such a structure. On the other hand, sidebands have been observed in the case of the blue bronze (described in the next chapter).

Even at the largest voltages, the linewidth plotted in Figure 15 exhibits an excess over the intrinsic value. This could represent a difference in dipolar broadening between sites. It is also possible that the source is crystals with low current due to poor electrical contacts. From the relative linewidth reduction, no more than approximately 10% of the sample can have such poor contacts. (The observed narrow-band noise in this sample, as well as saturation results described in the next section, indicate that CDW motion in most crystals is more or less uniform.) Clearly CDW motion encompasses the major portion of the sample.

The ^{93}Nb NMR signals include also the broad satellite lines. These satellites will not overlap the central lines, even for very rapid CDW translation, since the background EFG remains. (The EFG principal values and orientation modulation for the yellow site are well-known, as described above.) Thus, most neighbors remain inequivalent, and the (1/2,-1/2) coherence can be regarded in a "fictitious spin-1/2" sense [83]. In particular, the dipolar linewidth should change little from the rigid-lattice case, as observed.

In contrast to the dipolar linewidth, the spin-spin relaxation exhibits large changes in the motional narrowing regime, attributed to dynamical effects. The static T_2 decay of the yellow central line, at 77 K, is fit by a sum of functions which is 70% Gaussian and 30% exponential, with a signal reduction of 50% at $2\tau_D = 2.1$ ms. This is consistent with a J-coupling mechanism [55-57], causing spin exchange between equivalent neighbors. Motional narrowing will increase the number of equivalent neighbors, but only to the extent of the NQR situation, for which the T_2 has been measured (Figure 12). The observed rate is much faster than these expectations, in limited T_2 measurements of the yellow site with applied current.

The data, plotted in Figure 16, represent the spin echo amplitude *vs.* the time between the two rf pulses. A reduced T_2, with applied current, persists to the highest voltage applied, 2.0 V/cm, at which the narrow-band noise fundamental peak is observed at 80 MHz. The solid line in Figure 16 represents an exponential decay with $T_2 = 250\mu$s, intended to indicate the approximate time scale for the current-induced relaxation. While the T_2 reduction is similar to CDW phason relaxation [26], no comparable reduction in T_1 has been observed, so the dynamics that produce this effect must be quasi-static.

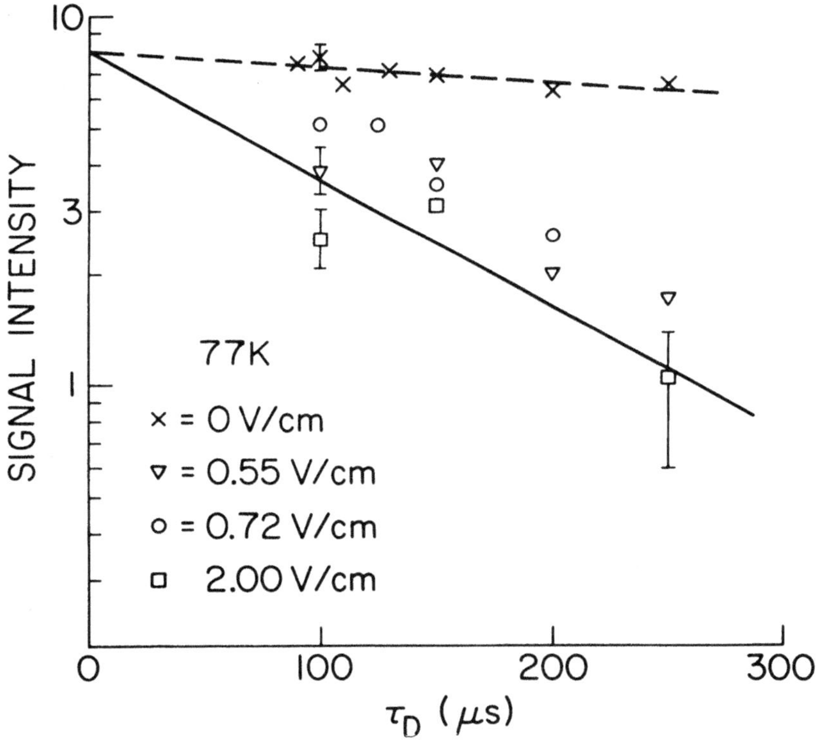

Fig. 16 Spin-echo decay in $NbSe_3$ as affected by sample voltage. Solid line is an exponentional decay with $T_2 = 250\mu s$, to indicate approximate time scales. [from Ross, Wang and Slichter [56]]

Rapid T_2 relaxation due to CDW motion will bias the motional narrowing data of Figure 15(b) by giving extra weight to any static portions of the sample. In fact, the echo width is observed to have some dependence on the delay time between pulses. This dependence, though, is shown to be consistent with a CDW motion inhomogeneity in the sample equal to the width of the observed narrow-band-noise peaks [55-57].

Furthermore, since the amplitude of the spin echo, extrapolated to zero delay time, is proportional to the number of spins observed, it is apparent from Figure 16 that the motionally narrowed signal represents all or most of the spins observed with no applied current. At least half of the sample is observed in the motionally narrowed signal. This rules out any model in which only small sections of the CDW are actually in motion, showing instead that the CDW motion involves the entire crystal.

While the observed motional narrowing is consistent with simple CDW translation at large currents, the behavior at intermediate currents is more complicated, and is found to agree with nonuniform CDW motion, such as the dislocation-motion theories described above. For uniform CDW translation, one expects that the NMR line will narrow when the CDW motion frequency equals approximately the static linewidth, which in our case is 30 kHz. This is the case for random fluctuations [2], and also for constant-velocity CDW motion [82], in which the sidebands already discussed vanish rapidly once they exceed the boundaries of the static line shape.

In contrast to these expectations, narrowing is not observed in $NbSe_3$ until the apparent CDW motion frequency far exceeds the static linewidth. At the voltage corresponding to the inflection point in Figure 15(b), the narrow-band noise fundamental was found to be slightly higher than 10 MHz. An estimate based on the excess CDW current yields a value approximately 5 times smaller (the discrepancy may result both from sample inhomogeneity and from uncertainties in sample cross-section). These CDW motion frequencies are much larger than the 30 kHz NMR linewidth.

Since the T_2 measurements indicate that essentially all regions of the sample participate in CDW motion, the discrepancy in motional narrowing was attributed to highly nonuniform CDW motion, consistent with both the T_2 and T_1 behavior. The CDW motion which fits these observations involves CDW advancement in a series of hops. This explanation relies on the periodic nature of the CDW-induced frequency modulation. For motion frequencies small compared to ω_L, the time-domain signal can be described,

$$S(t) = S_0 \left\langle \exp\left(i \int \Delta\omega(x,t)dt \right) \right\rangle, \tag{2.6}$$

where the angular brackets represent an average over x, and $\Delta\omega$ represents the local instantaneous NMR precession frequency (referenced to the observation frequency). For motion described as a sequence of hops, the integral is strongly weighted by the stationary time periods. Furthermore, if each hop involves a CDW phase advance by 2π (expected due to the nature of pinning), the signal will be nearly that of a stationary CDW, even though the average CDW motion frequency may exceed the classic narrowing criterion.

In this model, motional narrowing occurs when CDW motion goes over from hopping motion to smoother motion at voltages far exceeding the conduction threshold. A simple picture describes the situation; if each hop requires a time T_h, a crossover to smooth motion (and narrowing) will occur at an average hopping frequency of $1/T_h$. The narrow-band noise frequency, measured at the voltage when motional narrowing commences, implies that $T_h \approx 0.1\mu$s.

Comparison to models described in section 1.3 indicates that the hopping motion model may correspond to dislocation, or soliton, motion, based on strong pinning in some regions. In this case, T_h is the time for a dislocation to pass a fixed point. Weak pinning alone is unlikely to imply motion that is sufficiently discontinuous; see Figures 2 and 3. Noise-pulse observations in the blue bronze [51] show pulses of millisecond duration, a much longer time-scale than reported here, although screening by normal electrons in $NbSe_3$ may be responsible for reducing dislocation widths.

While a single time-scale was used to explain motional narrowing, the T_2 results imply that quasi-static components remain important even for average CDW motion frequencies near 100 MHz. A large low-frequency component to the spectral density is implied, and is presumably related to the broadband low-frequency electrical noise associated with CDW motion. The latter can be seen in Figure 14. The NMR results indicate that this low-frequency component is distributed homogeneously in the sample, instead of a small number of domains.

2.6.3. *Current-induced Saturation* Saturation of the yellow-site NMR signal induced by sample current gives an additional measure of the CDW motion [55-57]. The observations indicate CDW motion at 77 K having a local spectral density function roughly equivalent to the broadened "narrow-band noise" peaks, rather than a locally constant CDW velocity.

The observations are summarized in Figure 17. The measurements follow a common T_1 measurement scheme, in which the spin-echo repetition rate is varied to find the magnetization recovery. An effective single exponential $T_1 = 20$ ms gave a good fit to the static-CDW data. (The relaxation will be multiple-exponential, of the form of equation (2.5), but with undetermined coefficients.) The inset to Figure 17 shows the static-CDW recovery, and the recovery with a sample bias of 1.38 V/cm. The zero-bias measurements were fit to the 20 ms recovery curve measured with a larger sample. With the 1.38 V/cm sample bias, the recovery is faster (effective $T_1 = 11$ ms), and to a smaller equilibrium value.

All measurements were performed at 77 K, and using the sample described in section 2.6.1.. Sample bias was applied only during the recovery time between spin echoes. This eliminated ambiguities due to changes in T_2 caused when the current was on during the echo sequence. Data represent the signal-averaged spin echo amplitude.

The equilibrium signal, the main plot of Figure 17, represents the asymptote for the recovering magnetization (inset). In the absence of an external perturbation, the equilibrium signal will reflect only the temperature and the number of spins. The change in equilibrium, and the faster observed recovery with large bias, agree with a model in which CDW motion induces random spin flips, through an oscillating local EFG. A rate equation for such a process is,

$$\frac{dn}{dt} = -\left(\frac{n - n_o}{T_1}\right) - 2Wn, \tag{2.7}$$

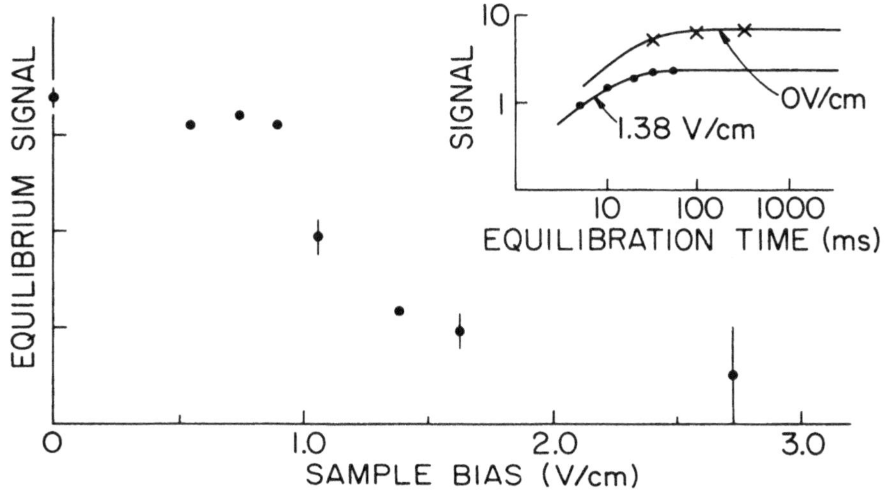

Fig. 17 Current-induced NMR saturation in NbSe$_3$. Inset shows recovery of the signal *vs.* repetition rate, with curves described in text. Main graph shows the asymptote extracted according to sample bias. [from Ross, Wang and Slichter [55]]

where n represents the magnetization density, with n_0 its equilibrium value in the absence of irradiation. The first term on the right represents a thermal T_1 process, approximated as a single exponential process, and W is the spin-flip rate induced by CDW motion. A solution to this equation is,

$$n(t) = n_{eff}\left[1 - \exp\left(\frac{-t}{T_{eff}}\right)\right], \tag{2.8}$$

where $n_{eff} = n_0/(1 + 2T_1W)$, and $T_{eff} = T_1/(1 + 2T_1W)$. The effective T_1, and the effective equilibrium magnetization, are reduced by the same factor. This is approximately valid for the current-induced recovery curves, such as shown in the inset of Figure 17.

The equilibrium magnetization of Figure 17 has a threshold near 1.0 V/cm above which the signal saturates. A continued reduction in signal is observed at all higher voltages. At the saturation threshold voltage, the narrow-band noise second-order harmonic reaches the NMR frequency, 88 MHz, whereas at the largest bias condition the narrow-band noise fundamental is centered near the NMR frequency. This is consistent with resonant saturation of the spins due initially to the $4k_F$ phonon component of the CDW. A $4k_F$ term is observed also in X-ray measurements [84].

The narrow-band noise second-order harmonic at this voltage has a width (FWHM) of 16 MHz, whereas the fundamental, at 45 MHz, has a width of 8 MHz.

Steady CDW motion would produce locally a very narrow spectral density, and at a given current only those crystals or domains with a motion frequency resonant with the ^{93}Nb spins will be effective in giving saturation. The small signal reduction implied by this situation is not observed, and so a locally fluctuating CDW motion is much more likely. (Saturation related to spin diffusion to only a few moving domains is short-circuited by the rapid T_1.) An approximate calculation of the spin-flip matrix elements [57], due to the known EFG axis modulation of the yellow site, gives a good accounting for observed saturation if the spectral width for CDW motion is that of the narrow-band noise peaks. Thus, these observations affirm that CDW motion encompasses the entire sample, and that the CDW motion spectrum is locally equivalent to that of the narrow-band noise.

2.6.4. *Pulsed CDW Displacements* A further NMR measurement [55-57] of the CDW response to applied voltages in $NbSe_3$ probed CDW displacements near the conduction threshold. The CDW electrical response includes very large dielectric constants ($\epsilon \sim 10^8$), which can be attributed to displacements of the charged CDW. In the rigid overdamped oscillator model, the CDW is displaced by 90° within its pinning barrier, just below the conduction threshold. From an NMR standpoint, the inhomogeneous line shape (Figure 8) results from nuclei in different portions of the CDW, and CDW displacement will change the NMR frequency of each site, which can be detected through pulsed NMR. The experiments detected much smaller frequency shifts than expected, which is attributed to an inhomogeneous CDW response.

A Carr-Purcell-Meiboom-Gill pulse sequence [85, 86] was used for these experiments, consisting of one 90° rf pulse followed by a series of 180° rf pulses. To improve the signal-to-noise ratio, echoes from an eight-echo sequence were added, using a weighting function equal to the T_2 decay function. As before, these measurements involved the yellow-site line at 77 K.

Results of this experiment are shown in Figure 18. The NMR frequency was in the center of the yellow line, and the data points are the integrated areas of the summed spin echoes. (The integrated area represents a narrow slice of the NMR signal, near the NMR frequency.) To the rf pulses were added one or two voltage pulses to the sample, at a fixed sample voltage of 150 mV, roughly the center of the broad conduction threshold in the sample. The inset pulse sequence in Figure 18 shows the timing for these pulses, along with the first echo of the Carr-Purcell train. Voltage pulses were applied between the 90° and 180° pulses, for each echo train, and between the first 180° pulse and the first echo, in odd-numbered echo trains. Echoes for one and two voltage pulses were signal-averaged separately.

The echo refocusing process relies upon spin precession that is the same before and after the spin-reversing 180° pulse. The reduction in signal, induced by the voltage pulse, implies a change in the quadrupole shift appearing only during the voltage

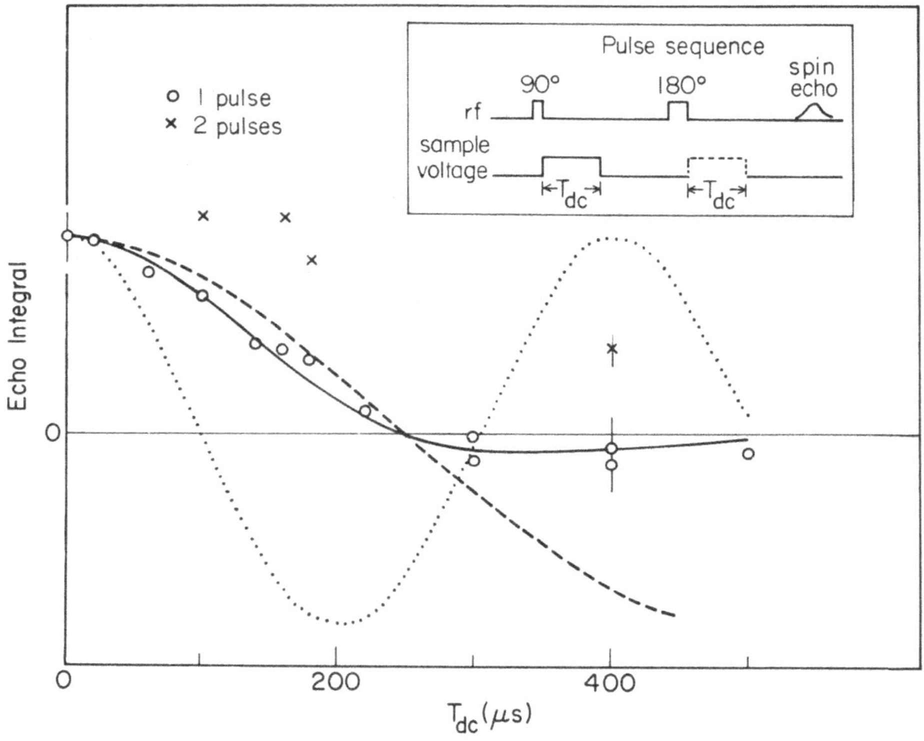

Fig. 18 Pulsed NMR study of dielectric displacements in $NbSe_3$. Inset: pulse sequence, showing only the first echo in a Carr-Purcell-Meiboom-Gill sequence. Main graph: echo integral *vs.* T_{dc}, with the observation frequency in the middle of the yellow-site central line. Curves: theoretical simulation with a phase advance of 5° (dotted curve), 2° (dashed curve), and a distributed phase advance about an average of 2° (solid curve). [from Ross, Wang and Slichter [55]]

pulse, due to a voltage-induced shift in the CDW position. Partial refocusing of the echo with two pulses indicates that the displacement of the CDW is not completely reproducible. The latter effect, however, may be due to some crystals in which the applied voltage is above threshold. In addition, under the experimental conditions a small amount of stimulated echo included in the signal average will reduce the relative refocusing for two voltage pulses (but not for one pulse).

The curves in Figure 18 were generated numerically to fit the one-pulse experiment, assuming constant CDW phase displacement during the voltage pulse, and immediate phase relaxation upon turning off the voltage. The numerical calculation for Figure 18 was based on a generic line shape fit, assuming that the observed linewidth was first-order in the charge density. The detailed line shape fit described in section 2.3. implies a phase displacement smaller by a factor of two than reported here, but this does not greatly affect the conclusions. The dotted and dashed curves in Figure 18 assume uniform CDW phase displacements of 5° and 2°, respectively. Much better agreement is given by the solid curve, which includes an average displacement of 2°, but with a large distribution, of half-width 2°. The distribution may be related to inhomogeneities within the sample, but the average value is much smaller than the 90° displacement expected for a rigid CDW.

It seems likely that CDW dislocations are required to explain the small CDW displacements measured in this experiment. In the quantum tunneling model, displacement within one domain can be smaller than the classical value since tunneling over the barrier is involved. Using parameters typical for I-V curve fits in $NbSe_3$, the phase displacement at threshold in this model is estimated [55-57] to be 13°, still too large to explain the NMR results. Whereas it was suggested [87] that unidirectional voltage pulses applied repetitively, for signal averaging, might lock the CDW into a strongly deformed state, in which the barriers for CDW motion are effectively steeper, modifications of the original experiments [56, 57] did not find such an effect. These modifications consisted of reversing the voltage pulse direction in successive pulse trains, and adding an additional pulse of several times the conduction threshold before the 90° rf pulse, with essentially unchanged results in each case.

The NMR measurements can be explained by large displacement in small regions of the sample only, such as within CDW dislocations. For instance, making the simplifying assumption that regions of the sample can be divided into those with a small dielectric response and those with a large dielectric response, the NMR response will show a fast decay followed by a slow decay, but a small fraction having a fast decay would be lost in the scatter of the data. It was shown [55-57] that both the NMR and electric polarizability measurements fit such a model, in which 4% of the sample has the classical 90° phase advance at threshold. In a detailed model, Tucker *et al.* [49] find a close quantitative agreement with the measured CDW displacement in a model, described above, in which large displacements occur at fixed strong-pinning sites.

2.7. Pressure and Impurity Effects

Both pressure and doping have a strong effect on the CDW transition in $NbSe_3$, lowering the CDW transition temperature and inducing a superconducting transition at temperatures below 3 K [11]. Wada *et al.* [59] have studied the effect of pressure and doping on the spin-lattice relaxation in $NbSe_3$, using both NMR and NQR, including limited ^{77}Se NMR measurements. They find that both conditions reduce the measured density of states, in conjunction with the enhanced superconductivity.

The pressure effect was observed in a ^{93}Nb NQR study, between 1 K and 30 K. The 4.2K NQR spectrum, measured between 2 and 5.5 MHz, is quite similar to that measured by Suits and Slichter above the 59 K transition (Figure 6). This is consistent with spectral changes only on the orange site at the lower CDW transition, since the frequency region contains only the red- and yellow-site resonances. Relaxation was measured at a fixed frequency of 3.3 MHz, a peak-intensity position which seems likely due to the narrow yellow line, seen at 3.25 MHz in Figure 6.

Pressure was applied hydrostatically, and the superconducting T_c was measured through the bulk susceptibility. The T_c exhibited a pressure dependence identical to that of Monceau *et al.* [88], giving an increase to $T_c = 2.4$ K at 6 kbar. The effect on the NQR relaxation rate was a monotonic decrease in $(T_1T)^{-1}$, reducing the rate to half the zero-pressure value at 5.5 kbar. Wada *et al.* estimate that this reduction in rate corresponds to a decrease in density of states from 0.14 to 0.12 states eV^{-1}.

Changes in T_1 due to doping were measured through NMR in powder samples, via the same techniques utilized by Wada, Aoki, and Fujita [58], for their study of undoped samples (section 2.5). The T_1 was measured at an NMR frequency of 10.5 MHz. Samples were doped with up to 5 atomic % tantalum, giving a superconducting transition increased from zero to 1.5 K, and CDW transitions decreased to 10 K and 85 K. The effect of such doping is a decrease in the relaxation rate, at all temperatures. A Korringa $(T_1T)^{-1}$ was extracted from the NMR in three temperature ranges: between T_s and T_{c2}, between T_{c1} and T_{c2}, and above T_{c1} (where T_s denotes the superconducting transition, and T_{c1} and T_{c2} denote the high- and low-temperature CDW transitions, respectively). In each range, the $(T_1T)^{-1}$ decreases by a factor 2 to 3 as the doping is increased to 5 atomic %. In addition, T_1 data for the ^{77}Se resonance (spin $I = 1/2$) are reported, which also decrease with Ta doping, from $(T_1T)^{-1} = 0.06\ s^{-1}K^{-1}$ in pure samples, to $(T_1T)^{-1} = 0.02\ s^{-1}K^{-1}$ for 5 atomic % doping (both values for the low-temperature regime).

The relaxation rate has a frequency dependence due to diffusion of electrons confined to chains, as described in section 2.5. While this complicates the analysis, Wada *et al.* show that the $(T_1T)^{-1}$ has a similar crossover frequency $(1/\tau_\perp)$ for doped and undoped samples at low temperatures. Furthermore, at the 10 MHz frequency where T_1 data were measured, the motion is in the 3-dimensional regime. Assuming that $\tau_\parallel$ is unchanged by doping, their analysis yields an electron density of states reduction by approximately one-half at doping of 5 atomic %. The fact that this is observed at all temperatures, and for both nuclei, gives convincing proof that this is a general feature of Ta doping.

A reduction in the density of states by doping and by pressure, both of which enhance the superconductivity, is contrary to standard BCS theory of superconductivity, in which a density of states increase tends to favor superconductivity. Wada *et al.* propose a model in which peaks in the density of states due to CDW gap formation may be responsible for the increase in superconducting transition temperature.

3. Monoclinic TaS_3

The monoclinic form of TaS_3 is nearly identical in many respects to $NbSe_3$. This material is not to be confused with the orthorhombic phase of TaS_3, which has also received much attention in the literature. Like $NbSe_3$, monoclinic TaS_3 has three structural chains, two containing paired chalcogen anions [89]. Two incommensurate CDW transitions are observed, with T_c = 240 K and 160 K (*vs.* 144 K and 59 K for $NbSe_3$). CDW conductivity effects are observed for each CDW, with threshold fields on the order of 0.5 V/cm. The lower-temperature transition is a complete metal-insulator transition, unlike $NbSe_3$. In addition, pre-transitional diffuse scattering is observed at room temperature in monoclinic TaS_3 [1], indicating more one-dimensional behavior in this material than $NbSe_3$.

Dai *et al.* [90, 91] report Mössbauer studies on substitutional Fe, in monoclinic $Ta_{1-x}Fe_xS_3$, with $0.0012 \leq x \leq 0.5$. They demonstrate CDW-induced changes in the EFG and isomer shifts, indicating electronic changes and CDW suppression due to Fe impurities. In addition, studies of $Ta_{0.8}Fe_{0.2}Se_3$ in a magnetic field reveal line shape changes attributed to a magnetic transition near H = 1.5 T. Multicrystalline samples were prepared by heating elemental powders under pressure. The monoclinic phase was indicated in X-ray measurements, and doping levels were verified through wet-chemical methods. The three chains are numbered as I, II, and III, according to increasing S-S bond lengths, so that III is the nominally insulating chain. This scheme is followed in the description here. Note that these labels differ from those of other groups; I, II, and III here correspond to III, I, and II, of Meerschaut *et al.* [89], respectively (or yellow, orange, and red, respectively, in Wilson's [5] convention).

3.1. ROOM-TEMPERATURE MEASUREMENTS

Room temperature Mössbauer measurements [90] are reported for eight stoichiometries between x = 0.0012 and x = 0.5. Data for all impurity levels could be fit to two pairs of Lorentzian doublets. While three doublets would be expected due to the three sites, it is suggested that one doublet is a superposition of chains I and II, which are quite similar electronically. The intensities of these lines are in the approximate ratio 3:2, for all impurity levels. This ratio corresponds to a slightly larger Fe concentration on chain III (the insulating chain) than on the other two chains.

A plot of the quadrupole splitting and isomer shift δ *vs.* doping level, at room temperature, is given in Figure 19. The data are computer fits assuming two doublets as described above. Both δ and ε increase with impurity concentration (where ε is the quadrupole splitting denoted ΔE in section 1.1.). Also, the chain-III isomer shift is larger than that of the other chains for all doping levels. The larger isomer shift on site III is consistent with a smaller charge density on this chain, as

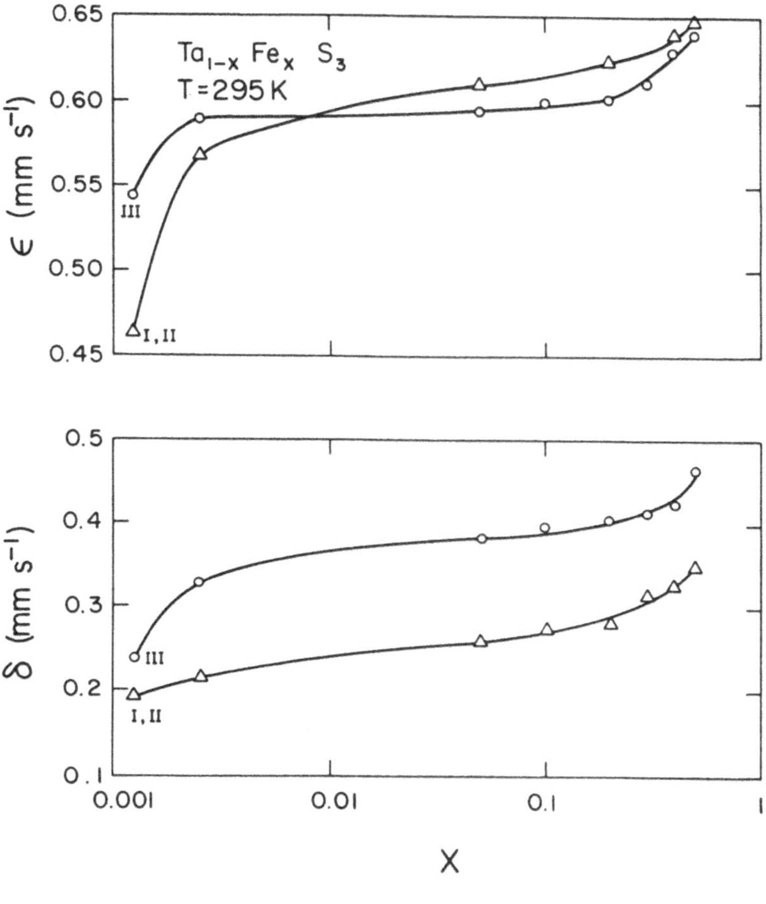

Fig. 19 ^{57}**Fe Mössbauer parameters for monoclinic $Ta_{1-x}Fe_xS_3$ *vs.* Fe doping, at room temperature. Data points are computer fits as described in the text. Signals labeled III are attributed to the insulating chain, and those labeled I, II to the two conducting chains. [from Dai *et al.* [90]]**

might be expected for impurities on the insulating chain, and this helps to confirm the site identification.

The increased δ with Fe concentration indicates that substitutional Fe acts as an acceptor. The natural valence states are Fe^{3+} and Fe^{2+}, whereas the Ta ions are nominally $Ta^{4.5+}$ and Ta^{5+} (this is Wilson's picture [5], as described for $NbSe_3$). Thus Fe will accept electrons from the bands on chains I and II, or from S^{2-} ions. A shift in δ with doping level implies that Fe impurities gradually deplete the available electrons, changing the average Fe valence state. A change in the number of Fe-Fe neighbors is suggested as the source of the increase in ε with doping.

3.2. CDW TRANSITIONS

In the same work, Dai *et al.* [90] report Mössbauer spectra for $Ta_{0.95}Fe_{0.05}S_3$ as the temperature is lowered through the two CDW transitions, and also report spectra at 77 K for materials with different Fe concentrations. The CDW phase transitions are apparent through broadening of the Mössbauer spectra, and at 77 K they find that the CDW is completely suppressed for x = 0.5.

In $Ta_{0.95}Fe_{0.05}S_3$, large changes in spectra are identified as CDW transitions at 215±5 K and at 153±2 K, suppressed from 240 and 160 K in the pure material. Note that this suppression is similar to the effect of Ta impurities in $NbSe_3$, described above. Spectra are shown in Figure 20. CDW-broadened Mössbauer spectra could not be fit to two doublets. Rather, the data were fit to four doublets for 210 K $\geq$ T $\geq$ 155 K, and five doublets for 150 K $\geq$ T $\geq$ 77 K. The individual doublets that comprise the fit are indicated by solid curves in Figure 20.

The four doublets for 210 K $\geq$ T $\geq$ 155 K are attributed to chains II and III with no CDW (one doublet each), plus a pair of doublets for chain I due to the peak and valley of the CDW on that chain (I_p and I_v in Figure 20). Similarly, the five doublets for 150 K $\geq$ T $\geq$ 77 K represent a splitting at chain II due to the development of a CDW on that chain. The site identification for these peaks follows from the analogous CDW structures identified in $NbSe_3$. The splitting of the signal from one chain into two doublets could signal a commensurate structure, or the edge singularities of an incommensurate line (such as in Figure 8); the resolution is not sufficient to distinguish these cases. Clearly, though, the broadened signals represent Fe sites distributed in CDW phase, implying that the pinning due to these impurities is not sufficiently strong to fix the phase at each site, for concentrations observed here.

The temperature-dependent quadrupole splittings and isomer shifts extracted from these fits are shown in Figure 21. An increase in δ with decreasing temperature, above the CDW transition, is attributed to the second-order Doppler shift. Just below the CDW transitions, the signal for the relevant chain splits into high- and low-δ components, consistent with low- and high-charge regions, respectively, for the CDW valleys and peaks. At 77 K, the signals from chains I and II become equivalent, and overlap. Discontinuities in the quadrupole splittings at the transition temperatures, including those of the non-CDW chains, are attributed to shifts in atomic positions due to changes in bonding.

Studies at 77 K of samples of different Fe concentration indicate that the CDW is further suppressed by increasing x. A sample with x = 0.50 had a sharp spectrum that could be fit to two doublets, like the room-temperature measurements. This implies that both CDW transitions must be below 77 K for that composition, and also tends to rule out magnetic ordering above 77 K (in zero field). Furthermore,

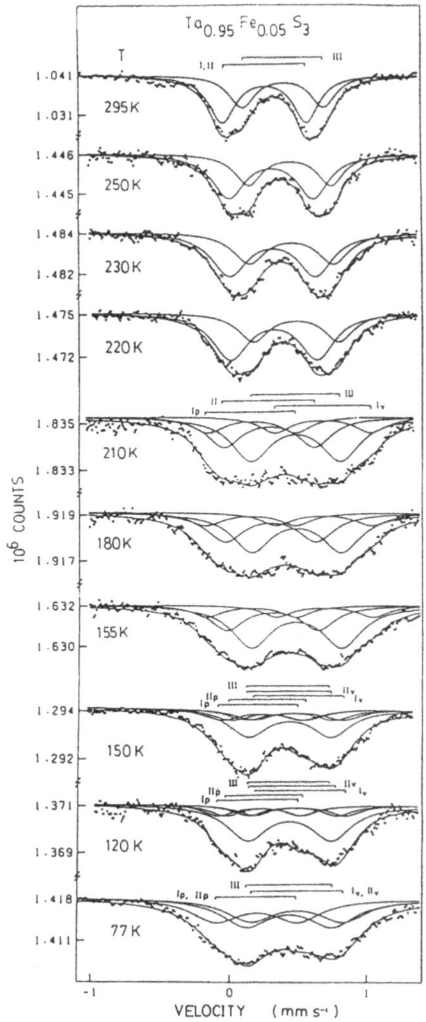

Fig. 20 ^{57}Fe Mössbauer spectra for monoclinic $Ta_{0.95}Fe_{0.05}S_3$ *vs.* temperature. Solid curves are theoretical fits, and their individual components as described in the text. [from Dai *et al.* [90]]

the 77 K spectrum for x = 0.40 could be fit to four doublets, indicating that 77 K lies between the two CDW transitions in that material.

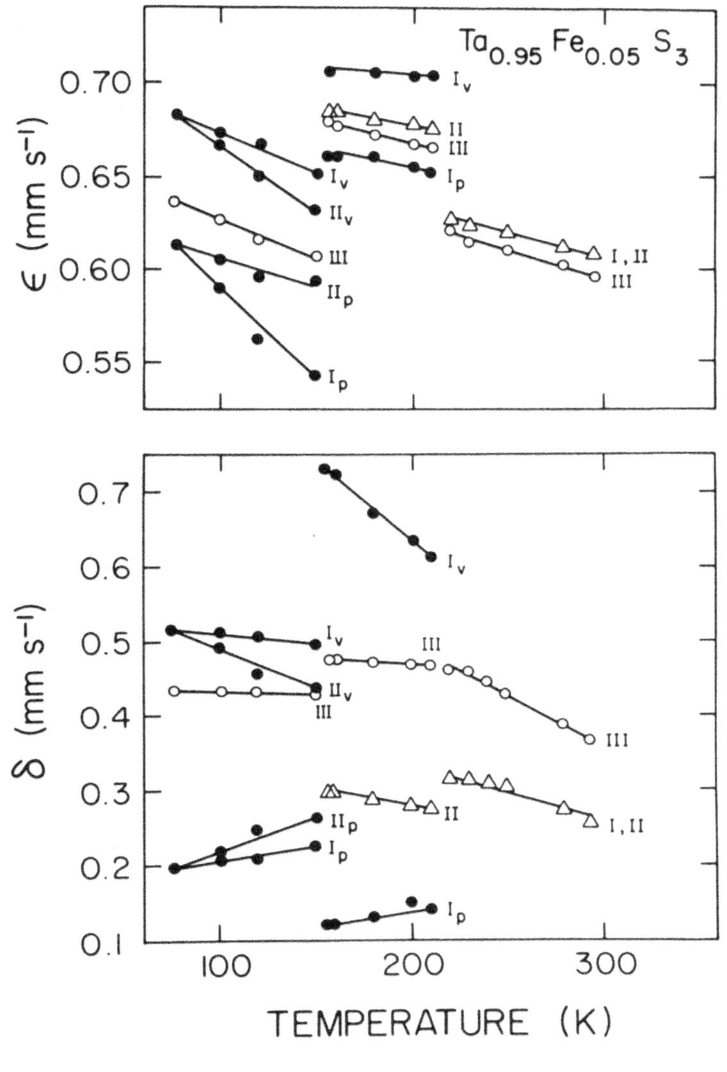

Fig. 21 ^{57}Fe Mössbauer parameters, extracted from spectra of Figure 20. Peak and valley components for the two CDW-bearing chains are labeled I_p and I_v, or II_p and II_v, respectively. [from Dai *et al.* [90]]

3.3. Magnetic Field Effects

Mössbauer studies of $Ta_{0.8}Fe_{0.2}S_3$ in a magnetic field were carried out in the same laboratory [91], and indicate a pronounced change in structure in a field of approximately 1.5 T. These measurements were carried out at 4.2K, and in fields between 0 and 2.5 T. The γ-ray beam was parallel to the applied field. Results are shown in Figure 22, and a large change is readily apparent for the highest magnetic fields.

Fits to these spectra were made assuming that the quadrupole splitting and isotope shifts for chains I and II were equal, as in the 77 K measurements. Furthermore, 3:3:4 intensities for sites I, II, and III were assumed, as before. In addition, a numeric calculation of the EFG's, based on ionic charge transfer, served as a guide. For H $\leq$ 1.5 T, successful fits were found with no hyperfine field. These fits are the solid curves in Figure 22. For 2.0 and 2.5 T, good fits required antiferromagnetic hyperfine fields on chains I and II, and a paramagnetic response for chain III. The best-fit hyperfine fields for chains I and II are 0.95 and 0.55 T, respectively, at an applied field of 2.0 T. For 2.5 T applied, the hyperfine fields change to 1.04 T (chain I), and 0.38 T (chain II).

Measurements of the magnetization for samples having x up to 0.2 were also reported, for $T = 1.5$ K. The measured magnetization gradually saturates at a field of approximately 1 T, also indicating some form of magnetic order. In addition, Mössbauer measurements for x = 0.05 were qualitatively similar to those for x = 0.2, developing hyperfine fields at approximately 1.0 T.

The magnetic order implied in these studies is attributed to an induced SDW component, by analogy with 2H-TaS_2 measurements [92], and possible indications of such structures in $NbSe_3$ [75]. It is not clear in this case whether the magnetic component may be induced by the impurities, or if this may be a property of pure TaS_3.

4. $FeNb_3Se_{10}$

A material structurally distinct from $NbSe_3$ and monoclinic TaS_3 is $FeNb_3Se_{10}$, discovered in attempts to dope $NbSe_3$ with Fe. The structure (Figure 23) exhibits two types of chains: a trigonal chain with three Se per Nb, very similar to the yellow $NbSe_3$ chain, and also an octahedral chain which contains Fe, Nb, and Se. This material [93] is in fact never obtained as $FeNb_3Se_{10}$; observed stoichiometries are $Fe_{1+x}Nb_{3-x}Se_{10}$, with $0.25 \leq x \leq 0.40$. The Fe atoms are segregated on the octahedral chain, where Fe and Nb locations are disordered. The disorder on this chain plays an important part in the observed properties, and attempts to grow ordered phases of this structure [94] have not been successful.

In $FeNb_3Se_{10}$ a single incommensurate CDW transition is observed, with T_c = 140 K. The near-equivalence to the $NbSe_3$ T_c, and the similarity of the trigonal chain

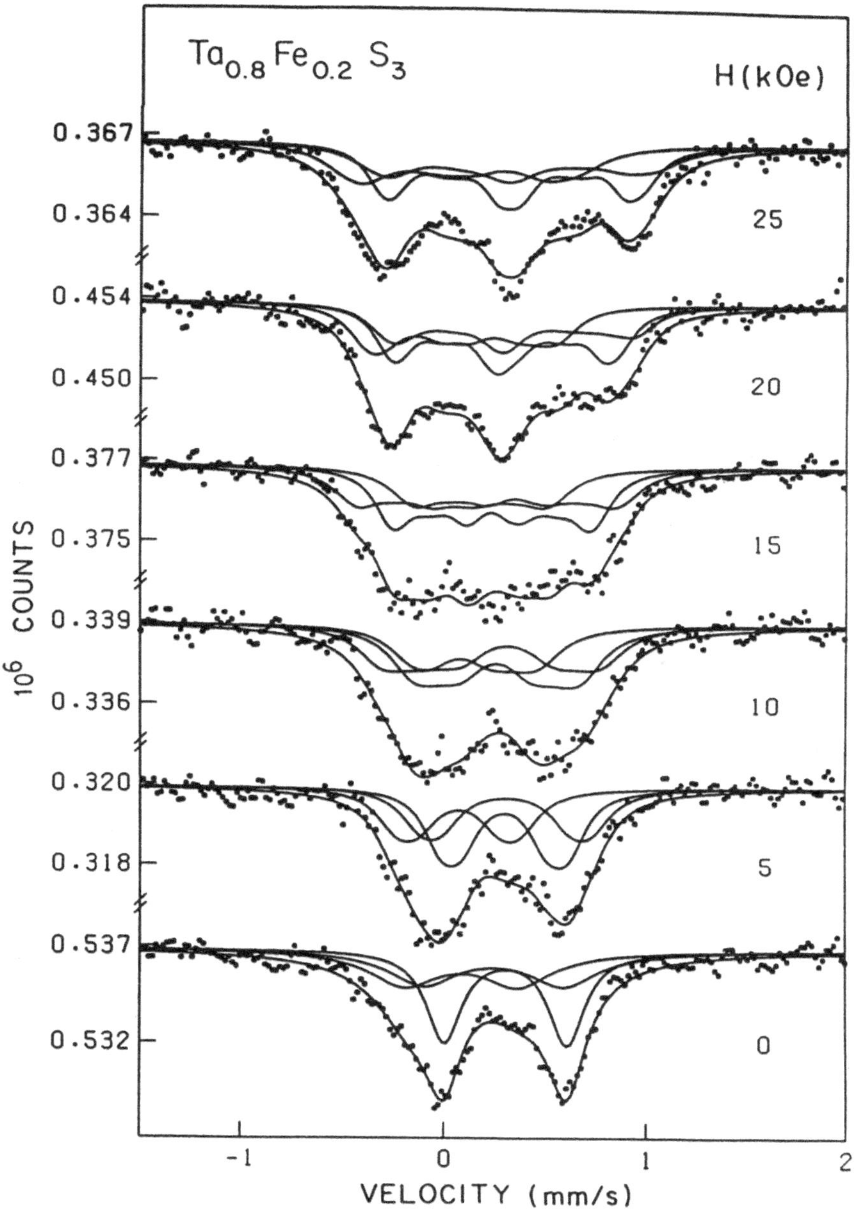

Fig. 22 ^{57}Fe Mössbauer spectra for monoclinic $Ta_{0.8}Fe_{0.2}S_3$, at 4.2K, in a sequence of magnetic field strenghts. The theoretical curves are described in the text. [from Dai *et al.* [91]]

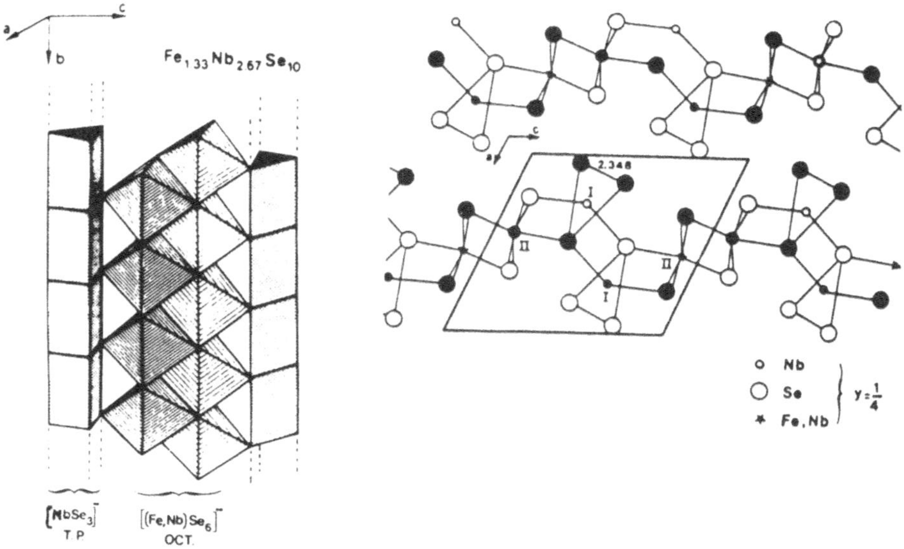

Fig. 23 $Fe_{1+x}Nb_{3-x}Se_{10}$ structure, shown in a view nearly perpendicular to the chains (left), and viewed along the chain axis (right). The trigonal Nb sites are labeled I, and the octahedral II site contains Nb and Fe. [from Ben Salam *et al.* [97]]

to the yellow chain in $NbSe_3$, suggests that the CDW resides on the trigonal chain in $FeNb_3Se_{10}$. Band-structure calculations [63, 95] confirm that the orbitals at the Fermi surface are based mainly on trigonal-chain d-orbitals. These calculations, however, have been performed for the ordered structure (x = 1), and the additional Fe certainly will change the electron count. No CDW conductivity effects have been observed in this material, and the disorder on the Fe-containing chains presumably dominates this behavior by strongly pinning the CDW.

Mössbauer studies on the Fe sites in $Fe_{1+x}Nb_{3-x}Se_{10}$ have been performed by several groups. In Figure 24 are shown the spectra for ^{57}Fe-enriched material, by Brill *et al.* [96]. Parameters extracted from these data are given in Table 4. For these measurements, the Fe content was approximately (x = 1/3), and the data are consistent with independent measurements for (x = 0.33) [94, 97] and (x = 0.25) [98]. The spectra are characterized by two pairs of doublets at room temperature, which coalesce into a single doublet at approximately 170 K.

The two sites observed at room temperature correspond presumably to different near-neighbor environments [94]. Assuming statistical distributions of metal

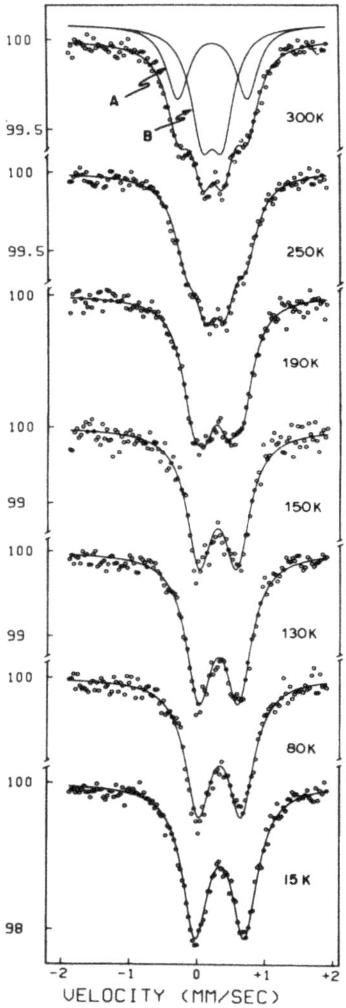

Fig. 24 Mössbauer spectra for $Fe_{1.33}Nb_{2.67}Se_{10}$, from Brill, Boolchand and Lemon [96]. Solid line fits correspond to the data in Table 4. The two sites observed at high temperatures are labeled A and B, as shown.

atoms in the octahedral chain for x = 0.33, one site may have zero, one, or two nearest-neighbor Fe atoms, with relative probabilities 1:4:4 respectively. A likely explanation is that spectral site A (Table 4) has one Fe neighbor, whereas site B is a

Temperature	Site	δ (mm/s)	ΔE (mm/s)	FWHM (mm/s)	Intensity ratio I_B/I_A
300 K	A	0.237(7)	1.021(8)	0.39(3)	1.35(8)
	B	0.229(7)	0.299(8)	0.39(3)	
190 K	A	0.298(8)	0.68(1)	0 34(3)	0.70(8)
	B	0.273(8)	0.25(1)	0.34(3)	
180 K	A	0 310*	0.538*	0.40(5)	0.24(5)
	B	0 290*	0.207*	0.40(5)	
160 K	-	0 292(8)	0 525(10)	0.45(3)	-
15 K	-	0 363(5)	0 724(5)	0.46(1)	-

* Parameters fixed in least-squares refinement of spectra

Table 4 Mössbauer parameters for $Fe_{1.33}Nb_{2.67}Se_{10}$, from the spectra of Figure 24. A and B refer to the two doublets observed at high temperatures. [from Brill, Boolchand and Lemon [96]]

superposition of the zero- and two-neighbor cases. Site A has the larger quadrupole splitting, which might be expected for the more asymmetric site having one Fe neighbor. The linewidths, however, indicate further disorder at these sites, perhaps due to next-nearest-neighbor interactions.

The isotope shifts at the two room-temperature sites are nearly equal, indicating equivalent ionic charges. Conflicting proposals have been given [94, 96] for the Fe valence configuration. Cava *et al.* [94] deduce that the Fe ions are high-spin, through magnetization measurements. The charge state may correspond to an intermediate valence; Ben Salem *et al.* [97] propose a valence of $Fe^{2.8+}$ from comparison to Mössbauer studies in similar materials.

The merging of the two doublets (Figure 24) upon lowering the temperature is apparently unrelated to the CDW transition, since the lines coalesce near 170 K, above the CDW transition temperature, 140 K. There is no sign of CDW-induced splitting at the transition; this is consistent with CDW formation on the trigonal chain, from which the Fe atoms are excluded. Brill *et al.* [96] have interpreted the line shape changes as a partial ordering of the Fe sites on the octahedral chain. Indeed, there is evidence [99] for a structural trimerization in this material, independent of the CDW ordering observed at 140 K. The merging of the Mössbauer spectra also coincides with a broad minimum in the magnetic susceptibility, which has been attributed to clusters of exchange-coupled Fe ions [94], and to Anderson localization of electrons on the disordered chain [97].

Additional information about the electronic state of this material was obtained through a lithium intercalation study by Ben Salem *et al.* [97]. This group prepared $Li_yFe_{1.33}Nb_{2.67}Se_{10}$, with Li concentrations between y = 0 and 2.5. Mössbauer, magnetic susceptibility, and EPR studies were used in this study to probe the electronic changes induced by intercalation. The Li NMR Knight shift was also measured, showing a small density of states at the Li sites which indicates electron transfer to the chains.

The intercalation discharge curves from this study show a set of broad plateaus, interpreted as filling of flat bands due to the octahedral chains. Thus the Fermi surface for the y = 0 material is found to lie in a region of trigonal-chain d-bands, in close agreement with band structure calculations [63, 95]. Mössbauer measurements, though, show that some of the Li-donated electrons directly reduce Fe ions, rather than joining metallic orbitals.

The Mössbauer measurements show a new doublet in the intercalated material, having $\Delta E = 2.67$ mm/s and $\delta = 1.12$ mm/s at room temperature, in addition to the less-shifted sites (as in Figure 24). This doublet position is nearly independent of y, and changes at 77 K to $\Delta E = 2.97$ mm/s. The new signal is attributed to a well-localized Fe^{2+} high-spin state, and the splitting is found comparable to results from a number of materials containing such states. This reduction process is found to take up approximately 30% of the electrons available from Li intercalation.

5. Tetrachalcogenides

A large number of transition metal tetrachalcogenide materials have been found to exhibit one-dimensional CDW's. The basic structural feature of these materials is chains composed of square or rectangular prisms, in contrast to the trigonal-chain materials described above. Examples are $NbTe_4$, $(TaSe_4)_2I$, and $(NbSe_4)_3I$. The tetratellurides exhibit rich CDW structures [100], but have not as yet been investigated through nuclear spectroscopy. In the iodine-containing materials, iodine ions separate the chains, giving typically more anisotropic metallic behavior, due to decreased communication among chains.

Figure 25 illustrates the $(NbSe_4)_3I$ structure [101]. The twisting chain structure is generic to the tetrachalcogenide materials. Conduction electrons reside to a good approximation in d_{z^2} bonds along the chain axis [7], giving one-dimensional conduction properties much like the trichalcogenides. This material has tetragonal symmetry, and only one type of chain is exhibited, but there are several inequivalent Nb sites along the chain.

Several materials of the form $(NbSe_4)_xI$ have been prepared, including x=2, 3, and $\frac{10}{3}$. Within each of these materials, the chains are all equivalent. Bond lengths indicate a nominal $Nb^{4+}((Se_2)^{2-})_2$ electronic configuration, so that one d-electron is available per Nb. As electron acceptors, the iodine ions determine the electron count in the chains, and therefore the CDW wavelength. Using this scheme for on-chain electron counts, the theoretical CDW wavelengths (measured in Nb-Nb distance along the chain) will be 4, 3, and $\frac{20}{7}$, for x=2, 3, and $\frac{10}{3}$, respectively [7].

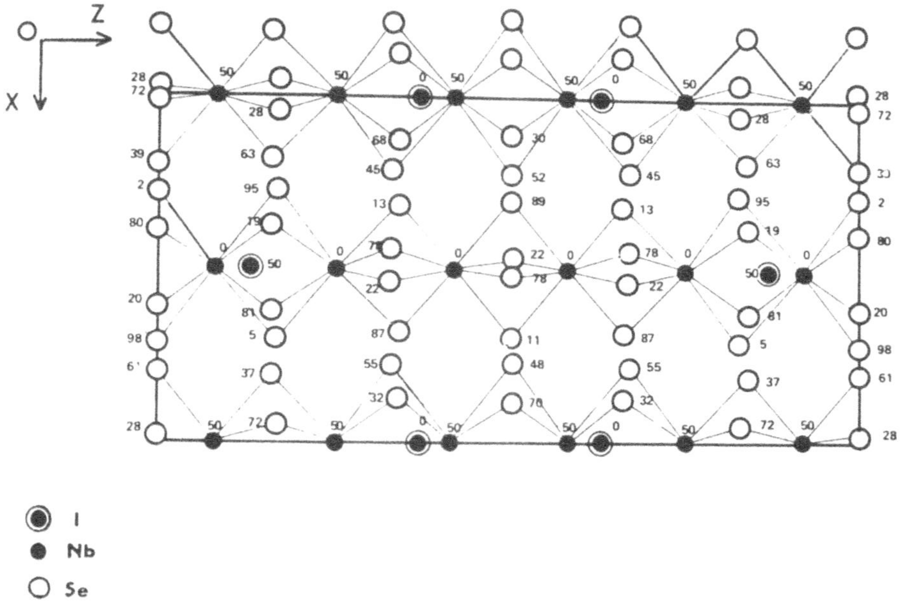

Fig. 25 Structure of $(NbSe_4)_3I$, viewed perpendicular to the chains. *y*-axis locations are indicated numerically (in percent of the tetragonal unit cell base, $a = 9.489$ Å). All chains are equivalent. [from Meerschaut, Palvadeau and Rouxel [101]]

5.1. $(NbSe_4)_3I$

As expected from the electron-counting argument described above, $(NbSe_4)_3I$ has a trimerized structure at room temperature [101], a structure that may be considered a CDW even though the distortion is observed at all temperatures. The room-temperature structure is shown in Figure 25. Although the unit cell is six Nb sites long, there is a reflection symmetry giving three inequivalent Nb sites. (A total of twelve Nb sites are contained in the unit cell, which includes two equivalent chains.) The Nb-Nb bond alteration goes as long-long-short (3.25 and 3.06 Å), giving clear indication that this commensurate Peierls distortion involves electrons condensed into Nb-Nb bonding orbitals. In addition to the trimerized configuration at room temperature, a structural transition is observed in $(NbSe_4)_3I$ at 274 K. This transition has been the focus of an NMR study by Butaud *et al.* [67].

The ^{93}Nb NMR studies [67] utilized a single crystal sample. The NMR frequency was fixed at 21 MHz, with the magnetic field rotated in a plane perpendicular to the chain axis. Figure 26 shows NMR results at 293 K and 270 K (above and below the phase transition), for the same crystal orientation. The figure shows the 1/2,-1/2 transitions at center, as well as some of the satellites. Two sets of lines are observed at the higher temperature, whereas three sets of lines are expected due to the trimerized structure. The two sites are identified as the two structural sites adjacent to short Nb-Nb bonds in the long-long-short sequence, and the third is perhaps excessively broadened.

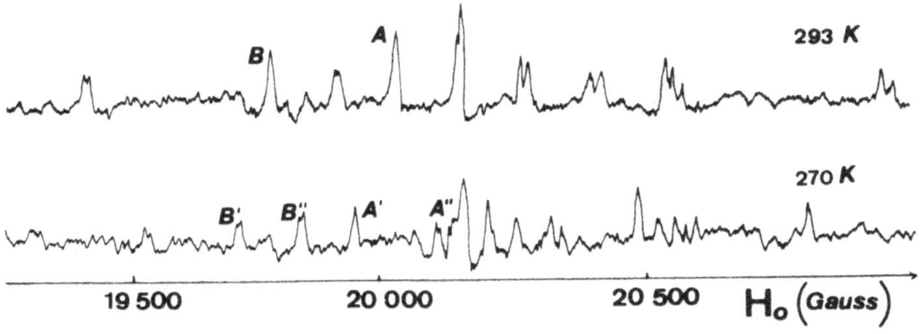

Fig. 26 ^{93}Nb NMR spectra of $(NbSe_4)_3I$, from Butaud *et al.* [67]. Field direction is identical for the two spectra. Peaks labeled A and B are (1/2, 3/2) transitions for two sites (related to each other by a $\pi/2$ rotation). Peaks B and A split below the 274 K transition into peaks B′ and B″, and A′ and A″, respectively.

The sites were identified by observing the orientation dependence of the signals, and assuming that the EFG's exhibit the symmetry of the 8 near-neighbor seleniums. Electrons in on-site Nb-Nb bonds will have a large EFG contribution, but this contribution will be axial, and independent of direction in the plane perpendicular to the chains. The nearest neighbors should thus well approximate the symmetry of the EFG. The two observed sites have the same EFG principal values, but different orientations of the principal axes. The principal values, indicated in Table I, were extracted assuming that the chain axis is a principal value (not strictly a symmetry requirement).

Upon lowering the temperature, the satellite lines were observed to disappear very close to the 274 K transition. In addition, a peak in the spin-lattice relaxation rate is found just above 274 K. Both are characteristic of fluctuations associated with a mode-softening structural transition. Below the transition, the satellites reappear, but split so that their number is doubled. This is illustrated in the 270 K trace of Figure 26. Thus the transition can be associated with a further symmetry reduction in $(NbSe_4)_3I$, apparently not a CDW-type transition. The symmetry change is characterized by previously equivalent pairs of sites becoming inequivalent, through a rotation of the selenium cages.

5.2. $(NbSe_4)_{10/3}I$

$(NbSe_4)_{10/3}I$ exhibits an incommensurate CDW transition at $T_c = 285$ K [102]. Its (undistorted) structure [103] is illustrated in Figure 27. Even in the absence of the Peierls distortion, the unit cell contains ten sites along the chain axis. While a CDW wavelength of $\frac{20}{7}$ is predicted by the electron-counting arguments described above, a CDW wavelength slightly greater that 4 lattice constants is observed. The difference can be attributed to valence-band energy shifts caused by significant inter-chain interactions [7]. $(NbSe_4)_{10/3}I$ has exhibited CDW conductivity effects, whereas $(NbSe_4)_3I$, with its commensurate CDW, has not. Non-ohmic conductivity and narrow-band noise have both been observed [11]. Threshold fields are on the order of 1 V/cm.

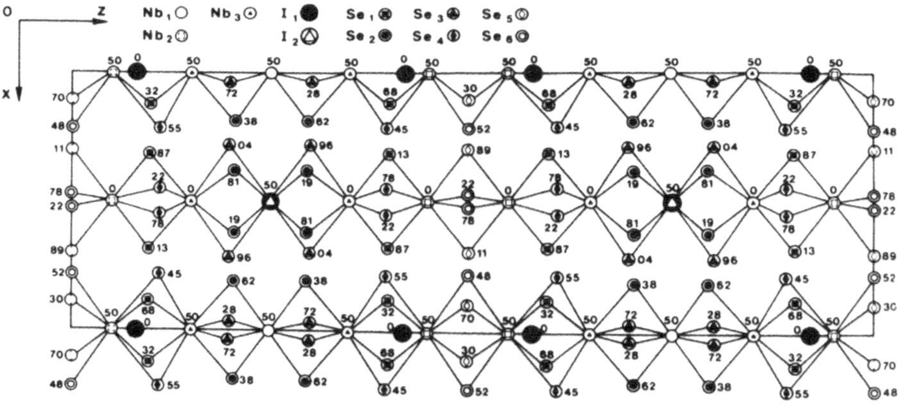

Fig. 27 Structure of $(NbSe_4)_{10/3}I$, viewed perpendicular to the chains. y-axis locations are indicated numerically, in percent of the tetragonal unit cell base, $a = 9.464$ Å. The chains are equivalent, and a total of three Nb sites are inequivalent by rotational symmetry. [from Meerschaut *et al.* [103]]

In a ^{93}Nb NMR study of $(NbSe_4)_{10/3}I$, Butaud *et al.* [104] have observed an incommensurate CDW line shape, and measured the temperature-dependence of the order parameter. These measurements utilized fixed NMR frequencies of 83 and 65 MHz (sweeping the field in the vicinity of 8.0 and 6.2 T, respectively). The sample was a single crystal of 1.8 mg. The external field H_0 was aligned perpendicular to the *c*-axis (chain axis).

Figure 28 shows field scans of the $(\frac{1}{2}, -\frac{1}{2})$ transitions, at 83 MHz, for temperatures between 310 K and 190 K. The NMR lines are a superposition of lines from the 10 inequivalent sites (above the transition). Separating the quadrupole and Knight shift contributions via their field-dependences, the line positions are found to be determined mainly by the quadrupole shift.

Although the NMR spectrum cannot be separated into components for each site, the expected CDW broadening is observed below 285 K, confirming an incommensurate CDW transition at this temperature. The two main peaks in Figure 28 can be observed to broaden, and a computer fit to the higher-field line shape gives a measure of the temperature dependence of the CDW order parameter, as plotted in Figure 29. Note that the order parameter shown in Figure 29 is extracted assuming that the linewidth depends linearly on the gap.

In limited measurements of spin-lattice relaxation, $1/T_1$ was found to be an order of magnitude larger in $(NbSe_4)_{10/3}I$ than for ^{93}Nb relaxation in $(NbSe_4)_3I$. Since at high temperatures the relaxation is dominated by fluctuating EFG's, the difference is ascribed to enhanced CDW phason fluctuations.

6. Conclusions

Like the organic conductors described elsewhere in this book, the one-dimensional inorganic conductors described here exhibit a wide variety of structural variations. These systems provide models for the behavior of low-dimensional metallic electrons, the charge transfer properties of transition metal systems, and the dynamics and phenomenology of CDW's.

Nuclear spectroscopy measurements have provided significant advances in the understanding of these materials. We have described measurements of the normal-state electronic properties, as well as the CDW structures and dynamics. Particularly important is the site-selective nature of these measurements, allowing a correlation of the electronic properties with structural features, such as the three chains in $NbSe_3$ and the octahedral Fe sites in $FeNb_3Se_{10}$. In addition, nuclear spectroscopy is a primary tool for identification of locally incommensurate CDW structures, as observed here in $(NbSe_4)_{10/3}I$ and $NbSe_3$. A new type of measurement was demonstrated in these materials, in NMR measurements of $NbSe_3$, in which the dynamics of electrically-triggered CDW motion was studied.

In this chapter, only those transition metal chalcogenides have been described that have yielded to nuclear spectroscopy investigations. Many interesting materials invite further investigation. For instance, $TaSe_3$ has a chain structure similar to $NbSe_3$, but it has exhibited a low-temperature resistance anomaly apparently

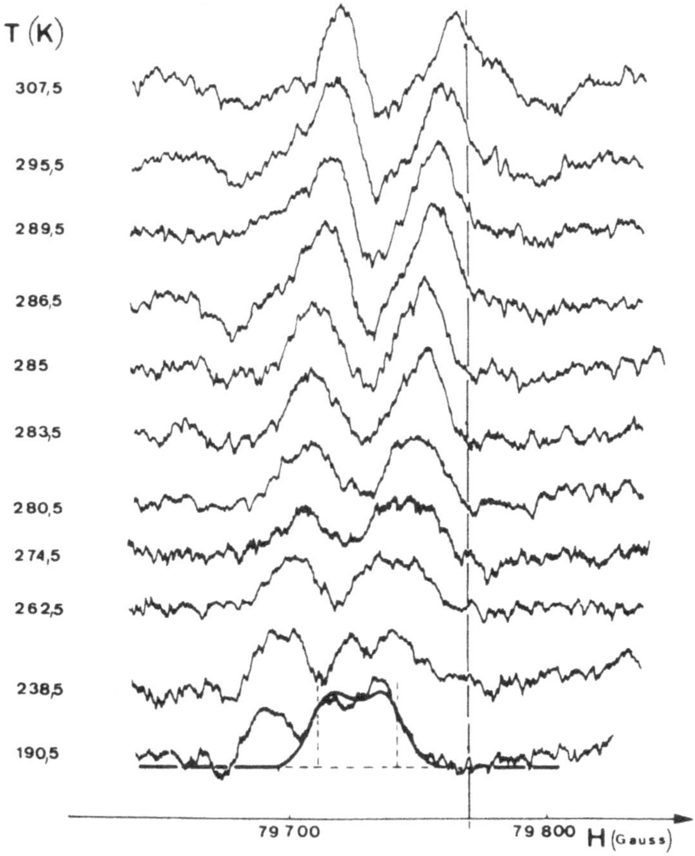

Fig. 28 ^{93}Nb NMR spectra of $(NbSe_4)_{10/3}I$, showing the (1/2, -1/2) transitions, from Butaud *et al.* [104]. Crystal orientation is such that the field is perpendicular to the chain axis. A total of 10 inequivalent sites contribute to the spectra. Below the CDW transition the main peaks are broadened, and the superposed curve at the lowest temperature is a fit to an incommensurate CDW line shape.

associated with its superconducting transition [105]. The chain-like material $HfTe_5$ exhibits a phase transition that is yet uncharacterized [106]. The magnetic-field-induced changes in CDW conductivity of $NbSe_3$ have been of great recent inter-

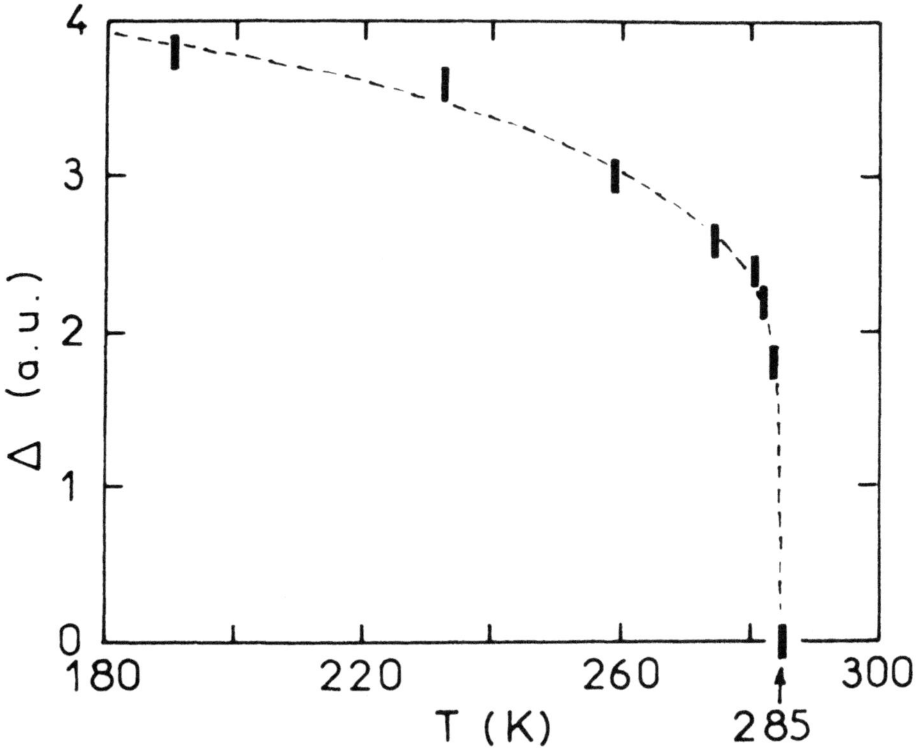

Fig. 29 The order parameter for the CDW transition in $(NbSe_4)_{10/3}I$, extracted from a fit to the NMR spectra of Figure 28. [from Butaud *et al.* [104]]

est [75, 76]. Furthermore, in several materials, CDW motion with a much larger threshold and little or no resistance has been detected in low-temperature transport measurements [107]. Our understanding of the complicated dynamics of CDW motion continues to be refined, and new surprises are certain to come in this field.

References:

[1] A. Meerschaut and J. Rouxel, in *Crystal Chemistry and Properties of Materials with Quasi-One-Dimensional Structures*, ed. J. Rouxel, p. 205, Reidel, Dordrecht, Holland, 1986.

[2] C. P. Slichter, *Principles of Magnetic Resonance*, Springer-Verlag, Berlin, 1980.

[3] H. Frauenfelder, *The Mössbauer Effect*, W. A. Benjamin, New York, 1962.

[4] J. L. Hodeau, M. Marezio, C. Roucau, R. Ayroles, A. Meerschaut, J. Rouxel, and P. Monceau, *J. Phys.* **C 11**, 4117 (1978).

[5] J. A. Wilson, *Phys. Rev.* **B 19**, 6456 (1979).

[6] R. E. Peierls, *Quantum Theory of Solids*, Oxford, 1955.

[7] P. Gressier, M.-H. Whangbo, A. Meerschaut, and J.Rouxel, *Inorg. Chem.* **23**, 1221 (1984).

[8] S. Kagoshima, H. Nagasawa, and T. Sambongi, *One-Dimensional Conductors*, Springer-Verlag, Berlin, 1988.

[9] N. P. Ong, *Phys. Rev.* **B 18**, 5272 (1978).

[10] F. Devreux, *Phys. Rev.* **B 13**, 4651 (1976).

[11] P. Monceau, in *Physics and Chemistry of Materials with Low-dimensional Structures, Part II*, ed. P. Monceau, p. 139, Reidel, Dordrecht, Holland, 1985.

[12] G. Grüner, *Rev. Mod. Phys.* **60**, 1129 (1988).

[13] R. Lacoe, H. J. Schulz, D. Jérome, K. Bechgaard, and I. Johannsen, *Phys. Rev. Letters* **55**, 2351 (1986).

[14] H. Fröhlich, *Proc. R. Soc. London* **A 223**, 296 (1954).

[15] R. E. Thorne, W. G. Lyons, J. W. Lyding, J. R. Tucker, and J. Bardeen, *Phys. Rev.* **B 35**, 6360 (1987).

[16] J. C. Gill, *Solid State Commun.* **39**, 1203 (1981).

[17] W. McMillan, *Phys. Rev.* **B 14**, 1496 (1976).

[18] Z. Z. Wang, H. Salva, P. Monceau, M. Renard, C. Roucau, R. Ayroles, F. Lévy, L.Guemas, and A. Meerschaut, *J. Physique Lett.* **44**, 311 (1983).

[19] K. Maki, in *Electronic Properties of Inorganic Quasi-One-Dimensional Materials, Part I*, ed. P. Monceau, p. 125 Reidel, Dordrecht, Holland, 1985.

[20] H. Fukuyama and P. A. Lee, *Phys. Rev.* **B 17**, 535 (1978).

[21] P. A. Lee and T. M. Rice, *Phys. Rev.* **B 19**, 3970 (1979).

[22] L. P. Gor'kov, *JETP Letters* **25**, 358 (1977).

[23] P. Monceau, J. Richard, and M. Renard, *Phys. Rev.* **B 25**, 931 (1982).

[24] J. McCarten, M. Maher, T. L. Adelman, and R. E. Thorne, *Phys. Rev. Letters* **63**, 2841 (1989).

[25] K. K. Fung and J. W. Steeds, *Phys. Rev. Letters* **45**, 1696 (1980).

[26] R. Blinc, *Phys. Reports* **79**, 331 (1981).

[27] B. H. Suits, S. Couturie, and C. P. Slichter, *Phys. Rev. Letters* **45**, 194 (1980).

[28] G. Grüner, A. Zawadowski, and P. M. Chaikin, *Phys. Rev. Letters*, **46**, 511 (1981).

[29] S. E. Barnes, and A. Zawadowski, *Phys. Rev. Letters* **51**, 1003 (1983).

[30] J. Bardeen, in *Physics and Chemistry of Materials with Low-dimensional Structures, Part I*, ed. P. Monceau, p. 105, Reidel, Dordrecht, Holland, 1985.

[31] J. Bardeen, *Phys. Rev. Letters* **55**, 1010 (1985).

[32] J. R. Tucker, *IEEE J. Quantum Electron.* **15**, 1234 (1979).

[33] J. H. Miller Jr., R. E. Thorne, W. G. Lyons, J. R. Tucker, and J. Bardeen, *Phys. Rev.* **B 31**, 5229 (1985).

[34] K. Seeger, A. Phillip, and W. Mayr, *Solid State Commun.* **50**, 223 (1984).

[35] L. Sneddon, M.C. Cross, and D.S. Fisher, *Phys. Rev. Letters* **49**, 292 (1982).

[36] R. E. Thorne, J. H. Miller, W. G. Lyons, J. W. Lyding, and J. R. Tucker, *Phys. Rev. Letters* **55**, 1006 (1985).

[37] D. S. Fisher, *Phys. Rev.* **B 31**, 1396 (1985).

[38] S. Bhattacharya, M. J. Higgins, and J. P. Stokes, *Phys. Rev. Letters* **63**, 1503 (1989).

[39] P. Sibani and P. B. Littlewood, *Phys. Rev. Letters* **64**, 1305 (1990).

[40] L. Sneddon, *Phys. Rev.* **B 29**, 719 (1984).

[41] G. Mozurkewich and G. Grüner, *Phys. Rev. Letters* **51**, 2206 (1983).

[42] X. J. Zhang, N. P. Ong, and J. C. Eckert, *Phys. Rev. Letters* **56**, 1206 (1986).

[43] P. B. Littlewood, *Phys. Rev.* **B 33**, 6694 (1986).

[44] P. Bak, *Phys. Rev. Letters* **48**, 692 (1982).

[45] J. A. Wilson, *J. Phys.* **F 12**, 2469 (1982).

[46] J. A. Wilson, Phil. *Trans. R. Soc. Lond* **A 12**, 2469 (1985).

[47] D. Feinberg and J. Friedel, *J. Phys. (Paris)* **49**, 485 (1988).

[48] N. P. Ong, G. Verma, and K. Maki, *Phys. Rev. Letters* **52**, 663 (1984).

[49] J. R. Tucker, W. G. Lyons, and G. Gammie, *Phys. Rev.* **B 38**, 1148 (1988).

[50] I. Tütto and A. Zawadowski, *Phys. Rev.* **B 32**, 2449 (1985).

[51] T. Csiba, G. Kriza, and A. Janossy, *Phys. Rev.* **B 40**, 10088 (1989).

[52] P. Monceau, M. Renard, J. Richard, and M. C. Saint-Lager, *Physica* **143B**, 64 (1986).

[53] F. Devreux, *J. Physique* **43**, 1489 (1982).

[54] B. H. Suits and C. P. Slichter, *Phys. Rev.* **B 29**, 41 (1984).

[55] J. H. Ross Jr., Z. Wang, and C. P. Slichter, *Phys. Rev. Letters* **56**, 663 (1986).

[56] J. H. Ross Jr., Z. Wang, and C. P. Slichter, *Phys. Rev.* **B 41**, 2722 (1990).

[57] J. H. Ross Jr., Thesis, University of Illinois at Urbana-Champaign, 1986 (unpublished).

[58] S. Wada, R. Aoki, and O. Fujita, *J. Phys.* **F 14**, 1515 (1984).

[59] S. Wada, M. Kaburagi, Y. Saito, and R. Aoki, *J. Phys.* **F 18**, 1231 (1988).

[60] A. Meerschaut and J. Rouxel, *J. Less Common Metals* **39**, 197 (1975).

[61] R. M. Fleming, D. E. Moncton, J. D. Axe, and G. S. Brown, *Phys. Rev.* **B 30**, 1877 (1984).

[62] D. W. Bullett, *J. Phys.* **C 12**, 277 (1979).

[63] D. W. Bullett, *J. Phys.* **C 15**, 3069 (1982).

[64] R. Hoffmann, S. Shaik, S. C. Scott, M.-H. Whangbo, and M. J. Foshee, *J. Solid State Chem.* **34**, 263 (1980).

[65] M.-H. Whangbo and P. Gressier, *Inorg. Chem.* **23**, 1305 (1984).

[66] N. Shima, *J. Phys. Soc. Japan* **51**, 11 (1982).

[67] P. Butaud, P. Ségransan, and C. Berthier, in *Proceedings, XXII Congress Ampère*, ed. K. A. Müller, R. Kind, and J. Roos, p. 40, The University, Zürich, 1984.

[68] C. Berthier, D. Jérome, and P. Molinie, *J. Phys.* **C 11**, 797 (1978).

[69] G. C. Carter, L. H. Bennett, and D. J. Kahan, ed., *Metallic Shifts in NMR*, Pergamon, Oxford, 1977.

[70] E. Ehrenfreund, A. C. Gossard, and F. R. Gamble, *Phys. Rev.* **B 31**, 1396 (1972).

[71] Y. Yafet and V. Jaccarino, *Phys. Rev.* **133**, 1630 (1964).

[72] J. D. Kulick and J. C. Scott, *Solid State Commun.* **32**, 217 (1979).

[73] A. M. Clogston, V. Jaccarino, and Y. Yafet, *Phys. Rev.* **134**, A650 (1964).

[74] C. Berthier, D. Jérome, P. Molinie, and J. Rouxel, *Solid State Commun.* **19**, 131 (1976).

[75] R. V. Coleman, M. P. Everson, H.-A. Lu, A. Johnson and L. M. Falicov, *Phys. Rev.* **B 41**, 460 (1990).

[76] T. M. Tritt, D. J. Gillespie, A. C. Ehrlich, and G. X. Tessema, *Phys. Rev. Letters* **61**, 1776 (1988).

[77] C. A. Balseiro and L. M. Falicov, *Phys. Rev.* **B 34**, 863 (1986).

[78] A. Narath, *Phys. Rev.* **162**, 320 (1967).

[79] C. Berthier and P. Ségransan, in *Low-Dimensional Conductors and Superconductors*, ed. D. Jérome and L. G. Caron, Plenum, New York, 1987.

[80] S. K. Lyo, *Phys. Rev.* **B 24**, 6253 (1981).

[81] Y. Obata, *J. Phys. Soc. Japan* **18**, 1020 (1963).

[82] M. Kogoj, S. Zumer, and R. Blinc, *J. Phys.* **C 17**, 2415 (1984).

[83] A. Wokaun and R. R. Ernst, *J. Chem. Phys.* **67**, 1752 (1977).

[84] R. M. Fleming, D. E. Moncton, and D. B. McWhan, *Phys. Rev.* **B 18**, 5560 (1978).

[85] H. Y. Carr and E. M. Purcell, *Phys. Rev.* **B 94**, 630 (1954).

[86] S. Meiboom and D. Gill, *Rev. Sci. Instrum.* **29**, 688 (1958).

[87] S. N. Coppersmith, *Phys. Rev. Letters* **57**, 1191 (1986).

[88] P. Monceau, J. Peyrard, J. Richard, and P. Molinie, *Phys. Rev. Letters* **39**, 161 (1977).

[89] A. Meerschaut, L. Guemas, and J. Rouxel, *J. Solid State Chem.* **36**, 118 (1981).

[90] S. Dai, X. Zhou, A. H. Morrish, and Z. Shau, *Phys. Rev.* **B 36**, 1 (1987).

[91] S. Dai, X. Z. Zhou, A. H. Morrish, Q. A. Paukhurst, and Z. Shau, *Phys. Rev. Letters* **61**, 1412 (1988).

[92] T. Butz, K.-H. Ebeling, E. Hagn, S. Saibene, E. Zech, and A. Lerf, *Phys. Rev. Letters* **56**, 639 (1986).

[93] S. J. Hillenius, R. V. Coleman, R. M. Fleming, and R. J. Cava, *Phys. Rev.* **B 23**, 1567 (1981).

[94] R. J. Cava, F. J. DiSalvo, M. Eibschütz, and J. V. Waszczak, *Phys. Rev.* **B 27**, 7412 (1983).

[95] M.-H. Whangbo, R. J. Cava, F. J. DiSalvo, and R. M. Fleming, *Solid State Commun.* **43**, 277 (1982).

[96] J. W. Brill, P. Boolchand, and G. H. Lemon, *Solid State Commun.* **51**, 9 (1984).

[97] A. Ben Salem, G. Fatseas, P. Molinie, and A. Meerschaut, *Synthetic Metals* **25**, 207 (1988).

[98] H. Gruber, E. Bauer, M. Reissner, and W. Steiner, in *Charge Density Waves in Solids*, ed. Gy. Hutiray and J. Sólyom, p. 125, Springer-Verlag, Berlin, 1985.

[99] R. Moret, J. P. Pouget, A. Meerschaut, and L. Guemas, *J. Phys. Letters* **44**, L93 (1983).

[100] Z. Y. Chen, M. B. Walker, and R. Morelli, *Phys. Rev.* **B 39**, 11 742 (1989).

[101] A. Meerschaut, P. Palvadeau, and J. Rouxel, *J. Solid State Chem.* **20**, 21 (1977).

[102] Z. Z. Wang, P. Mouceau, M. Renard, P. Gressier, L. Guemas, and A. Meerschaut, *Solid State Commun.* **47**, 439 (1983).

[103] A. Meerschaut, P. Gressier, L. Guemas, and J. Rouxel, *J. Solid State Chem.* **51**, 307 (1984).

[104] P. Butaud, P. Ségransan, C. Berthier, and A. Meerschaut, in *Charge Density Waves in Solids*, ed. Gy. Hutiray and J. Sólyom, p. 121, Springer-Verlag, Berlin, 1985.

[105] Y. Tajima, and K. Yamaya, *J. Phys. Soc. Japan* **53**, 495 (1984).

[106] S. Saibene, T. Butz, A. Lerf, and W. Biberacher, *Hyperf. Interact.* **34**, 255 (1987).

[107] L. Mihály and G. Y. Tessema, *Phys. Rev.* **B 33**, 5858 (1986).

[108] J. Shi and J.H. Ross, Jr., *Phys. Rev.* **B** *in press.*

Note added in proof Recent spin-lattice relaxation measurements [108] in aligned-crystal $NbSe_3$ give additional information regarding the per-site electron distribution. This includes a more detailed profile of the electrons contributing to each CDW. The red site is found to be non-insulating, although the orange and yellow site carry most of the metallic electrons, consistent with the Knight shifts described above.

NUCLEAR SPECTROSCOPY OF CHARGE DENSITY WAVES IN MOLYBDENUM BRONZES

CLAUDE BERTHIER, ANDRAŚ JÁNOSSY+, PATRICK BUTAUD AND PIERRE SEGRANSAN
Laboratoire de Spectrométrie Physique (associé au C.N.R.S.)
Université Joseph Fourier, Grenoble 1, B.P. 87
38402 St Martin d'Hères
France
+Central Research Institute for Physics, H 1525 Budapest 114
Hungary

1. Introduction

The transition metal bronzes are ternary compounds of generic formula $A_xT_yO_z$, where A is usually an alkali metal. The bronzes have been of interest for a number of research fields, ranging from intercalation chemistry, fast ionic transport, to physics of low dimensional conductors. Those of them which exhibit low dimensional electronic properties have been recently reviewed by M. Greenblatt [1]. The existence of charge density wave (CDW) instabilities has been firmly demonstrated for the molybdenum bronzes [2] which include the quasi two dimensional purple bronzes $A_{0.9}Mo_6O_{17}$ with A = Li, Na, K, Tl and the quasi one dimensional blue bronzes $A_{0.3}MoO_3$ with A = K, Rb, Tl. These latter are maybe the best studied among quasi one dimensional systems in which the charge density waves are depinned by a small electric field [3]. Compared to other systems, large crystals have been grown [4], enabling the determination of the crystalline structure and its deformations below the Peierls transition in detail [5]. The dynamic response to an electric field both at low and high frequencies has been thoroughly investigated [6]. The crystals grow with high purity and the CDW may be depinned at low electric fields so that various properties have been studied in the depinned state also. The microscopic methods, especially nuclear magnetic resonance, contributed considerably to the understanding of both the static and dynamic aspects. As we shall detail in this chapter, it provided an unambiguous test for the incommensurate plane wave nature of the CDW in a broad temperature range [7, 8, 9] and a direct proof for the collective motion [10, 11]. The phase velocity of the depinned wave was measured with high precision [12, 13]. Temporal fluctuations of the CDW phase have also been studied [13] but here a lot remains to be done.

T. Butz (ed.), Nuclear Spectroscopy on Charge Density Wave Systems, 177–227.

This chapter will review NMR, EPR and Mössbauer studies of the molybdenum blue bronzes. In the next section their physical properties are briefly recalled. NMR studies of $Rb_{0.3}MoO_3$ in the normal state (above the Peierls transition) are reviewed in section 3, section 4 deals with the NMR, EPR, and Mössbauer studies of the pinned CDW state, while section 5 describes the NMR study of thermal excitations. Finally, section 6 is devoted to the NMR study of sliding CDW's in $Rb_{0.3}MoO_3$.

2. Physical properties of the blue bronzes

Several reviews discuss the properties of charge density waves in quasi one dimensional systems in general (see e.g. [14, 15] and references therein). Also, exhaustive descriptions of the physical properties [16] and structural investigations [17] of blue bronzes have been published recently. In the following we shall review briefly aspects directly related to the microscopic studies.

2.1. CRYSTAL STRUCTURE

The three systems $K_{0.3}MoO_3$, $Rb_{0.3}MoO_3$, and $Tl_{0.3}MoO_3$ are isostructural and crystallize in the monoclinic form [18] (space group C/2 m). In describing the basic structural features we shall not distinguish between findings on the K and Rb compounds since their physical properties are very similar. The crystal structure of the unmodulated high temperature phase is shown in Figure 1. Sheets of (somewhat distorted) MoO_6 octahedra are separated by the alkali metal ions. Within each sheet groups of four octahedra chains form the quasi-one-dimensional conducting chains along the $\vec{b}$ monoclinic axis. Among the three inequivalent Mo sites (Mo 1, 2, 3) only two (Mo 2 and Mo 3) are included in the conducting chains, the Mo 1 are in octahedra lying somewhat aside from the chains. There are two crystallographically inequivalent Rb sites which serve as excellent NMR probes [7, 8, 9]. No NMR study is known of the molybdenum or oxygen sites. Electron spin resonance of Mo^{5+} ions [19, 20] at Mo 1 sites near unidentified defects and V^+ ions substituting Mo [21] have been reported together with Mössbauer studies [22] of Fe at unknown locations.

2.2. THE PEIERLS TRANSITION

One-dimensional fluctuations of the lattice along the $\vec{b}$ direction with a wave vector q_b incommensurate to the lattice have been observed already at room temperature [23]. By lowering the temperature a transverse order of the phase fluctuations develops and at 180 K a second order Peierls transition towards a three dimensionally ordered CDW state appears with $\vec{q} = (0, q_b, 1/2)$ in reciprocal lattice units. At

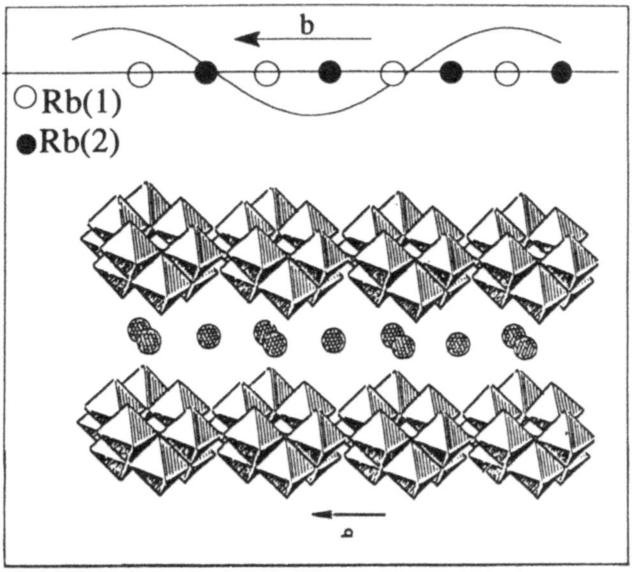

Fig. 1 Crystal structure of $K_{0.30}MoO_3$ showing the infinite slabs separated by the alkali ions and the infinite chains of MoO_6 octahedra parallel to the *b*-axis. (From [2].)

the transition q_b is somewhat less than 0.74 and the deviation from commensurability is obvious, but as the temperature is lowered q_b approaches the commensurate value of $q_0 = 3/4$ (Figure 2). Below 100 K the deviation of q_b from q_0 is comparable to the highest resolution attainable by X-ray [23b, 24] or neutron diffraction [25, 26]. Magnetic resonance shows [7, 8, 27, 28], however, that the system remains incommensurate with a plane wave character at temperatures much below 100 K (section 4.2.).

In one dimensional systems the CDW wavevector is equal to $2k_F$ with k_F the Fermi wavevector in the unperturbed state, and is directly proportional to the electronic density on the chains. Assuming the conduction electrons are provided by fully ionized alkali metal atoms, $q_0 = 3/4$ is obtained if each group of four strongly interacting chains of MoO_6 octahedra contain formally two "degenerate" chains. The origin of the deviation $\Delta q = q_b - q_0$ and its temperature dependence is not clear, it was suggested that it may follow from the band structure of the pure system [23b] but it may also be due to impurity electronic states in the gap [23b]. By doping $K_{0.3}MoO_3$ with vanadium the filling of the band changed and q_b was reduced [30].

The CDW is coupled to a periodic lattice displacement which has the same

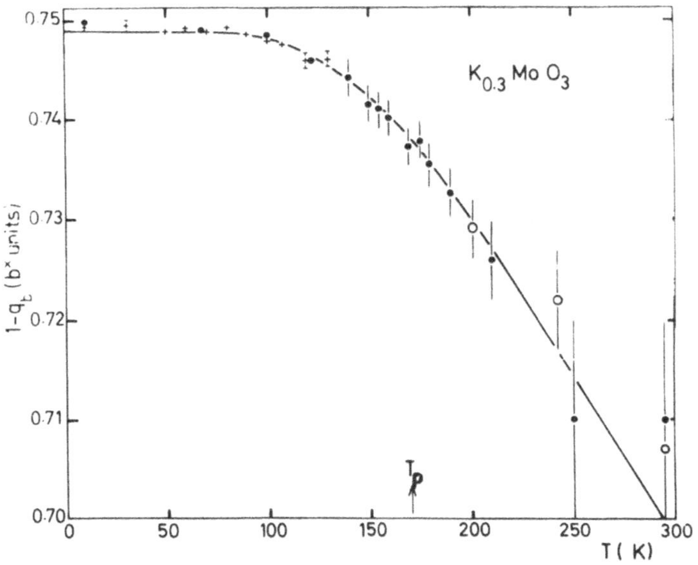

Fig. 2 Temperature dependence of the incommensurate component $1 - q_b(= 2k_F)$ of the satellite reflection (below T_P) and of the diffuse scattering (above T_P). (From [23].)

wavevector $\vec{q}$ but different amplitudes for each inequivalent atom at average position $\vec{R}_j$ within the unit cell. The displacements are not simply along the $\vec{b}$ direction, they are described by a spiral:

$$\Delta\vec{R}_j = \vec{\Delta}_{1j}\cos(\vec{q}\vec{R}_j) + \vec{\Delta}_{2j}\sin(\vec{q}\vec{R}_j)\,. \tag{2.1}$$

$\vec{\Delta}_{1j}$ and $\vec{\Delta}_{2j}$ have been measured by X-ray diffraction by Schutte and Boer [5] with high accuracy. The largest displacements are transverse to $\vec{b}$. A typical value for $|\vec{\Delta}_{1j}|$ is 0.03 Å. Higher harmonics with wavevector $2q_b$ along $\vec{b}$ have been searched for [23b] down to 60 K, these are certainly very small, the observed low intensity reflections correspond probably to second order reflections rather than a deviation from the plane wave modulation given by (2.1).

The 3D long range order developing at the Peierls transition leads to a static charge density wave i.e. a "pinned" state, with a phase $\varphi(\vec{R})$ determined by impurities and other lattice defects. The electronic charge density is described by:

$$\rho(\vec{R}) = \rho_0 + \rho_1\cos(\vec{q}\vec{R} + \varphi(\vec{R})) \tag{2.2}$$

if relatively short range amplitude variations are neglected. ρ_0 is the uniform density, in one dimension $\rho_0 = 2ek_F/\pi$. The CDW amplitude ρ_1 is proportional to the magnitude of the order parameter in mean field theory.

In estimating the phase coherence length two limits are usually considered [31]. Depending on the strength of the impurity potential relative to the elastic energy of the CDW the pinning is considered weak or strong. For strong pinning at each impurity site the CDW phase is fixed to the same value while for weak pinning the phase is determined by the overall action of many impurities within the coherence length. It is not clear which is the case for nominally pure blue bronze crystals. As expected from theory [31] the observed phase coherence length [32] is strongly anisotropic. Along $\vec{b}$ it is more than 1 μm and can not be determined by X-ray diffraction, in the transverse directions it is several tenths of a μm at low temperatures but shortens above 100 K.

2.3. DYNAMICS OF THE CDW

2.3.1. *Excitations at finite frequencies* Extensive inelastic neutron [25, 26, 33] and frequency dependent conductivity [6] studies have been performed on blue bronzes. Above T_P a soft mode is observed centered around $\vec{q} = (0, q_b, 1/2)$ with a relatively short but strongly anisotropic coherence length. As the temperature approaches T_P the energy of this mode decreases and the coherence length increases. At the same time an elastic peak develops also. The inelastic branch corresponds to quasi-1D fluctuations of the CDW while the elastic peak is probably due to CDW's pinned to impurities. Below the Peierls transition in the excitation spectrum two branches corresponding to the amplitude and phase modulation of the complex order parameter have been observed. The amplitudon branch has a gap of 1.7 THz [33], while the phasons have an acoustic dispersion relation. There is probably a small gap in the phason mode spectrum, a maximum in the microwave electrical conductivity as a function of frequency around 100 GHz is attributed to the pinning of the phase mode by impurities [6]. The maximum becomes sharper as the temperature is lowered giving rise to a drop in the d.c. and low frequency conductivity. This sharpening of the pinned mode is believed to be related to the increasing rigidity of the CDW with decreasing temperature, in other words with the increase of the phason velocity.

2.3.2. *The Fröhlich mode* The pinning potential is so small in some quasi 1D CDW systems that the CDW may be depinned above a small threshold electric field E_T. This sliding state, or Fröhlich mode [34] is characterized by a time dependent phase. Neglecting finite energy phasons and at electric fields much above threshold the phase increases constantly with time:

$$\varphi(t) = q_b v_d t \tag{2.3}$$

where v_d is the drift velocity. Apart from fluctuations the current density is

$$j_{CDW} = \rho_c v_d \tag{2.4}$$

where v_d depends on the field E in a non-linear way. The condensed electron density ρ_c is equal to ρ_0 at low temperatures. As discussed in section 6, v_d may be measured by NMR, and by simultaneously measuring j_{CDW} relation (2.4) has been verified.

The normal excitations over the Peierls gap contribute also to the total current and in a two fluid model

$$j = \sigma_n E + \sigma_{CDW}(E)E \tag{2.5}$$

σ_n and $\sigma_{CDW} = \rho_c v_d / E$ are the normal and CDW conductivities, respectively. The above expression (2.5) is not entirely correct since it assumes that the CDW current does not influence the normal current. Hall effect measurements indicate that this is not exactly so, the CDW current induces a normal current backflow [35].

The blue bronzes at room temperature are anisotropic metals. Deviations from the linear temperature dependence of the conductivity due to CDW fluctuations appear below 220 K. At the Peierls transition (T_P = 180 K) a sharp decrease of the normal conductivity signals the opening of the gap (Figure 3).

A conductivity non-linear in the electric field has been first reported by Fogle and Perlstein [36] at low temperatures in 1972. At this time it was not realized that the depinning of the CDW was observed. The system was reinvestigated by Dumas *et al.* [3] who showed that a Fröhlich conductivity occurs above a well defined threshold field in a broad temperature range. An example for a current voltage characteristics of a blue bronze crystal is shown in Figure 4, here the contributions of σ_n and σ_{CDW} are well distinguished.

Below the Peierls transition σ_n depends somewhat on thermal and electrical history [37, 38] probably due to the interaction of the normal current with irregularities of the CDW phase. $\varphi(R)$ depends on temperature (since q_b is also temperature dependent) and also on electric field, even below threshold. The result is that several phase distributions may have similar energies leading to metastable states. Above 40 K σ_n shows the characteristics of an intrinsic semiconductor with an activation energy [39] of E_a^n = 490 K. At liquid helium temperatures the normal conductivity is unmeasurably low.

Near threshold a detailed analysis of the depinning by Fisher [40] leads to the form:

$$\sigma_{CDW} = \sigma_0 (E - E_T)^\gamma \tag{2.6}$$

with $\gamma = 1/2$ for a finite number of defects and $\gamma = 3/2$ in the mean field approximation of the Fukuyama Lee Hamiltonian which describes the deformation of the CDW by impurities. The electric field dependence of the CDW current is hard to determine experimentally since the depinning does not occur simultaneously at a well defined E_T in the whole crystal. As shown [10] by NMR, the current distribution is inhomogeneous even when the field is high enough to depin the CDW in the whole crystal. An investigation of the electric field dependence of the narrow band noise [39] yields $\gamma \approx 1$ in contrast to the mean field theory. σ_0 was found to be activated $\sigma_0 \sim \exp\{-E_a/kT\}$ with an energy E_a of 750 K. A similar activation

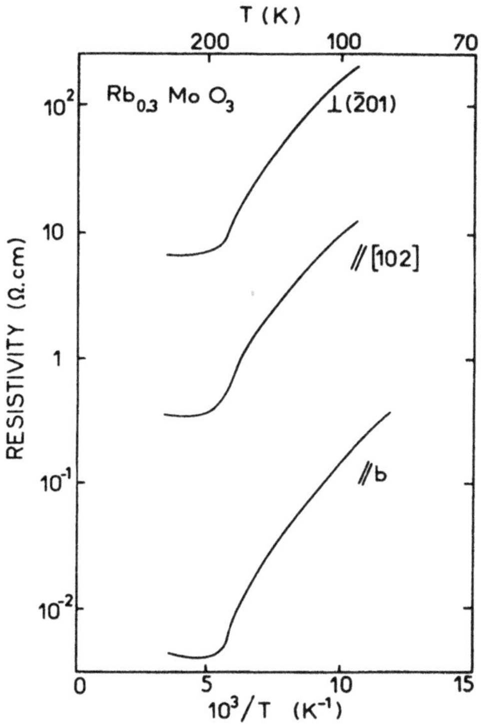

Fig. 3 Electrical resistivity (logarithmic scale) of $Rb_{0.30}MoO_3$ versus inverse temperature along b, along [102] and perpendicular to the (201) plane in which the layers of octahedra are lying. (From [2].)

energy was found [12] for the CDW current measured at a field fixed at a value several times E_t as shown in Figure 5.

The threshold field in nominally pure crystals varies somewhat from sample to sample and it has not been yet determined whether it arises from charged or neutral impurities or other defects, the crystal surfaces may also play a role since the phase coherence length is extremely long. It seems that at low temperatures (below 40 K) the mechanism for depinning the CDW may be very different from that at higher temperatures. At liquid helium temperatures a CDW conductivity which depends very strongly on E appears above a threshold of a few Volts/cm [41]. Most studies in the depinned state are confined to the high temperature regime where depinning occurs at much lower fields. A typical value of E_T is 80 mV at 40 K. (Relaxation between metastable states makes it difficult to measure E_T at lower temperatures).

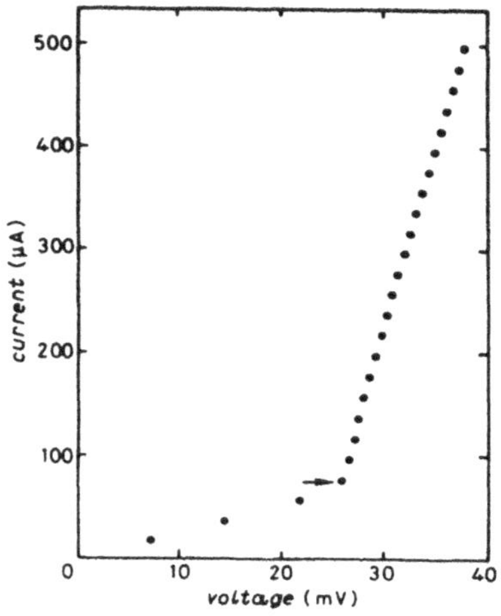

Fig. 4 D.C.current-voltage characteristics measured in a $Rb_{0.30}MoO_3$ single crystal at 77 K. The arrow indicates the onset of the non-linear conductivity. (From [50].)

The threshold field increases with increasing temperatures as shown in Figure 6. NMR studies up to present were conducted only in the higher temperature range above 40 K.

2.3.3. *Voltage oscillations* One of the most intriguing problems of the depinned CDW state is the origin of voltage fluctuations observed for crystals biased by a constant current source [42]. An example of low frequency oscillations is given in Figure 7.

The oscillation frequencies in some cases may be extremely well defined but generally the frequency spectrum is rather broad in blue bronzes. These oscillations reflect periodic fluctuations $f(t)$ of the phase superimposed onto the linear time evolution:

$$\varphi(t) = q_b v_d t + f(t) \tag{2.7}$$

which give rise to a corresponding fluctuation of the CDW current.[1] Various models have been proposed to describe the phenomenon. The simplest [44, 45] neglects all degrees of freedom except the average phase $\bar{\varphi}$ and visualizes the motion of the CDW in a potential $V(\bar{\varphi})$ arising from a single impurity. Since V has the periodicity of $\lambda = 2\pi/q_b$ along the chains, the motion of the CDW in a sufficiently large external

[1] Although the total current is constant the normal and CDW currents are oscillating with opposite signs.

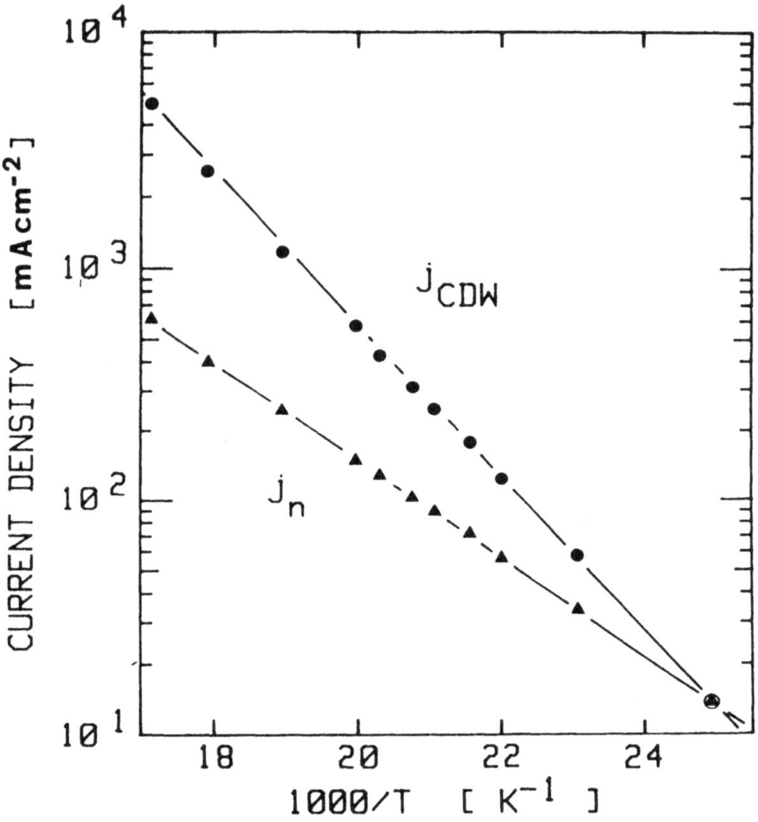

Fig. 5 CDW (circles) and normal (triangles) current densities at a field of 1.5 V/cm ($E/E_T \approx 15$) as a function of inverse temperature. The normal current is extrapolated from low-field data. Activation energies are 754 and 488 K for the CDW and the normal currents, respectively. (From [12].)

field will be periodic (although not simply sinusoidal) with the fundamental voltage noise frequency of:

$$\nu_n = q_b v_d \tag{2.8}$$

The extension of the above model to systems with several impurities leads to difficulties. If weakly coupled domains with dimensions of the coherence length are supposed, the voltage noise amplitude is expected to decrease with the square root of the actual dimensions of the crystal. The experimental evidences for this are controversial [46,47]. In blue bronzes it seems that the current oscillations may be

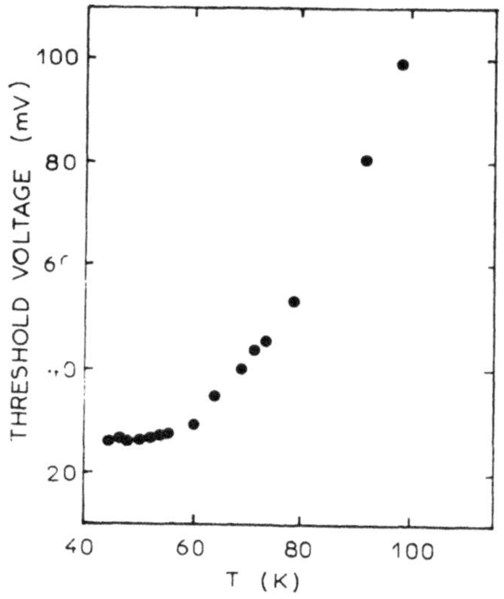

Fig. 6 Variation of the threshold voltage versus temperature T in $Rb_{0.30}MoO_3$. The sample length is 3.6 mm. (From [13].)

coherent along samples of several mm [48].

An alternative model proposed by Ong, Verma and Maki [49] suggests that the fluctuations originate at the boundaries between pinned and unpinned regions. The necessary conversion of normal to CDW electrons at such boundaries does not occur simultaneously on the whole surface but at CDW phase dislocation lines perpendicular to the conduction axis. The CDW slides forward by the transverse motion of dislocation lines which insert sheets of extra CDW wavelengths at one end and eliminate them at the other. The CDW current changes abruptly each time a dislocation line is created or annihilated. Recent experiments show that for fields near threshold low frequency voltage instabilities propagate along the sample and have distinct sources and sinks [50]. The sources and sinks are believed to be the locations within the crystal where dislocation loops perpendicular to the conduction axis are created and annihilated.

A common feature of the above models is that expression (2.8) holds, the basic assumption is that the periodicity of the phase fluctuations reflects the invariance of the CDW system to a translation by $\lambda = 2\pi/q_b$. Another model suggests [51] $\nu_n = \beta q_b v_d, (\beta \neq 1)$ but this seems to be ruled out by the NMR experiments described in section 6.

Together with the periodic oscillations non-periodic fluctuations of the phase are also observed. The peaks of voltage oscillation frequency spectra have finite widths, at 77K the best resolved peaks are still about 1 kHz broad. The frequency spectrum

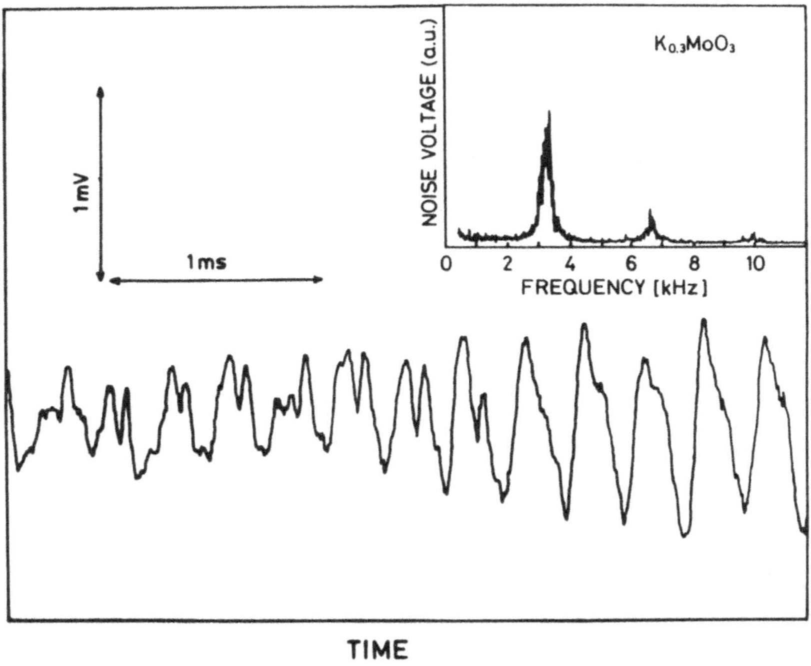

Fig. 7 Periodic voltage oscillations in a thin (about 5 μm thick) $K_{0.30}MoO_3$ single crystal. (T = 77 K). These oscillations are not transients, the trace is taken a long time after the on-switch of the current. (From [43].)

contains a broad band component which often has a large intensity at low frequencies. The voltage noise power spectra of a blue bronze crystal studied by NMR in the depinned state are shown in Figure 8 for various temperatures at a fixed external electric field. For fields much larger than threshold the spectra at various temperatures are similar, only the frequency scale changes. Thus, as the temperature is lowered the average current decreases exponentialy but the distribution of the current density remains the same.

3. NMR in the normal state

NMR studies of the blue bronzes have been concentrated on the ^{87}Rb (I = 3/2) nuclei of $Rb_{0.3}MoO_3$. K, Mo, and ^{17}O have either a low gyromagnetic ratio or a small natural abundance, which renders their observation much more difficult. The

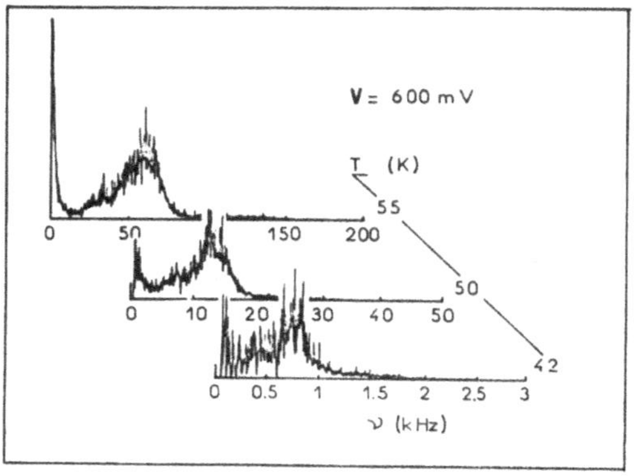

Fig. 8 Noise power spectra taken at $E = 1.67$ V/cm in a large crystal used also for NMR. The first harmonic component has been substracted from the raw data. The spectra are similar despite a change of a factor 60 in the frequency scale showing that at fixed field and varying temperature the CDW current distribution does not change much. The voltage power $|V(t)|^2$ changes inversely to the mean frequency so that the total intensity remains constant. (From [12].)

lack of NMR work on the Tl nuclei in $Tl_{0.3}MoO_3$ is likely due to the the poor quality of the crystals compared to those of $Rb_{0.3}MoO_3$ and $K_{0.3}MoO_3$. Before describing the NMR studies of the CDW state we shall briefly discuss results obtained in the normal state i.e. above the Peierls transition.

3.1. Electric field gradient and magnetic hyperfine shift tensors

A careful study of the Electric Field Gradients (EFG) and Magnetic Hyperfine Shift (MHS) tensors of Rb(1) and Rb(2) which are respectively ten (seven) fold coordinated to oxygen (Figure 1) has been performed by Douglass *et al.* [9]. They used a large single crystal (3x5x15 mm) which was rotated around the b-axis with respect to the applied magnetic field H_0. (Figure 9) The characteristics of the EFG tensors at 296 K, as well as the isotropic part of the MHS, are reported in Table I.

Table I

	V_{ZZ} (MHz)	V_{XX} (MHz)	V_{YY} (MHz)	η	K(%)
Rb(1)	12.62	-10.86	-1.76	0.72	0.015
Rb(2)	7.89	-5.97	-1.88	0.48	0.015

For Rb(2) the principal axes X, Y, Z lie, respectively, along the [1 0 2], $\vec{n}$, and $\vec{b}$ directions ($\vec{n}$ is the normal to the cleavage plane (-2 0 1)). For Rb(1) the X and

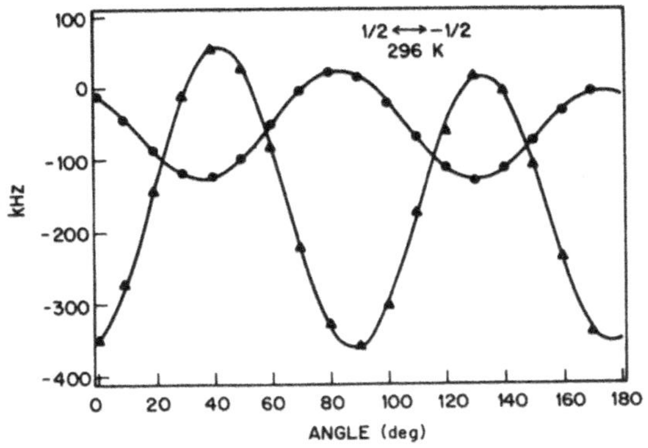

Fig. 9 Shift of the central transition of the sevenfold oxygen coordinated rubidium relative to RbCl as a function of rotation about the crystal *b*-axis at 47 kG and 296 K. Solid circles and triangles represent data for seven- and tenfold coordinated Rb sites, respectively. (From [9].)

Y axes are rotated by 42° around the Z (here $\vec{b}$) axis. In the notation used for metals, the magnetic hyperfine shift K is positive. No detailed analysis to separate an orbital (Van Vleck) contribution [52] from the contact interaction has been done yet.

The EFG tensor experienced by the Rb nuclei has two origins: one is the distribution of distant ionic charges, the other is the distribution of conduction electrons. Although the conduction band is primarily constructed from the Mo 4*d* orbitals, a hybridization of the Mo 4*d* wavefunctions with Rb orbitals is very likely; NMR studies of the sodium tungsten bronzes Na_xWO_3 [53] have shown that such a hybridization could play an important role. The effect of the distant charges is amplified by the antishielding Sternheimer factor which can be as high as 70 for Rb [54]. For on-site electrons, there is no antishielding amplification, but the small values of r in the expression for the main value of the EFG tensor, $q = V_{zz}/e$

$$q = \sum_{\vec{k}} \frac{\langle \psi_{\vec{k}} | P_2(\cos\theta) | \psi_{\vec{k}} \rangle}{r^3} \tag{3.1}$$

enhances their contribution.

At room temperature the full width at mid-height of the line corresponding to the (1/2, -1/2) transition was found to be 1.2 kHz. The spin-spin relaxation time T_2 was also measured by the spin-echo technique, it was found to be 4.6 ms at all temperatures, indicating clearly that the linewidth is inhomogeneously broadened by a small distribution of quadrupole coupling due to crystal imperfections.

3.2. QUASI-1D PROPERTIES

Above the Peierls transition the low-dimensionality of the conduction electron gas has several consequences which can be studied by NMR. One important feature is the strong temperature dependence of the electronic susceptibility $\chi(q)$ for $q = 2k_F$.[2] This susceptibility describes the charge perturbation $\delta\rho_q = \chi(q)\delta V(q)$ induced by an external perturbation potential $\delta V(q)$ and it diverges at the onset of the Peierls transition T_P. As $\chi(2k_F)$ is precisely what gives rise to the Friedel oscillations around charged impurities in metals [52], its divergence when approaching T_P from above is responsible for an anomalous temperature dependent screening of charged defects in low dimensional conductors. Damped charge oscillations grow around these defects with a range following the temperature dependence of $\chi(2k_F)$. The corresponding EFG disribution is responsible for a static inhomogeneous broadening of the NMR lines. This effect, which was best studied in $NbSe_2$ [55] was also observed in $Rb_{0.3}MoO_3$ [7, 27] for the (1/2,-1/2) transition of Rb(1).

Another signature of the quasi-1D character of an electron gas is the opening of a pseudo-gap well above the 3D Peierls transition. This is reflected in the NMR properties as a temperature dependence of the Knight shift and the spin-lattice relaxation rate which start to decrease well above T_P (see section 5). Although no precise data on the Knight shift are known for $Rb_{0.3}MoO_3$, the decrease of $(T_1T)^{-1}$ between 300 K and 200 K [27, 28] was found to scale with the temperature dependence of χ_s^2 where χ_s is the spin part of the macroscopic susceptibility [56].

The Knight shift deduced from nuclear spin-lattice relaxation measurements from the relationship $K = (\Re T_1 T)^{-1/2}$ with $\Re = \frac{4\pi k_B}{\hbar}\frac{\gamma_n^2}{\gamma_e^2} \approx 4{\cdot}10^5$ for Rb are in reasonable agreement with those reported in Table I.

4. Nuclear Spectroscopy of pinned CDW

A periodic modulation of the lattice associated with a CDW modifies both the magnetic and electric hyperfine couplings experienced by the nuclei in the crystal. As far as the magnetic part (i.e. the Knight shift tensor) is concerned, those contributions which are proportional to the density of states at the Fermi level $N(E_F)$, that is the contact, core-polarisation and dipolar ones, vanish exponentially below T_P as the Peierls gap opens. On the other hand the orbital contribution K_0 which arises from conduction electron states within the whole band is barely affected by the opening of the Peierls gap since usually kT_P is much smaller than the bandwidth. K_0 is a local quantity, the local electronic density at distance r from the nucleus contributes proportionally to $\langle r^{-3}\rangle$. In presence of a CDW, where the local

[2] For a 1D non interacting electron gas χ_0 for $q = 2k_F$ is proportional to $n(E_F)$ and has a logarithmic divergence at $T = 0$. A coupling between chains raises the transition to a finite temperature.

electronic density $\rho(r)$ becomes $\rho(r) + \delta\rho(r)$, this term is linearly coupled to $\delta\rho(r)$ and thus to the order parameter of the Peierls transition[3] [28].

As far as the quadrupole coupling is concerned, all the parameters of the EFG tensor (V_{zz}, η, the directions of the principal axes) become periodically modulated in the crystal. As mentioned in section 3.1, both the displacements of distant ions and the modulation of the conduction electron density will contribute to the EFG. If the last term is the dominant one, the modulation of the EFG can be expressed as a function of $\delta\rho(r)$; in the opposite case we do not expect a local relationship between the lattice distortion and the EFG modulation since the wavelength of the modulation in the blue bronzes is only a few interatomic distances. (This latter case applies to the modulation of the g-factor of ESR lines which is related to that of the crystal field.)

4.1. LINESHAPE IN PRESENCE OF AN INCOMMENSURATE CDW

4.1.1. *Nuclear magnetic resonance lineshape* The unusual character of NMR lineshapes in presence of an incommensurate modulation of the hyperfine field has been first recognized in CDW systems [57, 58]. Nevertheless the most extensive NMR studies of incommensurate phases have been performed in dielectric compounds [59]. We shall here restrict ourselves to the case relevant to the blue bronzes, that is a single $\vec{q}$ CDW, and we shall suppose that the local approximation discussed above is valid. Let us examine the theoretical NMR lineshape of a given $(m, m+1)$ transition of a single crystal below the onset of the CDW. Within the local approximation the resonance frequency for a nucleus at site $\vec{R}$ is:

$$\begin{aligned} \nu(\vec{R}) = \nu_0 + F\left[\delta\rho(\vec{R})\right] &= \nu_0 + \alpha\delta\rho(\vec{R}) + \beta\left[\delta\rho(\vec{R})\right]^2 \\ &= \nu_0 + \delta\nu(\vec{R}) \end{aligned} \tag{4.1}$$

where we have expanded $\nu(\vec{R})$ in powers of the real part of the order parameter. The coupling constants α and β depend on the direction of the magnetic field H_0 with respect to the principal axes of the EFG and (or) the orbital Knight shift tensor. In $Rb_{0.3}MoO_3$ the magnetic hyperfine tensor is so small that its modulation can be safely neglected. It is important to notice that even in the case of the (1/2,-1/2) transition (which is only affected to second order of perturbation by the quadrupole coupling $\Delta\nu^{(2)} = \frac{\nu_Q^2}{\nu_0} f(\theta, \varphi, \eta)$ [60]) $\alpha \gg \beta$ provided the modulation of the EFG is small compared to its unperturbed value. The resonance lineshape resulting from the distribution of the EFG is readily calculated:

$$g(\nu) = \text{const} \int\int_{\nu=\text{const}} \frac{dS}{|\vec{\nabla}(\delta\nu(\vec{R}))|} \tag{4.2}$$

[3] For similar reasons, one expects that in Mössbauer experiments the isomer shift will be linearly coupled to $\delta\rho(r)$.

where ν is measured from the resonance frequency in the absence of the CDW and the integral is performed over the surface S corresponding to $\delta\nu(\vec{R}) = \nu$. This lineshape has to be convoluted with the nuclear dipole-dipole homogeneous broadening. The major effect of incommensurateness is to give rise to an infinite number of inequivalent sites and to spread the intensity of the resonance line over a continuous range of frequency which can be as large as a few hundred kHz.

If we approximate the charge modulation by a plane wave, one has simply:

$$\delta\nu(R) = \nu_1 \cos(qR + \varphi_1) + \nu_2 \cos^2(qR + \varphi_2) \tag{4.3}$$

where ν_1 is linear in $\delta\rho$ and ν_2 is proportional to $(\delta\rho)^2$, and (4.2) reduces to $g(\nu) = \mathrm{const}\left| \left[\frac{d\delta\nu}{dx}\right]^{-1} \right|$ where $x = qR$.

If ν_2 is negligible, and as long as q is incommensurate

$$\begin{aligned} g(\nu) &= \mathrm{const}|1 - (\nu/\nu_1)^2|^{-1/2} && |\nu| < \nu_1 \\ g(\nu) &= 0 && |\nu| > \nu_1 \end{aligned} \tag{4.4}$$

As a result of the local approximation the lineshape is independent of the value of q, provided it is incommensurate. This is no longer true in the case where non-local interactions become dominant, like the electron-nuclear dipolar magnetic interaction in some SDW systems [61, 62]. Computer simulations of NMR lineshapes for various values of ν_1, ν_2, and q in the plane wave model are given in Figure 10.

4.1.2. *Temperature dependence of the order parameter* Since from (4.3) ν_1 is proportional to the modulus of the order parameter one can extract the temperature dependence of the order parameter from that of the lineshape. Two studies of this type have been performed in $Rb_{0.3}MoO_3$ [7, 8, 27]. Figure 11 shows the temperature dependence of the lineshape of the (1/2, -1/2) transition for Rb(1) in $Rb_{0.3}MoO_3$. The temperature dependence of the order parameter (Figure 12) was extracted by fitting the experimental lineshapes to the theoretical ones computed from (4.4) and convoluted with the appropriate broadening function. It compares well with the neutron [4] and X-ray results [23b] in which the intensity of the superlattice reflections is proportional to the square of the order parameter. A deviation from the mean field BCS theory is observed both by NMR [27, 28] and X-ray [23b]. From the NMR results a critical exponent $\eta = 0.32 \pm 0.03$ has been found [27, 28] in good agreement with the expected value $\eta = 0.35$ for a 3D phase transition with a two component order parameter. Close to the transition, within a temperature range of about 5 - 10 K, an anomalous behavior is observed corresponding to a partial motional narrowing of the line due to a thermally induced depinning of the CDW. A quite similar behavior was also observed in the insulating compound Rb_2ZnCl_4 [64] close to the paraelectric incommensurate phase transition.

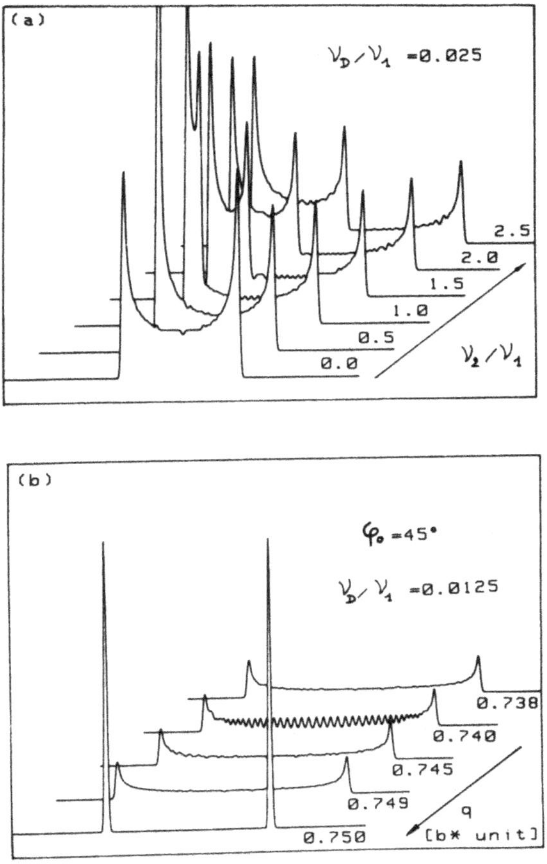

Fig. 10 Theoretical NMR lineshapes in the plane wave limit $\delta\nu(R) = \nu_1 \cos(qR+\varphi_1)+\nu_2 \cos^2(qR+\varphi_2)$
a) at various values of ν_2/ν_1, b) in the linear case ($\nu_2 = 0$) as a function of the wavevector q. The values $q = 0.75$ and $q = 0.74$ correspond to commensurate structures of order 4 and 50 respectively. All frequency distributions are convoluted with a Lorentzian of half width at mid-height ν_D (From [27, 28].)

4.1.3. *Electron Paramagnetic Resonance (EPR) in inorganic CDW systems* The concepts of the evaluation of ESR and NMR spectra of CDW systems are very similar. The modulation of the g-factor of defects in insulators with periodic lattice modulations not related to a Peierls instability has been extensively studied. The main difference is in the probes used, while NMR detects nuclei far from impuri-

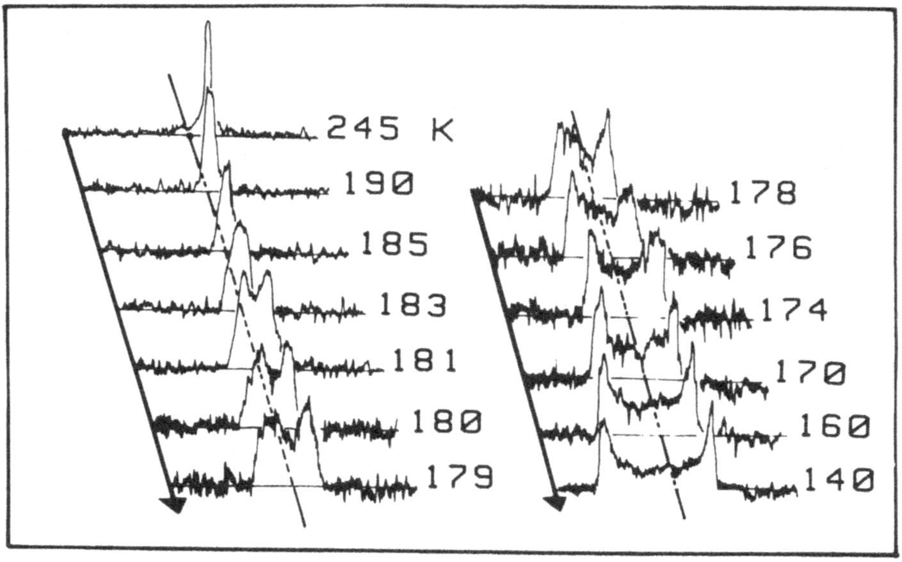

Fig. 11 Temperature dependence of the (-1/2,1/2) NMR line of Rb(1) in a $Rb_{0.30}MoO_3$ single crystal. (From [27, 28].)

ties, EPR on the other hand is originating at some particular defects. Conduction electron spin resonance - in contrast to some organic quasi 1D CDW systems - is not observed for inorganic crystals as the heavy ions relax the conduction electron spins rapidly via the phonon mediated spin orbit interaction [65].

To observe a resonance well localized uncompensated electrons are needed. These may be provided by doping with magnetic impurities or by the formation of a magnetic site near lattice defects.

An example for magnetic doping is the study of Mn doped $(TaSe_4)_2I$ by Shaltiel *et al.* [66]. Mn probably substitutes Ta in the chains. Mn^{2+} is an S = 5/2 state ion with only little mixing of non-zero momentum orbital states. To lowest order the Hamiltonian for the site of tetragonal symmetry is the same as for the nuclear quadrupole interaction of an $I = 5/2$ nuclear spin with an axially symmetric EFG. The oscillation of the crystalline field due to an incommensurate CDW broadens the line. Like the ^{87}Rb NMR in $Rb_{0.3}MoO_3$, the Mn^{2+} linewidth follows the order parameter as a function of temperature, (Figure 13) except that at T_P it does not fall to zero. One-dimensional low frequency or static fluctuations contribute to the linewidth in a broad temperature range above T_P.

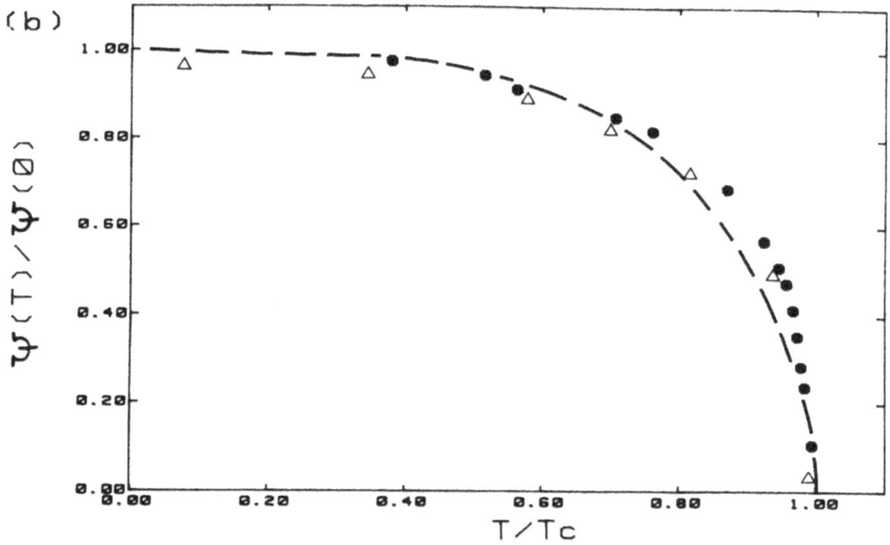

Fig. 12 Temperature dependence of the order parameter of the Peierls transition extracted from the temperature dependence of the Rb(1) NMR lineshape. (From [27, 28].) The open triangles represent the order parameter extracted from X-ray data. (From [23b].)

4.2. COMMENSURABILITY OF THE CDW WITH THE LATTICE

Below 100 K the CDW wavevector q_b deviates from $q_0 = 3/4$ less than the resolution of X-rays or neutron diffraction (Figure 2) and the question arises whether blue bronzes are commensurate or incommensurate. The theory of MacMillan [67] predicts for a bandfilling very close to a commensurate value an intermediate state called "nearly commensurate" where the commensurate regions are interrupted periodically by discommensurations in which the CDW phase changes by $2\pi/p$. The order of commensurateness is $p = 4$ for the blue bronzes. The width of the discommensurations α^{-1} depends on the rigidity of the CDW against the deformations by the lattice potential. (The nearly commensurate state is described by Walker in this book in detail.)

Contrary to diffraction methods, NMR and EPR are very sensitive to a commensurate incommensurate transition, the lineshapes depend qualitatively on the commensurateness of the system. As shown in the preceeding paragraph, for an incommensurate system a continuous range of Larmor frequencies occurs and the spectrum has characteristic singularities. On the other hand in a commensurate system only a limited number of discrete CDW phases are possible at the nuclear or electronic probe and the corresponding spectrum consists of a small number, k, of

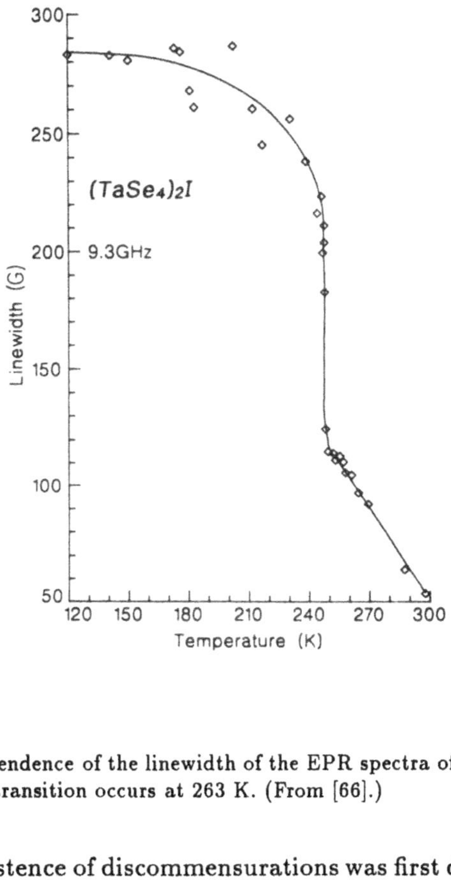

Fig. 13 Temperature dependence of the linewidth of the EPR spectra of Mn^{2+} in Mn:$(TaSe_4)_2I$ at $\nu \approx 36$ GHz. The Peierls transition occurs at 263 K. (From [66].)

discrete lines. The existence of discommensurations was first demonstrated by NMR in 2H-$TaSe_2$ [68]. A detailed description of the NMR lineshapes for dielectrics with incommensurate and commensurate lattice modulations has been given by Blinc *et al.* [59].

In the blue bronze, for a general magnetic field direction and taking into account the symmetry of the CDW displacements, $k = 8$ lines are expected at the ^{87}Rb or Mo^{5+} probes. For a magnetic field in the (010) plane there are $k = 4$ different sites. In the simple case when the Larmor frequency is simply related to the charge, as given by (4.1), the number of possible different sites are $k = 2$, 3, or 4 depending on the phase of the CDW relative to the lattice. Figure 14 shows the spectrum expected for $k = 2$. For any deviation from commensurability a continuous range of frequencies and an incommensurate lineshape occurs. A nearly commensurate system gives rise to a spectrum intermediate between the plane wave incommensurate lineshape and the discrete lines of a commensurate system.

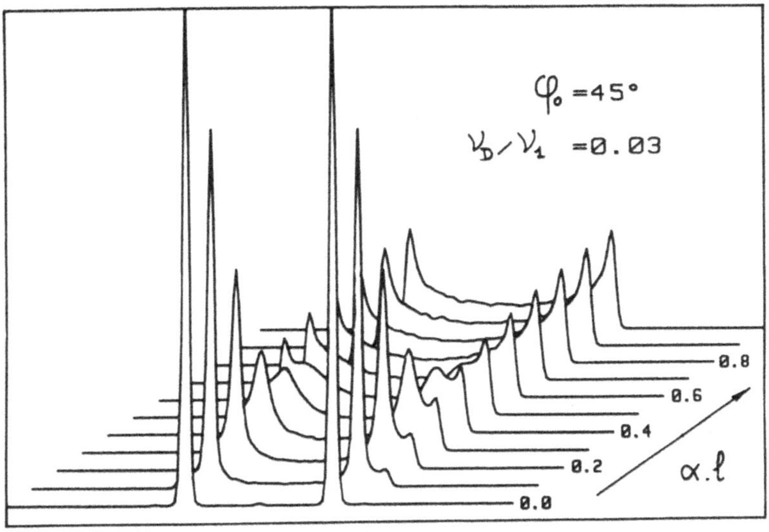

Fig. 14 Computer simulation of NMR lineshapes for a one dimensional soliton lattice ($q = 0.749$) as a function of the ratio of the soliton width α^{-1} to the soliton spacing l. (Eq. (4.6)) (From [27, 28].)

In presence of the lattice potential in the continuum limit the CDW phase is given by the Sine-Gordon equation

$$\frac{d^2\varphi}{dR^2} = \alpha^2 \sin(pR) \tag{4.5}$$

An approximate solution of this equation is a soliton lattice

$$\varphi(R) = \sum_{-\infty}^{+\infty} \frac{k}{p} \arctan\left\{\exp(-\alpha\sqrt{p}R - (n + \frac{1}{2})l)\right\} \tag{4.6}$$

where the parameters α and l correspond to the width and spacing of solitons. Expression (4.6) is valid for $\alpha l \ll 1$, i.e. in the dilute soliton limit.

A detailed analysis of the ^{87}Rb lineshape of $Rb_{0.3}MoO_3$ at 77 K at the Rb(1) and the Rb(2) sites based on (4.6.) shows that discommensurations, if they exist in this system at all, have a width α^{-1} of the order of their spacing, $l = b/p$. [27, 28, 13] (Figure 15). In other ^{87}Rb NMR experiments no evidence for an incommensurate to commensurate transition has been found for temperatures as low as 40 K.

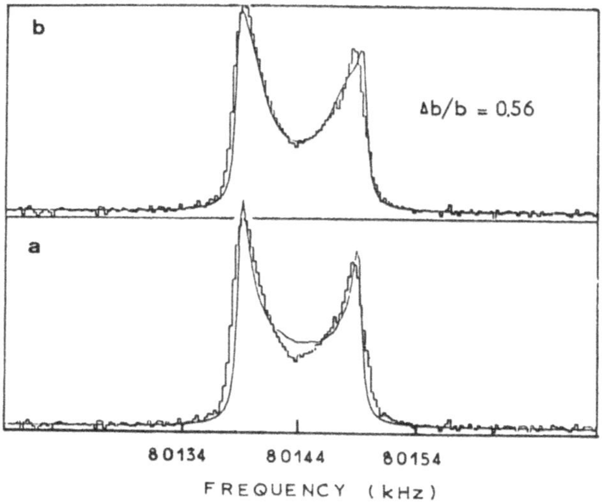

Fig. 15 Comparison of computer simulations to the experimental lineshape of ^{87}Rb in $Rb_{0.3}MoO_3$. a) plane wave limit, b) multisoliton lattice. (From [13].)

In nominally pure blue bronzes $K_{0.3}MoO_3$ and $Rb_{0.3}MoO_3$ some defects are formed during the growth of the crystals which give EPR lines easily detected at low temperatures. Defect lines from different sites are distinguished by their g tensors. A spectrum of two types of defects is shown in Figure 16. From the observed hyperfine structure the defects are identified as electrons localized to form Mo^{5+} ions probably at the off-chain Mo(1) sites [63]. A small overlap is observed with neighbouring on-chain sites. Normally Mo(1) sites are occupied by non magnetic Mo^{6+} ions, however some oxygen vacancies may be charge compensated by the observed Mo^{5+} ions. A detailed analysis of the EPR of this type of defects in $K_{0.33}MoO_3$ (red bronze) in terms of a d^9 electron structure in a distorted octahedral environment has been given by Bang and Sperlich [19]. The red bronze has a structure similar to that of blue bronze but it is a normal semiconductor with no CDW. Accordingly, each defect type gives rise to one narrow line with a hyperfine structure. The blue bronze spectrum is much more complicated. In addition to the hyperfine structure, the spectrum corresponding to each defect consists of a few narrow lines superimposed onto a broad one. The broad and narrow lines are due to the same type of defect since their g factor anisotropies are nearly the same. The former are attributed to defects lying in commensurate regions while the broad component arises from incommensurate regions [20].

Out of the eight lines expected in an arbitrarily oriented magnetic field for a $q_b =$

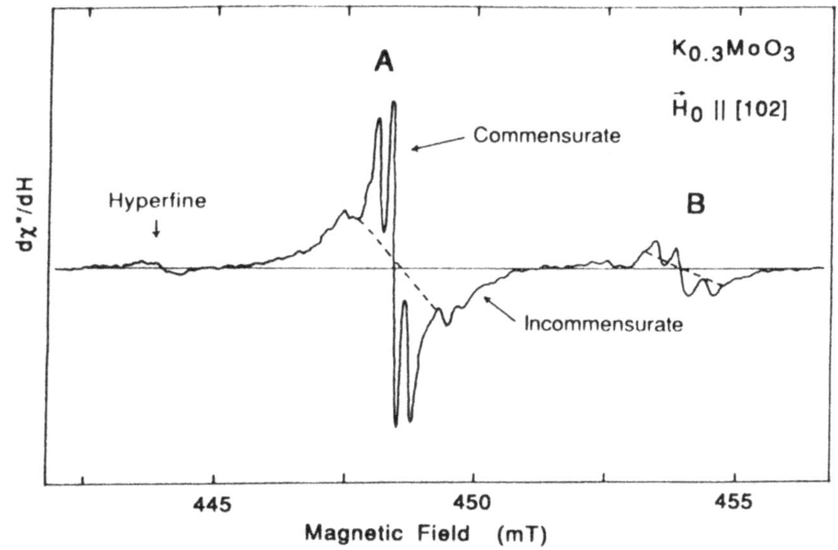

Fig. 16 ESR absorption derivative of $K_{0.3}MoO_3$ at T=41 K. Lines A and B correspond to two types of defects. The dotted line indicates a decomposition into an inhomogeneously broadened component and a set of narrow lines due to regions of incommensurate and commensurate CDW, respectively. One component of the ^{95,97}Mo hyperfine structure is indicated (another component partially overlaps line B.) (From [20].)

3/4 commensurate CDW weakly pinned to Mo^{5+} defects, six are clearly resolved. For fields oriented perpendicular to $\vec{b}$ three of the four narrow lines are resolved with the center line having a larger intensity (Figure 16). The expected number of lines for a general type of modulation (eq. (2.2)) does not depend on the particular phase of the CDW thus it is believed that not all lines are resolved. As expected, (see section 5), the relaxation rate in the incommensurate regions is significantly faster than that in the commensurate ones. The intensity of the commensurate component decreases relative to the incommensurate one as the temperature increases. A rough analysis showed a commensurate fraction of 5 % at 53 K and 20 % at 25 K.

There is some controversy between the EPR data and the ^{87}Rb NMR results since by the latter method no commensurate fraction has been detected down to at least 40 K. One possibility is that commensurate regions form preferentially around charged impurities. At low temperatures the average incommensurability is so small that the charge screening of a single defect is sufficient to shift k_F (and thus q_b) to a commensurate value in a region of several hundred unit cells. It is, however, hard to reconcile this idea with the observed weak pinning (see section 4.3); the phase of a CDW perturbed by a single defect is well determined. Another possibility is that the defects giving rise to the EPR are not homogeneously distributed [21]. In this case the average commensurate fraction of the full crystal may be less than that deduced from the EPR intensity. Further studies of the question would be desirable.

4.3. ESR AND MÖSSBAUER STUDIES OF THE STRENGTH OF THE CDW IMPURITY PINNING

The phase of the static CDW is determined by impurities and charged impurities are expected to be particularly effective pinning centers (section 2). If the pinning is strong enough to locally overcome the deformation potential of the CDW - i.e. if one deals with the strong pinning limit - a single narrow ESR line is expected independently of whether the system is commensurate or incommensurate. This is certainly not the case for the Mo^{5+} defects since several commensurate lines and a broad incommensurate line are detected. Thus, Mo^{5+} does not pin strongly the CDW (at least below 90 K) despite the charge localized onto it. On the other hand, there are no obvious singularities of the incommensurate line, like those observed for the ^{87}Rb NMR. This could signal that at the impurities not all phases have equal probabilities, i.e. the pinnig strength is intermediate between weak and strong.

The Mössbauer study of ^{57}Fe doped into CDW systems may give insight to the local environment at impurity sites. The analysis is in principle similar to that of NMR or EPR lineshapes. The EFG interacts with the quadrupole moment of the ^{57}Fe nucleus in the excited state ($I = 3/2$) and splits the Mössbauer line into a doublet. Decreasing the temperature below T_P a gradual change of the spectrum is expected due to the change of the EFG as the periodic lattice deformation develops. The change depends qualitatively on whether the CDW is pinned by the iron ions strongly or weakly. In the former case only a change in the splitting of the quadrupole doublet is expected, while if the pinning is weak (and the CDW is incommensurate) a broadening occurs.

In the study of Veuillen *et al.* [22,69] six lines corresponding to three Fe^{3+} sites were observed in the Mössbauer spectra of ^{57}Fe doped $K_{0.3}MoO_3$. The positions of the iron impurities within the crystal structure could not be identified. A broadening of one of the lines below 120 K has been observed (Figure 17). This extra linewidth increases strongly with decreasing temperature and does not follow the order parameter which below 120 K is practically constant. It has been suggested [22] that a change in the pinning regime takes place. The pinning is strong at high temperatures, hence no broadening, but below 120 K it becomes gradually weak according to this interpretation. Such a change of regime is expected from the theory of pinning of Lee and Rice [31]. The broadening is surprisingly large, it is at least 20 % of the splitting of the doublet i.e. the change of the EFG tensor is not negligible compared to the EFG itself. On the other hand, according to NMR studies at the Rb sites the EFG changes only by a few percent. Further studies are

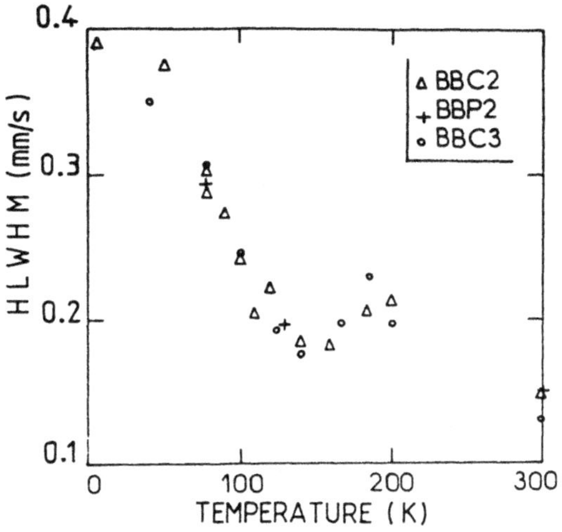

Fig. 17 Half Line-Width at Half Maximum of one of the Mössbauer lines of ^{57}Fe doped $K_{0.30}MoO_3$ for three different samples. (BBC2, BBC3 are single crystals and BBP2 is a powder sample obtained by crushing single crystals). (From [69].)

needed to clarify the Mössbauer results.

4.4. Defect density waves

It has been suggested by Lederer *et al.* [70] that mobile charged defects could partially order because of their interaction with the CDW. The defect density wave has a wave-length equal to that of the CDW and is a source of pinning. A time dependent broadening of the NMR line of $Rb_{0.3}MoO_3$ just below the Peierls transition [7] has been interpreted as the increase of the CDW order parameter, related to the development of a defect density wave, on the time scale of a few hours. This phenomenon is well known in the incommensurate dielectrics [71].

4.5. Metastable states and EPR

A rapid change of temperature or the decrease of the electric field under threshold leaves the CDW phase distribution in a metastable state which relaxes towards equilibrium in a non-exponential way [37, 72]. This relaxation is best monitored by the change of the low electrical field resistance [37] which within a broad range of times varies proportionally to log t. The reordering of the phase may take place as a continuous evolution through states of varying CDW deformations as discussed in detail by Littlewood [73] or by a rearrangement of phase dislocations as suggested by Maki [74]. EPR is in principle useful for following the variation of metastable states since it gives direct information on the phase around defects.

A temporal variation of the EPR spectrum of $K_{0.3}MoO_3$ crystals has been observed after sudden changes of the temperature [21, 63], and also after an electric field had been applied [75] by Dumas *et al.*, but a systematic study of the metastable states by EPR is still lacking. After quenching $K_{0.3}MoO_3$ crystals from room temperature to 4 K the EPR spectrum kept changing for more than an hour [75]. A hysteresis in the temperature dependence of the linewidth and intensity was observed above 50 K. At temperatures below 25 K the analysis of the data is complicated by the very long spin relaxation times. The high dielectric polarizability of $K_{0.3}MoO_3$ and at higher temperatures the high conductivity, modifies the penetration of microwave fields into the crystal and this presents a further difficulty. Small changes in the microwave conductivity due to metastable effects may modify the EPR intensity.

5. Dynamics in the pinned CDW

Spin lattice relaxation rate (SLRR) studies in CDW systems have been mainly concerned with the modification of the density of states at the Fermi level as the gap opens [28]. Contributions to the SLRR by the critical fluctuations of the CDW around the Peierls transition were found to be within the error bars in 2D systems [55] and barely observable in $NbSe_3$ [76]. However, the contributions from both critical fluctuations and excitations of the condensate below T_P (phasons and amplitudons) have been observed in $Rb_{0.3}MoO_3$ [27, 28]. This study illustrates the power of NMR techniques to investigate CDW dynamics, for details we refer to the extensive work on the incommensurate dielectrics [59]. Let us give a brief account of the theory. The spin lattice relaxation rate is due to the usual Korringa relaxation (which rapidly vanishes below T_P), and to the fluctuations associated with the CDW (critical fluctuations around T_P and excitations of the condensate below T_P). We have the general expression:

$$\frac{1}{T_1(R)} = AT\chi^2_{spin} + B\int\langle\delta\nu(R,t),\delta\nu(R,0)\rangle e^{i\omega t}dt \tag{5.1}$$

where χ_{spin} is the measured magnetic susceptibility and $\delta\nu(R,t)$ is the time dependent part of $\delta\nu(R)$ which in the local approximation can be expanded as a function of $\delta\rho(R)$ (eq. (4.1)). The second term in the right hand side of eq. (5.1) is thus determined by the correlation function of the order parameter. A and B are two constants related to the hyperfine fields. Below T_P the phase and the amplitude fluctuations of $\delta\nu(R,t)$ become decoupled. In order to evaluate the quantum correlation function in eq. (5.1) $\delta\nu(R,t)$ has to be replaced by the expression:

$$\sum_q \left(A_q(t) + A_q^+(t)\right)\cos(2k_F R + \varphi_0) + \left(\varphi_q(t) + \varphi_q^+(t)\right)\sin(2k_F R + \varphi_0) \qquad (5.2)$$

where $A_q(\varphi_q)$ is the anihilation operator of the amplitudon (phason) mode with wave vector q. Note the phase shift $\pi/2$ between the two modes which gives rise to an inhomogeneous distribution of the relaxation rate whithin the spectrum [59] and also a characteristic time dependence of the magnetisation $M_z(t)$ [27]. The relative weight of the contributions of the phasons and the amplitudons to the spin-lattice relaxation rate varies with $X = \cos(2k_F R + \varphi_0)$, i.e. the position in the line. The phasons are inefficient in relaxing nuclei contributing to the singularities of the spectrum but fully affect those at the center of the line. This has been experimentally observed in incommensurate dielectrics [59, 77] and in $Rb_{0.3}MoO_3$ [27].The calculation of Zumer and Blinc [77] has been generalized [27] to the more realistic case of an anisotropic dispersion relation:

$$\omega_\alpha^2(\vec{q}) = (\omega_\alpha^0)^2 + \kappa_\parallel q_\parallel^2 + \kappa_\perp q_\perp^2; \qquad \alpha = A, \varphi \qquad (5.3)$$

where ω_α^0 is the pinning frequency.

One has to distinguish between two relaxation processes:

i) the direct process, which corresponds to the emission (absorption) of only one excitation at the Larmor frequency. In presence of a gap ($\omega_L \ll \omega_\alpha$) this process is efficient only if the modes are overdamped (overlap the Larmor frequency). In this case

$$\frac{1}{T_1}\alpha\nu_1^2\left[X^2 J_\varphi(\omega_L) + \left(1 - X^2\right) J_A(\omega_L)\right] \qquad (5.4)$$

where the spectral density J_α can be expressed as:

$$J_\alpha = \frac{kT}{\omega_L}\sum_{\vec{q}}^{\vec{\Lambda}} \frac{\Gamma_q \omega_L}{\left(\omega_\alpha^2(\vec{q}) - \omega_L^2\right)^2 + \Gamma_q^2\omega_L^2} \qquad (5.5)$$

where $\vec{\Lambda}$ is a cut-off parameter, and Γ_q a damping factor which in the following is assumed equal to Γ for all q. In the limit where $\kappa_\parallel \Lambda_\parallel^2 \gg \kappa_\perp \Lambda_\perp^2$, ${\omega_\alpha^0}^2 \gg \Gamma\omega_L$ one obtains:

$$J_\alpha = M^{-1}\kappa_\perp \kappa_\parallel^{-1/2}\frac{\Gamma}{\omega_\alpha^0} \qquad (5.6)$$

which does not depend on ω_L and M is the effective mass of the excitation [14].

For a more important damping $\Gamma\omega_L \gg {\omega_\alpha^0}^2$

$$J_\alpha = M^{-1}\kappa_\perp \kappa_\parallel^{-1/2}\left(\frac{\Gamma}{\omega_L}\right)^{-1/2} \qquad (5.7)$$

we have a different dependence on Γ and in this case we have a dependence on the NMR frequency. With an anisotropic dispersion relation, the intermediate case $\Gamma\omega_L \gg \kappa_\perp \Lambda_\perp^2$ leads to the result:

$$J_\alpha = M^{-1}\kappa_{\|}^{-1/2}\frac{\Lambda_\perp^2\Gamma^{-1/2}}{\omega^{3/2}} \tag{5.8}$$

ii) another process to consider is the Raman process which corresponds to the scattering of excitations by the nuclei from wave vector q to q' with an energy change. The Raman process involves terms bilinear in A_q and φ_q only and can be written as [59]

$$\frac{1}{T_1} = \nu_2^2\left[X^2 J_{AA}(\omega_L) + \left(1 - X^2\right) J_{\varphi\varphi}(\omega_L)\right] \tag{5.9}$$

where the two-phason (amplitudon) spectral density can be written as:

$$J_{\alpha\alpha} = \frac{1}{M^2}\int_{\omega_\alpha^0}^{\omega_c} \frac{[\mathcal{D}(\omega)]^2}{\omega^2}\frac{\exp\left(\frac{\hbar\omega}{kT}\right)}{\left(1 - \exp\left(-\frac{\hbar\omega}{kT}\right)\right)^2}. \tag{5.10}$$

If one assumes a density of states for the phasons (amplitudons) calculated with an isotropic dispersion

$$\mathcal{D}_\alpha(\omega) = \kappa^{-3/2}\omega\left(\omega^2 - \omega_\alpha^{0\,2}\right)^{1/2} \tag{5.11}$$

one finds:

$$J_{\alpha\alpha} \propto \frac{1}{M^2\kappa^{1,5}}(kT)^3\int_{x_0}^{x_1}\left(x^2 - x_0^2\right)\frac{e^x}{\left(1 - e^x\right)^2}dx \tag{5.12}$$

where $x_0 = \frac{\hbar\omega_\alpha^0}{kT}$ and x_1 is the cut-off component. At low temperatures one obtains a characteristic behavior

$$J_{\alpha\alpha} \propto \frac{1}{M^2\kappa^{1,5}}kT\exp\left(-\frac{\hbar\omega_\alpha^0}{kT}\right) \text{ for } \hbar\omega_\alpha^0 \gg kT\,. \tag{5.13}$$

Let us consider the experimental results in $Rb_{0.3}MoO_3$. In Figure 18 the temperature dependence of $(T_1T)^{-1}$ for ^{87}Rb is shown. For $T > 200$ K the spin lattice relaxation rate is dominated by the Korringa process and scales with the squared experimental spin susceptibility χ_{spin}^2. In the temperature range 170 - 200 K a peak is observed corresponding to the critical fluctuations around 181 K. Between 170 and 100 K the spin-lattice relaxation rate is nearly flat, below 100 K it decreases exponentially. Several explanations are possible for this decrease:

i) If an incommensurate-commensurate transition sets in the phason gap may increase by an order of magnitude, leading to a sudden decrease of the SLRR,

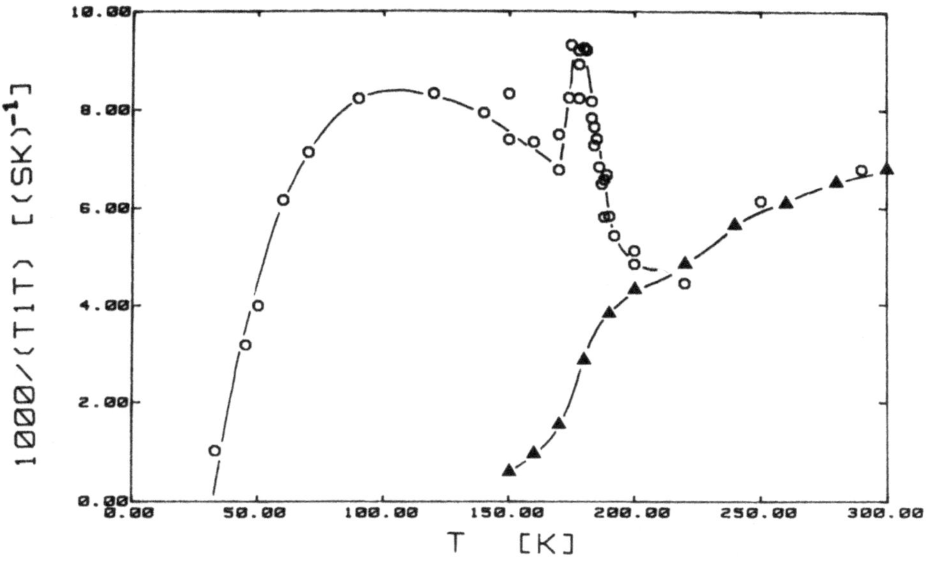

Fig. 18 ^{87}Rb spin-lattice relaxation rate as a function of temperature in $Rb_{0.30}MoO_3$. The full triangles are the squared experimental spin susceptibilities [56] reduced to an appropriate scale. (From [27, 28])

as observed in Rb_2ZnCl_4 [59]. However the lineshape at low temperature in $Rb_{0.3}MoO_3$ is in contradiction with the existence of commensurate domains.

ii) The overdamping of the phason modes, that is the value of Γ could decrease at low temperature, in relation with the decreasing number of free carriers available for screening. This leads to a decrease of the SLRR and this explanation is in agreement with recent conductivity measurements [6] which show a rapid decrease of the CDW conductivity $\sigma(\omega)$ at 80 MHz (the Larmor frequency of the NMR experiment) and low temperature. Contrary to $\sigma(\omega)$ which strongly depends on the longitudinal phason mode at $q = 0$, NMR averages over all $\vec{q}$ components and this leads to an apparent activation energy of 130 K of the SLRR.

iii) As far as the Raman processes are concerned, a thermally activated behavior is also expected at temperatures less than the excitation gap.

6. NMR in the sliding CDW state

6.1. Introduction

The possibility of depinning a CDW by an electric field was first discussed by Fröhlich in 1954 in an attempt to understand superconductivity by a mechanism involving the electron-phonon interaction. The Fröhlich conductivity was considered as an academic exercise on an unrealistic one-dimensional model until 1973 when Bardeen [78] suggested that it might explain the unusually high conductivity of the organic quasi-one-dimensional salt TTF-TCNQ at the Peierls transition. A non-linear current later unambiguously identified as the Fröhlich mode was reported in 1976 by Monceau *et al.* [79] in the inorganic salt $NbSe_3$. The unusual transport and optical properties of $NbSe_3$ and some other quasi 1D conductors led to the general acceptance of the existence of the sliding mode but the most convincing experimental proof has been provided by NMR. After some unsuccesful attempts in blue bronze [80] the sliding mode was observed in 1986 in $NbSe_3$ [81] and in $Rb_{0.3}MoO_3$ [11, 10]. In $Rb_{0.3}MoO_3$ it was shown that the voltage noise accompanying the CDW current is indeed related to the phase velocity of the motion [12] which could be measured [12, 13] with great precision. As explained below the sliding of the CDW gives rise to a periodic fluctuation of the hyperfine field which narrows most of the inhomogeneous spectrum of the incommensurate pinned CDW into a single line.

The first observations of the motional narrowing of ^{87}Rb NMR in blue bronze are shown in Figure 19a and b. The analysis of the lineshape is complicated by the inhomogeneity of the sliding CDW current due to an inhomogeneity of the current injection at the contacts. In the early experiments in large parts of the crystal the CDW remained pinned even at high applied electric fields. By improving the preparation of electrode contacts the CDW could be depinned in the full crystal and sidebands to the central line due to the periodicity of the hyperfine fluctuations were resolved [12, 13, 82].

6.2. Theory of NMR under CDW current

6.2.1. *NMR spectra for a sliding CDW* In the following we assume that for a depinned CDW the resonance frequency at site R is determined by the instantaneous phase $\varphi(R,t)$ of the CDW. Thus

$$\delta\nu(R,t) = \nu_1 \cos\left(qR + \varphi_1\left(R,t\right)\right) + \nu_2 \cos^2\left(qR + \varphi_2\left(R,t\right)\right) \tag{6.1}$$

In the simplest case the phase velocity is constant in time and uniform in space and

$$\delta\nu(R,t) = \nu_1 \cos(q(R - v_d t) + \varphi_1) + \nu_2 \cos^2(q(R - v_d t) + \varphi_2) \tag{6.2}$$

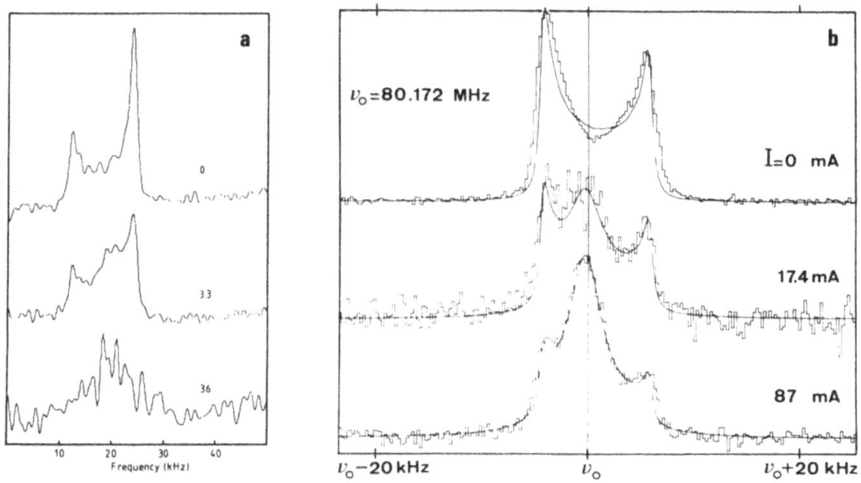

Fig. 19 NMR lineshapes of ^{87}Rb in $Rb_{0.3}MoO_3$ at various current levels probing the motion of the CDW by the motional narrowing of the incommensurate pattern observed for $I = 0$; a)From [11]; b) Solid lines are calculated fits taking into account the fraction of the crystal in which the CDW stays pinned. (From [10].)

with

$$qv_d = d\varphi/dt = 2\pi\nu_d. \tag{6.3}$$

The simple uniformly sliding plane wave approximation may not be entirely correct in describing the CDW motion. As suggested by Lee and Rice [31] the current may be transported by a series of loop dislocations. Far from the dislocation center the change of the amplitude of the order parameter may be neglected and the variation of φ is non-uniform but still periodic. The situation is similar to the motion of a discommensuration lattice. Although the dislocation loops or discommensurations may be widely spaced, the periodicity of the oscillations of the EFG at a given site is the same as for a plane wave CDW provided the same (average) current is flowing. For discommensurations with large wall widths or far from phase slip dislocation lines the main term is the time averaged phase winding rate:

$$\overline{\frac{d\varphi}{dt}} = v_{dc}\delta q = 2\pi\nu_d \tag{6.4}$$

and corrections are small [13].

For equal currents, the average phase winding rate of a sliding plane wave and that of a discommensuration lattice are equal. Near CDW dislocation lines, if dislocations are narrow, the harmonics in the expression of $d\varphi(R,t)/dt$ may be important in the calculation of the NMR spectra. In this case the assumption of a constant amplitude of the order parameter is, however, unjustified. The periodic generation and anihilation of phase dislocations at the boundaries or the uneven motion due to the impurity pinning potential also give rise to harmonics but, except for very small currents, the amplitude of the current oscillations is small compared to the average CDW current and does not influence much the NMR spectrum.

In the following we discuss the NMR spectra for a plane wave CDW. We first consider a drift velocity which is constant in time. The effect of non-periodic fluctuations of φ is deferred to section 6.2.3.

We first return to (6.2) where $d\varphi/dt = 2\pi\nu_d$ is uniform in space and time. The correlation function of the transverse magnetization $<I_x(t)I_x(0)>$ is:

$$G(t) = \exp\{-2\pi\nu_0 t\} <I_x(t)I_x(0)>= \sum_{R_i} \exp\left\{2i\pi \int_0^t \delta\nu(R_i,t')dt'\right\} \cdot \quad (6.5)$$

Neglecting the quadratic term and assuming an incommensurate CDW one finds

$$G(t) = \sum_{n=-\infty}^{n=+\infty} \left[J_n\left(\frac{\nu_1}{\nu_d}\right)\right]^2 \exp\{-2i\pi n\nu_d t\} = J_0\left(\frac{2\nu_1}{\nu_d}\sin(\pi\nu_d t)\right) \cdot \quad (6.6)$$

The NMR line shape is the Fourier transform of $G(t)$

$$\widetilde{G}(\nu) = \sum_{n=-\infty}^{n=+\infty} \left[J_n\left(\frac{\nu_1}{\nu_d}\right)\right]^2 \delta(\nu - n\nu_d)\,. \quad (6.7)$$

These expressions were first derived by Kogoj *et al.* [83].

It is natural that in (6.6) the correlation function $G(t)$ is periodic in time since we have considered up to now a deterministic periodic local hyperfine field modulation. The NMR spectrum (6.7) is composed of a central line ($n = 0$) and sidebands at $\nu = \pm n\nu_d$. The central line is analogous to the usual motional narrowing (except that for the moving CDW it has zero width). The sidebands are analogous to those observed in High Resolution NMR in solids where the sample is spining at a magic angle to average out the dipolar fields. Just as the position of the sidebands is a direct measure of the angular velocity of the rotor in Magic Angle Spinning, the position of the sidebands in $\widetilde{G}(\nu)$ for a uniformly sliding CDW measures the CDW phase velocity $v_d = \nu_d\lambda$. The amplitude of the sidebands decreases rapidly with increasing v_d since $J_n(x) \sim x^n$ for small x (Figure 20a), for $\nu_d > \nu_1$ nearly all the intensity is in the central line, (90% for $\nu_d/\nu_1 > 2$). For $\nu_d \geq \nu_1 \gg \nu_2$ the only effect of the quadratic term is a shift by $\nu_2/2$ to higher frequencies.

In the actual crystals the current is not homogeneous in space and instead of a single drift velocity we have to consider a distribution $P(v_d)$. We have in mind a

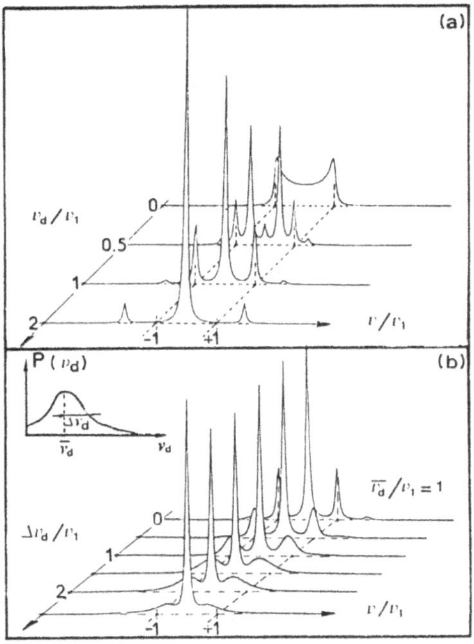

Fig. 20 a) Theoretical NMR lineshapes in presence of a CDW sliding with a uniform velocity $v = \nu_d\lambda$. The local hyperfine shift is $\nu(R) = \nu_1 \cos(qR + 2\pi\nu_d t + \phi_0)$;
b) Variation of the NMR lineshape as a function of the width of a Gaussian velocity distribution $P(\nu_d) = \exp - \left\{ \frac{(\nu_d - \overline{\nu_d})}{2(\Delta\nu_d)^2} \right\}$ at fixed value of $\overline{\nu_d}/\nu_1$. (From [13])

crystal with a large number of domains, in each domain the current is homogeneous and slides with a velocity v_d. The form of the central line $n = 0$ is not affected by a velocity distribution since it is independent of v_d in contrast to the sidebands which are broadened (Figure 20b).

Both the central and sideband lines are broadened by the dipolar interaction between neighbouring nuclei characterized by the spin-spin relaxation rate T_2^{-1}. Perturbation of delocalized electrons, strains around defects and mosaicity are also of importance, the linewidth above the Peierls transition is inhomogeneous and much larger than expected from the dipolar interaction alone. The dipolar contribution to the linewidth is somewhat decreased by a static CDW, neighbouring Rb nuclei at sites i and j become "unlike" in this case since their corresponding EFGs are different, i.e. the spin flip-flop terms $I_i^+ I_j^-$ become ineffective in inducing dipolar transitions. A sufficiently rapid motion of the CDW leads to an effective hyperfine field which is the same for all nuclei and causes the flip-flop term to reappear.

6.2.2. *Spin-echo in presence of a sliding CDW* Spin-echo is the traditional way to measure the homogeneous spin-spin correlation time T_2. It enables the distinction between homogeneous and inhomogeneous contributions to the NMR linewidth. The sliding of the CDW dramatically changes the echo and this offers a method to measure in a direct way the drift velocity v_d.

In the static case the form of the echo centered at 2τ (where τ is the delay between the $\pi/2$ and π r.f. pulses) is given in the plane wave approximation (neglecting ν_2) by:

$$G_e(t) = J_0(2\pi\nu_1(t-2\tau))e^{-t/T_2} \tag{6.8}$$

and the echo amplitude has the usual exponential form; except that the onset of the CDW motion modifies the amplitude [13, 82]:

$$I(2\tau) = J_0\left(\frac{4\nu_1}{\nu_d}\sin^2(\pi\nu_d\tau)\right)e^{-2\tau/T_2}. \tag{6.9}$$

The echo amplitude oscillates as a function of τ with a period $2\pi/\nu_d$. These oscillations are more pronounced for low drift velocities, if $\nu_d < 1.6\nu_1$ the amplitude becomes even negative in some intervals of τ. The effect is somewhat smeared by a distribution of ν_d:

$$I(2\tau) = \int J_0\left(\frac{4\nu_1}{\nu_d}\sin^2(\pi\nu_d\tau)\right)\cdot P(\nu_d)d\nu_d \tag{6.10}$$

but as we show in section 6.3. the effect remains well observable.

Further information on the CDW motion is gained from an analysis of the shape of the echo for $t \geq 2\tau$ [13]. The Fourier transform of the echo is similar to that of the FT of the free induction decay in the sense that it is composed of a central line and sidebands at $\pm n\nu_d$, but the relative intensities of these components differ and depend on τ. For example if $\nu_1/\nu_d = 0.8$ and $\nu_d\tau = 1/2$ the intensity of the sidebands is increased by 40 % and under special conditions the intensity of the central line may even be zero.

6.2.3. *Stochastic fluctuations of the velocity; spin-spin relaxation time* Up to now we have considered a constant or a periodically modulated CDW velocity. We have seen that this leads to a periodic correlation function for the transverse magnetisation $G(t)$ and in this case the damping is due to the nuclear magnetic dipole-dipole interaction characterized by the relaxation time T_{2d}. In the presence of a static CDW T_{2d} was found equal to 4.6 ms in $Rb_{0.3}MoO_3$; it is slightly shorter if the CDW is sliding with a fast constant velocity since then all Rb(2) nuclei become "like" (as explained above). T_{2d} for a hyperfine distribution motionally narrowed by a CDW sliding with constant velocity is expected to be similar to that measured above the Peierls transition.

However, random fluctuations of the CDW phase may drastically increase the damping of $G(t)$ and have a much more pronounced effect than the slight change of the dipolar interaction due to the CDW motion. In general we would like to describe the effect of a time and position dependent phase $\varphi(R,t)$ with a periodic and a non periodic component. The calculation of G(t) when both the random and constant phase variations are taken into account has been reported in Ref. [13] for two limiting cases:

At high velocities, one assumes that the amplitude of the random phase fluctuations is small and that its variation may be decoupled from the spatial dependence, *i.e.* $\varphi(R,t) = \varphi_0 + \delta\varphi(t) + 2\pi\nu_d t$. Supposing $|\delta\varphi(t)| \ll \pi/2$ for all t and assuming a Gaussian stationary process, one finds, for times $t \gg \tau_c$ the correlation time of the phase, the usual result [84]:

$$G(t) = \exp\left(-2\left(2\pi\nu_1\right)^2 \overline{\delta\varphi}^2 \tau_c t\right) \tag{6.11}$$

while for $t \ll \tau_c$:

$$G(t) = \exp\left(-2\left(2\pi\nu_1\right)^2 \overline{\delta\varphi}^2 t^2\right) \tag{6.12}$$

We emphasize that the above results were obtained by employing severe approximations and in particular we assumed that the drift velocity is fast. The calculation is limited to electric fields $E \gg E_T$ where fluctuations due to the pinning are small compared to the constant winding of the phase. An approximation which is more adequate to fields close to threshold is the following: the amplitude of the fluctuations is no longer considered to be small. One assumes the motion to be coherent for short times ($\varphi = 2\pi\nu_d t$) but with a probability Γ per unit time that the motion changes drastically. One assumes that nuclei at positions where the motion changes suddenly contribute no more to the echo. Thus the number of nuclei refocused at time 2τ (for a pulse sequence $\pi/2-\tau-\pi$) is $N(2\tau) = N(0)e^{-2\Gamma\tau}$ and the amplitude of the echo is

$$J_0\left(\frac{4\nu_1}{\nu_d}\sin^2\left(\pi\nu_d\tau\right)\right)e^{-2\Gamma\tau}. \tag{6.13}$$

In both cases the lineshape resulting from a coherent motion of the CDW has to be convoluted by a function taking into account broadening from the phase fluctuations as well as the static imperfections and the dipolar relaxation rate.

6.3. DETERMINATION OF THE CDW PHASE VELOCITY

As mentionned in section 6.1., the first NMR experiments under applied electric field in $Rb_{0.3}MoO_3$ showing the motional narrowing of the hyperfine structure associated to the static CDW successfully demonstrated the motion of the CDW on the microscopic level. As we have seen in the preceeding section this technique enables one to aim at a number of more ambitious goals.

The observation of the sidebands predicted by the theory of the lineshape provides a direct measurement of the oscillation frequency of the nuclei, the so-called phase winding rate $\nu_d = \frac{1}{2\pi}\frac{d\varphi(R)}{dt}$ which is directly related to the CDW velocity $v_d = \lambda\nu_d$. The simultaneous measurement of the non-linear current density j_{CDW} and the CDW velocity allowed the verification of the long-assumed relationship $j_{CDW} = nev_d$ where n is the total number of electrons in the band affected by the opening of the Peierls gap. In the case of the blue bronze, assuming that each alkali atom gives one electron to the otherwise empty conduction band, the theoretical value predicted for λne is 12.5 kHz/ A·cm^{-2}.[4] Moreover, the observation of these sidebands provides a direct evidence of the coherence of the CDW motion. Another important goal was to compare directly the voltage noise frequency ν_n with the local oscillating frequency of the nuclei ν_d since different ratios have been predicted depending on the models [44, 45, 51].

In order to be able to relate j_{CDW}, ν_d and ν_n three types of experiments were performed **in-situ** in the NMR coil at various temperatures: I-V characteristics, measurement of the noise spectrum, and recording of the NMR lineshape of the (1/2,-1/2) transition of Rb(2) under current. The size of the single crystals was chosen as a best compromise between reducing as much as possible the inhomogeneity of the current distribution in the sample and ensuring a reasonable NMR signal to noise ratio. (Typical dimensions were a few mm along the b-axis with perpendicular cross-sections of about 0.2 x 1 mm^2.). Greatest care was taken in the design of the electrical contacts which were copper plated on polished sections perpendicular to the b-axis. In spite of this, the voltage noise spectrum as a function of frequency never consisted of sharp well defined peaks (except for electric fields close to the threshold, E_T), but rather of a broad Gaussian disribution with a low frequency tail as shown in Figure 8. The final goal is to record the change of the NMR lineshape as a function of the CDW velocity in the particular range where $\nu_d = v_d/\lambda$ is of the order of ν_1 where the sidebands are most intense (see Figure 20, section 6.2.). One way for this could be to measure under a series of electric fields at fixed temperature. However, in order to keep the same inhomogeneous current distribution in the sample whatever its mean value is, it was decided to fix the ratio E/E_T and to sweep the mean current density $\overline{j_{CDW}}$ *i.e.* the mean CDW velocity $\overline{v_{CDW}}$ by varying the temperature using the fact that it is thermally activated (section 2.2.2, Figure 5). To ensure a constant velocity distribution during each scan, a constant current was fed into the sample so that small fluctuations of the temperature would only induce a change in the voltage. In Figure 21 seven NMR spectra (obtained by Fourier Transform of the free induction decay) recorded in the temperature range 40 to 60 K at a fixed value of $E/E_T = 15$ are shown. At 40 K, although the current voltage characteristics and the noise spectrum show that the CDW is depinned, the CDW velocity is too small to change the lineshape. As j_{CDW} increases with increasing temperature the edges of the spectrum broaden and diminish and a peak emerges at the center frequency of the static spectrum.

---

[4] The value of the wavelength λ of the CDW is equal to $\pi/k_F = 4b/3 = 10$ Å and $n = 4.8 \cdot 10^{21}$ cm^{-3}.

On further increase of j_{CDW} (by increasing T) the central peak becomes the most prominent part of the spectrum. At the intermediate current of $j_{CDW} = 0.405$ A/cm^2 additional steps arise at the sides of the spectrum outside the range of the static line.

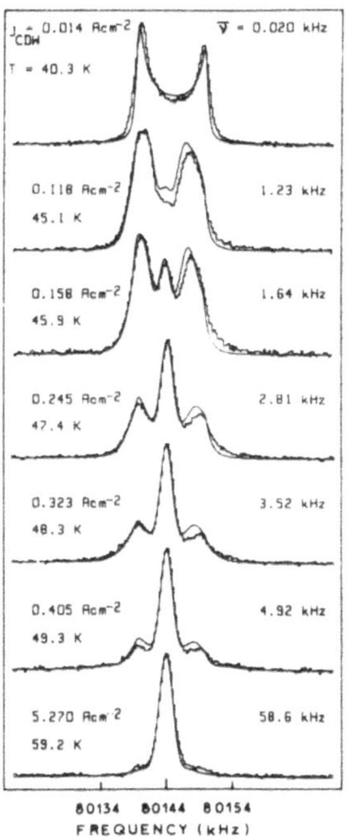

Fig. 21 Variation of the NMR lineshape and computer-fitted curves as a function of the CDW current j_{CDW}. The field is fixed at $E = 1.5$ V/cm ($\sim 15E_T$) and the temperature is varied to change j_{CDW}. For the computer simulated spectra the relative CDW velocity distribution is determined from the experimental noise spectrum at 59.2 K. For each simulated spectrum only one parameter, the average local field oscillation frequency $\overline{\nu_d}$, was fitted. (From [13].)

The solid curves have been calculated in the following way: the sample was assumed to consist of a large number of domains each having a uniform CDW velocity. The spatial distribution $f(v/\bar{v})$ of the CDW current was taken to be proportional to the experimental noise spectrum at 59 K(after removal of the first harmonic). Under the assumption that $\nu_n = \nu_d$ this provided a master curve for the phase winding rate distribution $P(\nu/\overline{\nu_d})$. The NMR spectra were calculated by folding (6.7) (the exact lineshape for a given ν_d) with $P(\nu/\overline{\nu_d})$ and adjusting

$\overline{\nu_d}$ to obtain the best agreement with experimental data. For very low velocities corresponding to $\nu_d = v/\lambda < 0.17$ kHz, the dynamic lineshape given by (6.7) was replaced by the static asymmetric one. The reason was that the dynamic histograms calculated for these low frequencies are indistinguishable from the static ones calculated for $\nu_2 = 0$. As can be seen the agreement is fairly good, the main features of the spectra are quite well reproduced for average values of ν_d ranging from 0.020 kHz to 58.6 kHz. The ratio $\overline{\nu_d}/j_{CDW}$ was found equal to 11 ± 1 kHz/A·cm^{-2} by averaging over fits to NMR spectra at seven temperatures. This value is in close agreement with the value of 12.5 kHz/A·cm^{-2} expected for the conduction electron density of $Rb_{0.3}MoO_3$.

We now review experiments which have demonstrated the coherence of the CDW motion. The observation of sidebands in the NMR spectra, predicted by eq. (6.6), was first reported in Ref. [10] and studied in more details by Butaud *et al.* [13]. Figure 22 shows for a small temperature range the variation of the lineshape of Rb(2) at a fixed electric field ($V = 600$ mV $\approx 24V_T$). Arrows indicate the position of the peak in the voltage noise spectra at the same temperatures. The $p = \pm 1$ sidebands situated about at the frequency of the peak of the noise spectra are well observable; they shift to higher frequencies as the average current is increased. Note a component of the spectrum resembling the static line which hides the observation of sidebands for $\nu_d/\nu_1 \approx 1$ where their intensity is in fact large. The optimum condition for observing the sidebands is $\nu_d/\nu_1 \approx 1.5$ where, hovewer, their intensity is only 10 % of the total spectrum. The wings of the spectra at about $\pm\nu_1$ only resemble a static line but in fact they have been shown [13] to arise from regions of the crystal where the CDW is sliding, albeit slowly. These regions are also the source of the low frequency upturn in the noise frequency spectrum.

Figure 23 shows the spectrum recorded at 49 K and the computer simulation of the NMR lineshape obtained by folding (6.7) with a distribution of the winding rate frequency $P(\nu)$ equal to the experimental noise spectrum recorded at the same current and temperature. The agreement is fairly good. It must be stressed that for this spectrum there is no fitting parameter apart from the width of a Gaussian broadening function which is known approximatively from the high temperature data. The best proof of the equivalence of $\nu_d = \nu_n$ is that the sidebands are correctly reproduced. An assumption of $\nu_n = 2\nu_d$ would lead to calculated sidebands displaced from the center by half the observed ones and they would be masked by the central part of the line. Better simulations of the line shape, and particularly of the wings at about $\pm\nu_1$ (Figure 22), were obtained by increasing the weight of the low frequency part of the distribution $P(\nu)$. More details concerning the fitting procedure can be found in Ref. [13].

Maybe the best and more direct proof of the temporal periodicity of the CDW phase is given by the variation of the amplitude of the echo $I(2\tau)$ as a function of the delay τ, as shown for $T = 49$ K in Figure 24. A deep minimum in $I(2\tau)$ is observed at precisely that delay time for which $2\pi\nu_p\tau = \pi/2$ where ν_p is the peak frequency of

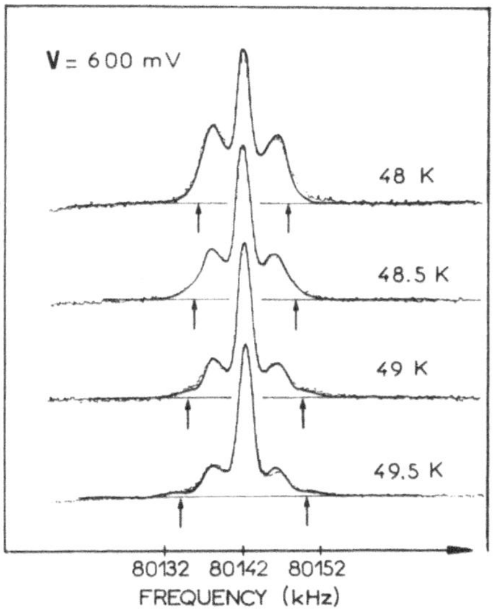

Fig. 22 NMR spectra giving evidence for the NMR sidebands $n = \pm 1$ in $Rb_{0.3}MoO_3$. The simulated spectra were computed by folding the theoretical lineshape given by (6.7) with a distribution $P(\nu_d)$ deduced from the experimental noise spectra taken at the same electric field (1.67 V/cm) and temperature. The arrows indicate the position of the peak in the experimental noise spectra shown in Figure 8.

the voltage noise spectrum at the same temperature and field. A damped periodic behaviour is observed with a period of ν_p^{-1} and a damping constant related to the inhomogeneity of the current *i.e.* the width of the noise peak. At large values of 2τ the oscillations are completely damped and $I(2\tau)$ decreases exponentially with a constant time T_2 resulting from the dipolar spin-spin interaction and stochastic fluctuations of the CDW phase. The solid line was calculated from (6.9) where $P(\nu_d)$ was obtained directly from the noise spectrum measured at the same applied voltage and temperature. The calculated and measured $I(2\tau)$ curves agree very well. An assumption of $\nu_n = 2\nu_d$ would clearly be incompatible with the measurement.

A similar experiment at 43 K and 600 mV was conducted also. Under these conditions the free induction NMR spectrum is only slightly modified by the motion of the CDW (Figure 25). Although the oscillations in $I(2\tau)$ are more complicated than at 49 K they are in complete agreement with the theory. The comparison of $I(2\tau)$ with the noise spectrum shows clearly the equivalence of ν_d and ν_n.

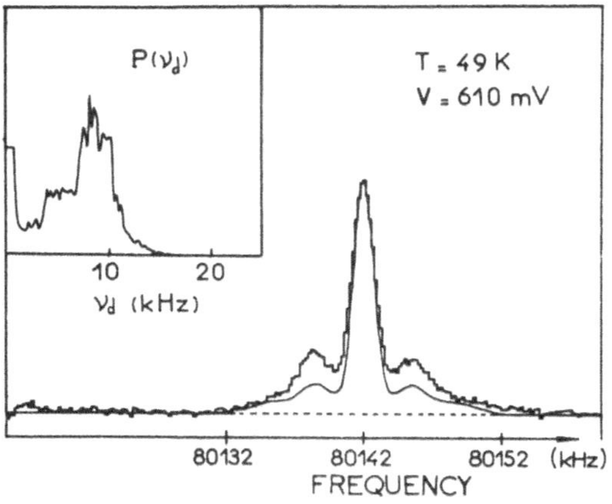

Fig. 23 NMR spectrum at 49 K and its computer simulation from the experimental noise spectrum shown in the inset. The sidebands $n = \pm 1$ are fairly well reproduced, but the difference in the quality of the fit compared to that of Figure 22 results from the levelling out of the amplitude of the low frequency tail of the noise spectrum which could not be measured. (From [13].)

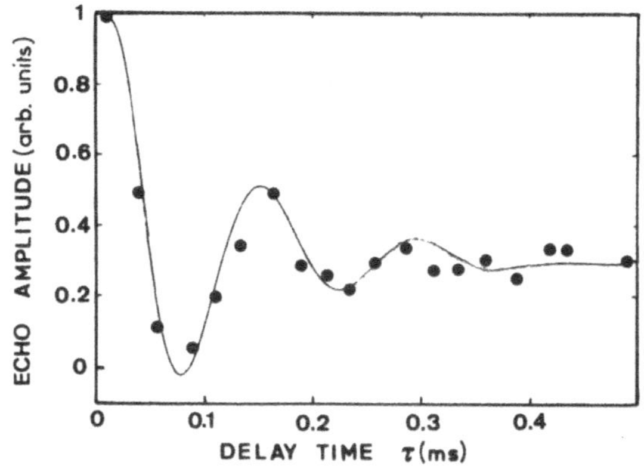

Fig. 24 Experimental variation of the amplitude of the spin echo $S(\tau)$ as a function of the delay time τ in $Rb_{0.3}MoO_3$ at $E/E_T \sim 20$ and 49 K. The solid curve is the convolution of the theoretical dependence $J_0(4\nu_1/\nu_d \sin^2(\pi\nu_d\tau))$ with the experimental distribution $P(\nu_d)$ obtained from the noise spectra. (From [13].)

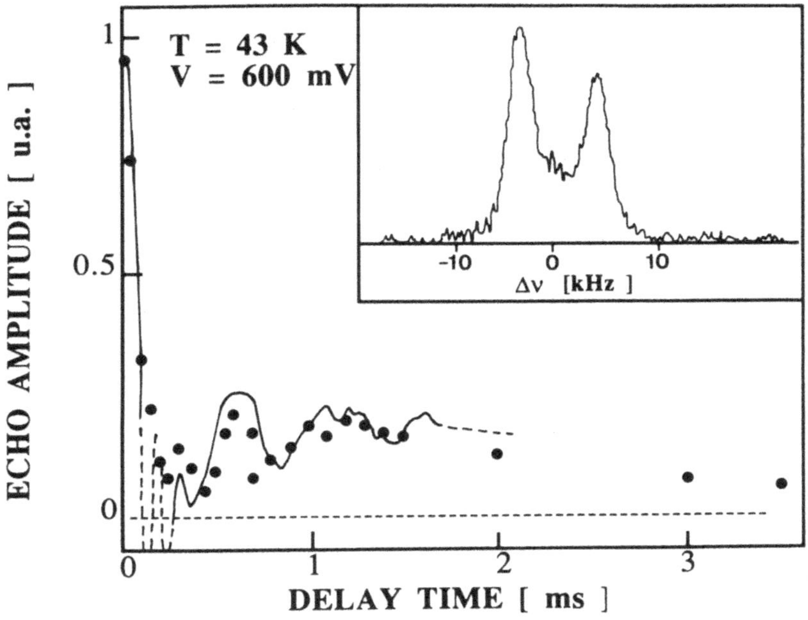

Fig. 25 Experimental variation of the amplitude of the spin echo as a function of the delay time τ in $Rb_{0.3}MoO_3$ at $E/E_T \sim 20$ and 43 K. Although the velocity of the CDW is too small to modify the lineshape obtained from the Fourier transform of the free decay, it produces strong oscillations of $S(\tau)$. The solid curve is calculated as for Figure 24. (From [85].)

6.4. INHOMOGENEITY OF THE THRESHOLD FIELD FOR DEPINNING

In the early NMR experiments showing the motional narrowing of the NMR line associated to the depinning of the CDW [11,10] the observed lineshapes, even at the highest applied electric field, were always the superposition of a Gaussian line onto the pattern corresponding to the pinned CDW. These lineshapes were fitted to the expression:

$$G(\nu) = f_S G_S(\nu) + (1 - f_S) G_D(\nu - \nu_2/2) \tag{6.14}$$

where f_S corresponds to the fraction of the crystal in which the CDW is pinned, $G_S(\nu)$ is deduced from a fit to the theoretical expression of the lineshape with no applied current, and $G_D(\nu)$ is a Gaussian.

In the subsequent experiments [10, 13, 82], with improved electrical contacts at fields well above the threshold ($E/E_T \approx 15 - 20$) the static fraction in the range 40 to 60 K was reduced to zero. (As explained in the preceding section, even in this case in some area of the crystal the CDW is moving much more slowly than the average velocity). But for values of the applied electric field closer to the threshold, ($E/E_T \approx 2-3$), this static fraction was always present, as shown in Figure 26 where f_S is plotted at various electric fields and temperature. At 64 K, a clear increase of the static fraction is observed as the field is decreased towards E_T. On the other hand for E/E_T between 2.5 and 3, the tendency of a decrease of the static fraction with an increase of temperature above 64 K is observed.

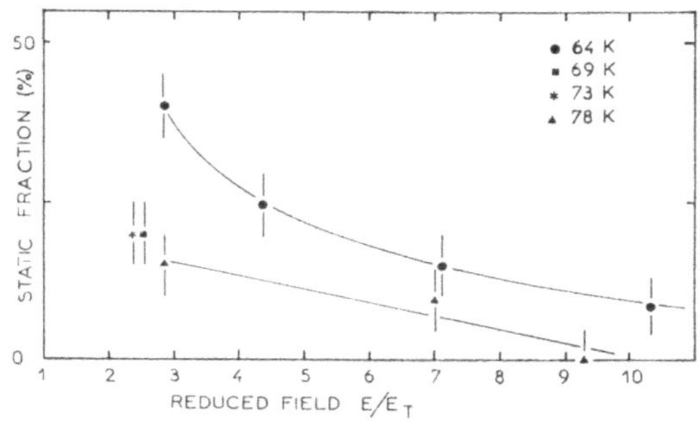

Fig. 26 Variation of the static fraction f_S (in which the CDW is pinned) as a function of E/E_T at various temperatures. (From [85].)

The careful determination of the static fraction f_S, which can only be determined from NMR experiments, proved to be crucial for several reasons: before the direct measurement of the CDW velocity by NMR, the correlation between the experimental values of j_{CDW} and the noise frequency ν_n was the only possibility to check the relationship $j = nev$. However, the value of j_{CDW} was defined as I/A, where A is the cross-section of the sample (perpendicular to the b-axis), ignoring the possibility that an appreciable amount of the CDW stays pinned in the crystal. By using the value of f_S obtained from NMR to determine an effective cross-section $A' = A(1 - f_S)$, one has been able to correct the value of $\overline{\nu_n}/\overline{j_{CDW}}$ which led to a reasonable agreement with the theoretical value [10, 13]. Note (section 6.3) that at higher fields and lower temperatures there is no static fraction but an appreciable amount of the crystal in which the CDW velocity is very low and which generates the low frequency tail of the noise power spectra. Taking this into account in the detemination of $\overline{\nu_n}$ led to $\overline{\nu_n}/\overline{j_{CDW}}$ equal to 12.5 kHz $A^{-1}{\cdot}cm^2$, a remarkably good agreement with theory.

At high temperature and low fields (79 K and $E/E_T = 2.3$), the comparison of f_s with the ν_n/j_{CDW} ratio led to the unexpected conclusion that at high temperatures large amplitude phase fluctuations occur in regions where there is no or negligible CDW current. This may be an indication for an electric field assisted thermal depinning. Thermal depinning of the phase ("floating phase") has been observed in incommensurately modulated dielectrics [59].

A quite different model for the variation of the static fraction with field and temperature has been suggested by Tucker *et al.* [86]. It was assumed that the current flows in a rather particular way: it advances by 2π jumps due to the appearance and disappearance of CDW phase slips. It has been argued that as long as the average drift frequency ν_d is smaller than the dielectric relaxation frequency ν_0 the 2π jumps of the phase are not significantly smeared and the NMR spectrum is not narrowed by the motion. To support this view they considered the results of Ross *et al.* [81, 87] on $NbSe_3$ where the condition $\nu_1 < \nu_d < \nu_0$ seems to be satisfied. These $NbSe_3$ data show significant motional narrowing only for $E/E_T > 3$ where $\nu_d = 15$ MHz while the hyperfine width is about $\nu_1 = 30$ kHz. ν_d is somewhat larger than $\nu_0 = 10$ MHz. However Ross [87] mentions that the noise spectra recorded in the range $E/E_T \leq 3$ lead to anomalously high values of ν_n/j: this is a strong indication that the CDW is moving only in a small part of the sample.

In blue bronze ν_0 is activated [88] and to satisfy the condition $\nu_0 < \nu_d < \nu_1$, as discussed by Tucker *et al.* [86], the temperature must be raised above $T = 65$ K and the electric field must be kept sufficiently low, $E/E_T \leq 3$. The data of Figure 26 show that for $T \geq 69$ K and E/E_T between 2.3 and 3 the static fraction is about $f_s = 20$ %. These data do not rule out the model of Tucker *et al.* [86] for $E/E_T < 2$ but certainly do not support it for larger values. The current is inhomogeneous in the sample and a fraction of 20 % in which the CDW remains static is reasonable and explains the data without sudden jumps by 2π of the phase. Earlier measurements by Ségransan *et al.* [10] gave a static fraction of 40 % at 78 K and $E/E_T \approx 3$ showing that this value depends on sample and electrode contact quality.

The question of whether one can avoid a large static fraction of CDW in the sample close to the threshold field ($E/E_T < 3$) is still open. Nevertheless, the results of Butaud *et al.* [13, 82] indicates that the motional narrowing is not dependent on the ratio ν_0/ν_d as suggested by Tucker *et al.* [86]. Contrary to their suggestions the static fraction is decreased at high temperature by thermal depinning or some other mechanism.

6.5. TEMPORAL FLUCTUATIONS

In the absence of temporal fluctuations and instabilities of the CDW velocity the correlation function of the transverse magnetization $G(t)$ and the variation of the spin-echo amplitude $S(\tau)$ would be periodic and only damped by the usual spin-spin interaction. Under current the spin-spin relaxation is, however, shorter than the dipolar relaxation time, T_{2d} (equal to 4.6 ms in blue bronze) due to the non-periodic variation of the phase. (see section 6.2.3.)

Let us first discuss the example shown in Figure 27. The fast decay (characteristic time T_{2F}) corresponds to nuclei in domains where the CDW is sliding (or at least fluctuating) while the slow decay has a time constant equal to T_{2d} and corresponds to domains with a static CDW. The static fraction f'_s may be deduced from the T_2 measurements by fitting the amplitude of the echo to the expression:

$$G_e(t) = f'_s \exp(-t/T_{2d}) + (1 - f'_s)G_D(t) \tag{6.15}$$

where $G_D(t)$ may be an exponential defining T_{2F} or a Gaussian if there is a distribution of correlation times. The values of f'_s and f_s deduced from the lineshape of the free precession are in reasonably good agreement. The Fourier transform of the echo lineshape at 73 K and delay time $2\tau = 1$ ms shows that at longer times the static component becomes dominant if the CDW is not depinned in the entire sample. This explains the failure of Douglass *et al.* [80] to observe any change in the echo lineshape in $Rb_{0.3}MoO_3$ with a voltage well above threshold. They used an echo delay time of $2\tau = 1$ ms for which the contribution of the dynamic central line vanishes (due to non-periodic fluctuations of the phase) and observed only a reduction of the intensity of the static line.

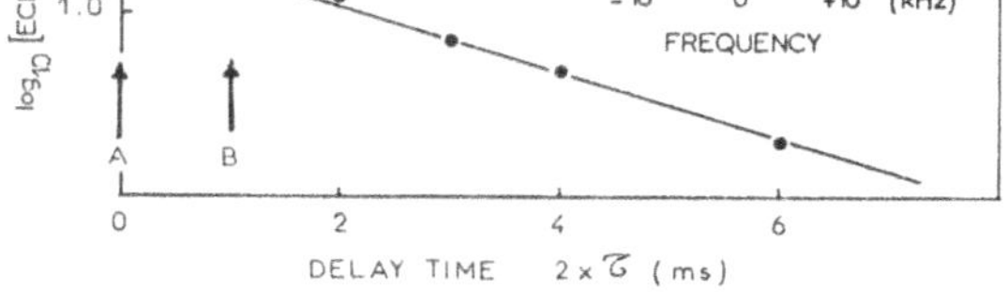

Fig. 27 Variation of the amplitude of the spin-echo as a function of the delay time τ at $T = 73$ K and $V = 110$ mV ($\approx 3V_T$). The decay with the long time constant corresponds to nuclei in regions of the crystal were the CDW is pinned, whereas the fast decay corresponds to nuclei in regions were the CDW is sliding. The inset shows the Fourier transforms of the half of the spin-echo for delay times $\tau = 0$ and 500 μs, indicated by the arrows A and B. The weight of the static contribution increases with τ. (From [13].)

The spin-spin relaxation time data, T_{2F}, measured at various temperatures and fields are plotted in Figure 28. The rapid decrease of T_2 with increasing temperature is the most obvious feature. The data must be considered with some caution since they are not taken at a fixed value of $E/E_T = \alpha$. The inset shows the variation of T_{2F} as a function of α at 77 K where the dependence is strong. At 63 K a much less pronounced field dependence is observed.

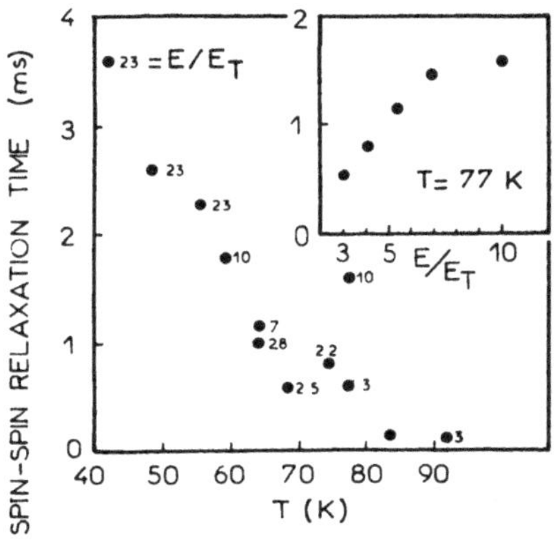

Fig. 28 Variation of the dynamic spin-spin relaxation time T_{2F} versus temperature. The number on the right of each point is the ratio E/E_T. The inset shows in more detail the field dependence of T_{2F} at 77 K, sketched by the two dotted lines in the main figure. (From [13].)

If the model of section 6.2.3. with rare and large phase fluctuations is assumed then the product $n = T_{2F}\nu_d$ corresponds to the number of coherent oscillations. ($T_{2F} = \Gamma^{-1}$ for $T_{2F} \ll T_{2d}$. At 63 K and $E/E_T \approx 3$ we find $\nu_d = 25$ kHz, $T_{2F} = 1$ ms and $n \approx 25$.) Thus the CDW remains coherent for about 25 oscillations at any given nucleus.

Contrary to the 63 K measurement, the relaxation at 77 K was found to be strongly dependent on E/E_T, T_{2F} decreases sharply with decreasing field. This may reflect that the motion is more chaotic for fields near threshold than for large ν_d. It is related to the problem of how large is the part of the sample in which the CDW is really static for small fields. As the field approaches E_T from above the apparent ν/j ratio increases showing that the fraction in which a CDW current with a steady component flows decreases. On the other hand both the lineshape analysis and the relaxation shows that in most of the sample the CDW remains fluctuating at high temperatures and low fields. This may imply that there are regions of the sample for which the average CDW current is zero ($\overline{\dot{\varphi}(t)} = 0$) but where large amplitude fluctuations occur, fast enough to narrow the NMR line and increase the relaxation rate.

The fast decaying component of the relaxation at high temperature and low value of E/E_T is not exponential but Gaussian like, characteristic of a spatial distribution of Γ. This is related to the inhomogeneity of the current distribution. In regions where the field is just sufficient to depin the CDW, but no steady current flows, the phase fluctuations have a large amplitude and are random, while in regions where a steady current flows the phase changes with a constant rate and fluctuations are less important.

Similar results are found from the analysis of the spin-echo amplitude $S(\tau)$ which, like the appearance of sidebands in the spectrum gives ample evidence for a coherent motion in high fields and low temperatures. The coherence of the motion leads to oscillations in $S(\tau)$ which are damped by the dipolar interaction, the distribution of the CDW velocity, and stochastic phase fluctuations (Figure 24). At high fields the static fraction is negligible. By measuring the total relaxation time T_2 at spin-echo delay times for which the oscillations in $S(\tau)$ are already damped the contribution of fluctuations is found to be $\Gamma = T_{2F}^{-1} = (T_2')^{-1} - (T_{2d})^{-1}$. A shortening of the spin-spin relaxation rate due to phase fluctuations is however still observed. From this for $E/E_T \approx 20$ at 43 K and 47 K $n_c = 25$ and 45 were obtained [13], respectively.

The above analysis is based on the large amplitude fluctuations model which, however, is not justified at fields much above threshold where the opposite limit of small amplitude fluctuations with a short correlation time is more reasonable. For this one found (section 6.2 4.)

$$(T_{2F})^{-1} = (2\pi\nu_1)^2 \overline{\delta\varphi^2}\tau_c \tag{6.16}$$

from which $\overline{\delta\varphi^2}\tau_c = 6 \cdot 10^{-8}$ s and $1.6 \cdot 10^{-7}$ s for 43 K and 47 K, respectively.

From the relaxation rate T_{2S} the intrinsic linewidth of the voltage noise oscillations $P(\nu)$ may be estimated. This width decreases with temperaure, at about 50 K from T_{2F} a width of 100 Hz was estimated. This is much less than the inhomogeneous width observed by the noise frequency analyser, T_{2F} corresponds to the fluctuations of the phase winding rate at a fixed position.

7. Conclusion

The molybdenum blue bronzes ($Rb_{0.3}MoO_3$, $K_{0.3}MoO_3$) are quasi-one dimensional conductors in which the CDW state has been extensively studied by microscopic techniques like Mössbauer, EPR, and mostly NMR. The temperature dependence of the order parameter of the Peierls transition has been carefully measured, a deviation from BCS theory has been found in agreement with neutron and X-rays results. The Rb NMR lineshapes at 77 K show that there is no incommensurate-commensurate transition down to this temperature, although the wavevector of the

Peierls transition approaches the commensurate value $q_b = 3/4$ within the resolution of X-ray and neutron techniques below 100 K. At lower temperatures, the EPR lineshape indicates the developement of small commensurate areas in the crystal, (5% at 50 K), however these were not found by NMR down to 40 K. As far as the strength of the pinning is concerned, the various impurities or charged defects used as local probes by EPR and Mössbauer techniques were found to act as weak pinning centers, at least far below the Peierls transition. The excitations of the pinned CDW (phasons and amplitudons) have also been studied by ^{87}Rb nuclear spin-lattice relaxation, like in incommensurate dielectrics, the relaxation was found to be dominated by phasons below the onset temperature of the incommensurate distortion. In spite of the absence of an incommensurate-commensurate transition, the relaxation rate was found to decrease exponentially below 100 K; this behaviour seems to be related to the drastic decrease of the a.c. conductivity in the same temperature range, a more complete study remains to be done.

The most spectacular and maybe the most important contribution to the microscopic study of the CDW in the molybdenum blue bronzes has been the NMR study of the Fröhlich mode. Transport, noise, and NMR measurements on the same single crystal - **in situ in the NMR coil** - have proved to be a very powerful way to investigate the dynamics of the sliding CDW. Beyond the evidence for the CDW motion as a bulk property which follows from the motional narrowing of the NMR spectrum a number of other questions have been clarified in the blue bronze $Rb_{0.3}MoO_3$ which is likely to be representative for the whole class of materials in which the CDW can be depinned by the application of an electric field.

From the sidebands of the motionally narrowed NMR and the oscillations of the spin-echo amplitude the bulk phase winding rate ν_d is directly measured. They also prove unambiguously that at any given nucleus in the crystal the sliding motion is coherent within the time scale of the NMR experiment. Comparison with the noise spectra recorded at the same applied electric field and temperature establishes that the bulk winding rate of the phase ν_d is equal to the "noise" frequency ν_n with an accuracy better than 10%. This rules out all theoretical models which predict a ratio ν_n/ν_d different from one.

Experimental NMR spectra obtained at a fixed high value of the ratio E/E_T in the temperature range 40 to 60 K have been fitted to the theoretical lineshapes calculated under the assumption that the noise spectra correspond to the static spatial distribution of the CDW velocities in the crystal. The accuracy of the fit demonstrates the validity of this assumption; it also allows to verify the relationship $j = nev_{CDW}$ with an accuracy better than 10 %, thus establishing that **all the electrons condensed below the Peierls gap are involved in the Fröhlich mode.**

It follows from the interpretation of the joint NMR and noise experiments that the low frequency increase of the noise spectra corresponds to a low frequency tail in the velocity distribution. This means that in some parts of the crystal, due to a stronger interaction with defects, the sliding motion of the CDW is very slow.

NMR experiments in $Rb_{0.3}MoO_3$ can be fully interpreted by considering the crystal as a collection of independent domains. An incommensurate CDW is sliding with a nearly uniform velocity in each domain. However, one cannot exclude that the "domains" are in fact separated by the paths of localized CDW defects, *e.g.* dislocation loops. Except for fields very close to the threshold, nuclei far from the paths of dislocations will experience a relatively constant phase drift. For values of the electric field close to the threshold ($E/E_T < 3$) the CDW stays pinned in some parts of the crystal. This effect, which is proven by NMR experiments, is strongly dependent on the quality of the contacts and may lead to an anomalously high value of ν_n/j_{CDW}. As explained in section 6.4. we believe that this problem is not restricted to the blue bronzes, but is also present in $NbSe_3$ and in our view is the origin of the anomalous features of the NMR data in $NbSe_3$ reported by Ross *et al.* [81, 87]. The temperature dependence of this static fraction is rather surprising, it decreases with increasing temperature, leading to the unexpected conclusion that in some parts of the crystal, where there is no or negligible CDW current, the fluctuations of the phase are large enough to motionally narrow the NMR line.

Finally, we have addressed the problem of the stochastic fluctuations of the CDW velocity within a time scale smaller than the NMR window, *i.e.* the spin-spin relaxation time. While the width of the sidebands of the NMR lineshape results from an inhomogeneous static distribution of velocities within the crystal, the width of the central line or more precisely the spin-spin relaxation rate as measured by spin-echo is affected by the stochastic temporal fluctuations of v_{CDW}. The correlation function of the transverse nuclear magnetization, which is periodic in the case of a periodic modulation of the local hyperfine field, becomes damped when stochastic fluctuations are present. From the analysis of the spin-spin relaxation rate the number of coherent oscillations may be determined. In the crystal studied, this number was found to be between 25 and 45. These values are related to the homogeneous linewidth of the voltage noise spectrum.

A number of problems are left open which would deserve new experiments: *e.g.* the dynamics of the CDW when the applied electric field is in the vicinity of the threshold is poorly understood yet, especially when thermal fluctuations become important. A continuous incommensurate to commensurate transition has been reported by EPR but not confirmed by NMR. The study of phason and amplitudon excitations by the dependence of the SLRR within the spectrum is of interest also. The most obvious extension of this work will be the study of the undamped sliding CDW regime recently observed at very low temperature [89]. In this review we dealt with charge density wave systems. The ideas presented have, however, much in common with the NMR spectra of type II superconductors in a magnetic field. Delrieu [90] calculated and observed in 1973 the lineshapes of the Tl NMR in $Pb_{0.75}Tl_{0.25}$ in which the vortex lattice was set in motion by a current.

References

1 M. Greenblatt, in *Low-Dimensional Electronic Properties of Molybdenum Bronzes and Oxides*, ed by C. Schlenker, Kluwer Academic Publishers, Netherlands, p.1 (1989).

2 C. Schlenker, J. Dumas, C. Escribe-Filippini and H. Guyot, in *Low-Dimensional Electronic Properties of Molybdenum Bronzes and Oxides*, ed. by C. Schlenker, Kluwer Academic Publishers, Netherlands, p.87 (1989).

3 J. Dumas, C. Schlenker, J. Marcus and R. Buder, *Phys. Rev. Lett.* **50**, 757 (1983).

4 M. Sato, H. Fujishita, and S. Hoshino, *J. Phys.* **C 16**, L877 (1985).

5 W.J. Shutte and J.L. de Boer, to appear in *Acta Crist.* B.

6 G. Mihaly, T. W. Kim, and G. Grüner, *Phys.Rev.* **B39**, 13009 (1989).

7 P. Butaud, P. Ségransan, C. Berthier, J. Dumas, and C. Schlenker, *Phys. Rev. Lett.* **55**, 523 (1985).

8 K. Nomura, K. Kume, and M. Sato, *Solid State Commun.* **57**, 611, (1986).

9 D.C. Douglass, L.F. Schneemeyer, and S.E. Spengler, *Phys. Rev.* **B36**, 1831 (1987).

10 P. Ségransan, A. Jánossy, C. Berthier, J. Marcus, and P. Butaud, *Phys. Rev. Lett.* **56**, 1854, (1986); *Physica* **143B**, 120 (1986).

11 K. Nomura, K. Kume, and M. Sato, *J.Phys.* **C 19**, L 289 (1986).

12 A. Jánossy, C. Berthier, P. Ségransan, and P. Butaud, *Phys. Rev. Lett.* **59**, 2348 (1987).

13 P. Butaud, P. Ségransan, A. Jánossy, C. Berthier, *J. de Physique (Paris)*, **59** (1990).

14 G. Grüner, *Reviews of Modern Physics*, **60**, 1129 (1988).

15 P. Monceau in *Properties of Inorganic Quasi-One Dimensional Materials*, Vol. II, ed. by P. Monceau, p. 139 (Reidel, Dordrecht, Holland, 1985).

16 For rewiews see: C. Schlenker and J. Dumas in *Crystal Chemistry and Properties of Materials with Quasi-One Dimensional Structures*, ed. by J. Rouxel, p.135 (Reidel, Dordrecht, Holland, 1986); C. Schlenker in *Low Dimensional Conductors and Superconductors*, NATO Advanced Study Institute Series B Physics Vol.155, eds. D. Jérome and L.C. Caron, Plenum Press NY, p.477 (1987).; and Ref. [2].

17 J.P Pouget, in *Low-Dimensional Electronic Properties of Molybdenum Bronzes and Oxides*, ed by C. Schlenker, Kluwer Academic Publishers, Netherlands, p.87 (1989).

18 J. Graham and A. D. Wadsley, *Acta Crist.* **20**, 93 (1966).

19 G. Bang and G. Sperlich, *Z. Phys.* **B22**, 1 (1975).

20 A. Jánossy, G.L. Dunifer, and J.S. Payson, *Phys. Rev.* **B38**, 1577 (1988).

21 J. Dumas, B. Laayadi, and R. Buder, *Phys. Rev.* **B40**, 2968 (1988).

22 J.Y. Veuillen, R. Chevalier, R. Marcus, and C. Schlenker, *Physica* **143B**, 186 (1986).

23 a)J.P. Pouget, S. Kagoshima, C. Schlenker, and J. Marcus, *J. Physique Lett.* **44**, L113 (1983); b) J.P. Pouget, C. Noguera, A.H. Moudden, and R. Moret, *J. Physique* **46**, 1731 (1985).

24 R.M. Fleming, L.F. Schneemeyer, and D.E. Moncton, *Phys. Rev.* **B31**, 899 (1985).

25 M. Sato, H. Fujishita, S. Sato, and S. Hoshino, *J. Phys.* **C18**, 2603 (1985).

26 J. P. Pouget, C. Escribe-Filippini, C. Hennion, B. Currat, A.H. Moudden, R. Moret, J. Marcus, and C. Schlenker, *Mol. Cryst. Liq. Cryst.* **121**, 111 (1985).

27 P. Butaud, *Thesis*, Grenoble (1987).

28 C. Berthier and P. Ségransan in *Low Dimensional Conductors and Superconductors*, NATO Advanced Study Institute Series B Physics Vol.155, eds. D. Jérome and L.C. Caron, Plenum Press NY, p.455 (1987).

[30] S. Girault, A.H. Moudden, J.P. Pouget, and J.M. Goddard, *Phys. Rev.* **B38**, 7980 (1988).

[31] P.A. Lee, and T.M. Rice, *Phys. Rev.* **B19**, 3970 (1979).

[32] R.M. Flemming, R.G. Dunn, and L.F. Schneemeyer, *Phys. Rev.* **B31**, 4099 (1985).

[33] J.P. Pouget, S. Girault, A. H. Moudden, B. Hennion, C.Escribe- Filippini and M. Sato, *Physica Scripta* **T25**, 58 (1988).

[34] H. Fröhlich, *Proc. R. Soc. London Ser.* **A223**, 296 (1954).

[35] L. Forro, J.R. Cooper, A. Jánossy, and K. Kamaras, *Phys. Rev.* **B34**, 9047 (1986).

[36] W. Fogle and H. Perlstein, *Phys. Rev.* **B6**, 1402 (1972).

[37] G. Mihaly and L. Mihaly, *Phys. Rev. Lett.* **52**, 149 (1984).

[38] R.J. Cava, R.M. Fleming, E.A. Rietman, L.F. Schneemeyer, and R.G. Dunn, *Phys. Rev. Lett.* **53**, 1667 (1984).

[39] A. Jánossy, G. Kriza, S. Pekker, and K. Kamaras, *Europhys. Lett.* **3**, 1027 (1987).

[40] D. Fisher, *Phys. Rev.* **B31**, 1396 (1985).

[41] L. Mihaly and G.Y. Tessema, *Phys. Rev.* **B33**, 5858 (1986).

[42] R. Fleming and C.C. Grimes *Phys. Rev. Lett.* **42**, 1423 (1979).

[43] A. Jánossy, C. Berthier, and P. Ségransan, *Physica Scripta*, **T19B**, 578 (1987).

[44] G. Grüner, A. Zawadovski, P.M. Chaikin, *Phys. Rev. Lett.* **46**, 511 (1981).

[45] P. Monceau, J. Richard, and M. Renard, *Phys. Rev.* **B25**, 931 (1982).

[46] G. Mozurkewich,M. Maki, and G. Grüner, *Solid State Commun.* **48**, 453 (1983).

[47] T.W. Jing and N.P. Ong, *Phys. Rev.* **B33**, 5841 (1986).

[48] A. Jánossy, G. Mihaly, S. Pekker, and S. Roth, *Solid State Commun.*, **61**, 33 (1987).

[49] N.P. Ong, G. Verma, and K. Maki, *Phys. Rev. Lett.* **52**, 663 (1984).

[50] T. Csiba, G. Kriza, and A. Jánossy, *Europhys. Lett.* **9**, 163 (1989).

[51] J. Baıdeen, *Phys. Rev. Lett.* **45**, 1978 (1980);*Phys. Rev. Lett.* **55**, 1010 (1985).

[52] J. Winter, *Magnetic Resonance in Metals*, Clarendon Press, Oxford (1971).

[53] B.R. Weinberger, *Phys. Rev.* **B17**, 566 (1978).

[54] T.P. Das and E.L. Hahn, *Solid State Physics, Suppl. I*, eds F. Seitz and D.T. Turnbull, Academic Press, New York (1958).

[55] C. Berthier, D. Jérome, and P. Molinie, *J. Phys.* **C 11**, 797 (1978).

[56] D.C. Johnston, *Phys. Rev. Lett.*, **52**, 2049 (1984).

[57] C. Berthier, D.Jérome, P. Molinie, and J. Rouxel, *Solid State Commun.* **19**, 131 (1976).

[58] J.A.R. Stiles and D.L.G. Williams, *J. Phys.* **C 9**, 3941 (1976).

[59] R. Blinc, *Phys. Rep.* **79** 331 (1981); R. Blinc, P. Prelovsek, V. Rutar, J. Seliger and S. Zumer in *"Incommensurate phases in dielectrics"* eds. R. Blinc and A.P. Levanyuk, Elsevier Publishers B.V. (1986), chap. 4.

[60] J.F. Baugher, P.C. Taylor, T. Oja, and P.J. Bray, *J. Chem. Phys.*, **50**, 4919 (1969).

[61] J.M. Delrieu, M. Roger, Z. Toffano, A. Moradpour, and K. Bechgaard, *J. de Physique* (Paris) **47**, 839 (1986).

[62] T. Takahashi, Y. Maniwa, H. Kawamura, and G. Saito, *J. Phys. Soc. Japan* **55**, 1364 (1986).

[63] J. Dumas, R. Buder, J. Marcus, C. Schlenker, and A. Jánossy, *Physica* **143B**, 183 (1986).

[64] R. Blinc, F. Milia, B. Topic, and S. Zumer, *Phys. Rev.* **B29**, 4173 (1984).

[65] Y. Yafet, in *Solid State Physics* **14**, ed. by F. Seitz and D.T. Turnbull, (Academic Press, New York), p. 1 (1963).

[66] D. Shaltiel, A. Grayevsky, V. Zevin, and G. Grüner, *Phys. Rev.* **B38**, 10075 (1988).

[67] W.L. McMillan, *Phys. Rev.* **B12**, 1187 (1975); *Phys. Rev.* **B14**, 1496 (1976).

[68] B.H. Suits, S. Couturie, and C.P. Slichter, *Phys. Rev. Lett.*, **55**, 253 (1985).

[69] J.Y. Veuillen, R. Chevalier, J. Marcus, C. Schlenker, *Solid State Commun.* **63**, 587 (1987).

[70] P. Lederer, G. Montambaux, and J.P. Jamet, *Mol. Cryst. Liq. Cryst.* **121**, 99 (1985).

[71] G. Errandonéa, J. Schneck, J.C. Tolédano, A. Litzer, H. Savary, J. Aubree, J.M. Kiat, and G. Calvarin, *Ferroelectrics*, **53**, 247 (1984).

[72] J.C. Gill, *Solid State Commun.* **39**, 1203 (1981).

[73] P. Littlewood, *Phys. Rev.* **B33**, 6694 (1986).

[74] K. Sasaki and K. Maki, *Phys. Rev.* **B33**, 2665 (1986).

[75] J. Dumas, B. Laayadi, and J. Marcus, *Synthetic Metals*, **29**, F487 (1989).

[76] S. Wada, R. Aoki, and O. Fujita, *J. Phys.* **F 14**, 1515, (1981).

[77] S. Zumer and R. Blinc, *J. Phys.* **C 14**, 465 (1976).

[78] J. Bardeen, *Solid State Commun.* **13**, 357 (1973).

[79] P. Monceau, N.P. Ong, A. M. Portis, A. Meerschaut, and J. Rouxel, *Phys. Rev. Lett.* **37**, 602 (1976).

[80] D.C. Douglass, L.F. Schneemeyer, and S.E. Spengler, *Phys. Rev.* **B32**, 1813 (1985).

[81] J.H. Ross Jr., Z. Wang, and C.P. Slichter. *Phys. Rev. Lett.* **56**, 663 (1986).

[82] P. Butaud, P. Ségransan, A. Jánossy, C. Berthier, *Synthetic Metals*, **29**, F499 (1989).

[83] M. Kogoj, S. Zumer, and R. Blinc, *J. Phys.* **C 17**, 2415 (1984).

[84] A. Abragam, *Principles of Nuclear Magnetism*, Clarendon Press Oxford (1961).

[85] P. Butaud, P. Ségransan, A. Jánossy, C. Berthier, unpublished results.

[86] J.R. Tucker, *Phys. Rev. Lett.* **60**, 1574 (1988); J. R. Tucker, W. G. Lyons, and G. Gammie, *Phys. Rev.* **B38**, 1148 (1988).

[87] J.H. Ross Jr., Z. Wang, and C.P. Slichter, *Phys. Rev.* **B41**, 2722 (1990); J.H. Ross Jr, Thesis, University of Illinois (1986).

[88] R. J. Cava, R. M. Flemming, P. Littlewood, E. A. Rietman, L. F. Schneemeyer, and R.G. Dunn, *Phys. Rev.* **B30**, 3228 (1984).

[89] G. Mihaly and P. Beauchêne, *Solid State Commun.* **63**, 911 (1987).

[90] J-M. Delrieu, *J. Phys.* **F 3**, 893 (1973).

NMR STUDIES OF THE QUASI ONE-DIMENSIONAL INORGANIC COMPLEX SALT $K_2Pt(CN)_4Br_{0.3}3.2H_2O$ (KCP)

DETLEF BRINKMANN
Physik-Institut, Universität Zürich
8001 Zürich
Switzerland

OTMAR KANERT
Institut für Physik, Universität Dortmund
4600 Dortmund
Germany

1. Introduction

Highly anisotropic or quasi-one-dimensional (Q1D) conductors have received much interest since the pioneering work by Krogmann [1] who suggested that mixed-valency planar d^8 metal complexes should exhibit one-dimensional metallic properties. Numerous reviews on the subject of Q1D-conductors have been written in the past and we only refer here to the two by Toombs [2] and by Kagoshima *et al.* [3] where references to the earlier literature can be found.

Among the Q1D conductors $K_2Pt(CN)_4Br_{0.30}3.2H_2O$ (KCP) is the most prominent example. The compound comprises parallel linear metal (Pt) strands with an intrastrand Pt-Pt separation of about the same size as in Pt metal while the interstrand distance is about three times the Pt-Pt separation in the chain (see also section 2). Highly anisotropic electrical conductivity at room temperature for dc and at optical frequencies demonstrates the Q1D character of KCP [4]. The conductivity along the Pt chains is about 10^5 times larger than perpendicular to the chains [5,6].

KCP is not a conductor in the usual sense, since it becomes insulating at low temperatures. Due to strong electron-phonon coupling a Peierls distortion [7] or a Kohn anomaly [8] is observed in X-ray [9] and neutron scattering [10] experiments at room temperature. An extension of these measurements to low temperature revealed a central peak, which implies there is a one-dimensional periodic lattice distortion (PLD) in addition to the Kohn anomaly [11]. Above 200 K there is no correlation between the distortion of neighboring chains, while below about 140 K evidence for the onset of long-range three-dimensional order is found [11]. The

T. Butz (ed.), Nuclear Spectroscopy on Charge Density Wave Systems, 229–273.

onset is accompanied by a critical slowing-down of lattice fluctuations [12]. It appears, however, that complete three-dimensional ordering is never achieved for the Peierls distortion in KCP. The correlation length for the order perpendicular to the Pt chains was found to rise rapidly below 120 K to a constant value of about $3 \cdot a$ where a denotes the chain separation [11]. Those and other observations are adequately explained by charge-density-wave (CDW) structures connected with a PLD of the $Pt(CN)_4^{2-}$ complexes with a displacement amplitude of the order of $10^{-2} \cdot d$ where d is the Pt-Pt distance along the chain [13]. These CDW's seem to order three dimensionally below about 120 K, although complete order is never established down to 5 K. Electrical conductivity data [14] suggest a Peierls gap, 2Δ, of about 0.16 eV whereas from optical data 0.2 eV was concluded [15].

NMR is shown to be a powerful tool for studying microscopic details of the structure and dynamic properties of KCP because the substance contains a fairly large number of appropriate nuclei able to act as local probes inside and outside the conducting Pt chains. While investigations using ^{195}Pt NMR are directly related to the static and dynamic properties of the conducting chains, the study of the "out of chain" nuclei such as ^{1}H, ^{13}C, ^{14}N, and ^{81}Br probes other features which may be related to the behavior of the conducting chains. Typically from "out of chain" NMR experiments one obtains crystallographic characteristics of KCP such as different sites for the water molecules or the Br ions in KCP and of the onset and thermodynamics of their localized motion with rising temperature [16,17,18].

2. Properties of KCP

2.1. Lattice structure

KCP forms a tetragonal lattice structure of the space group P 4 mm [18]. The lattice parameters of the unit cell shown in Figure 1 are $a = 9.88$ Å and $c = 5.76$ Å at room temperature. The main feature of KCP is the columnar stacking along the c-axis of square planar $(CN)_4$ ligands comprising each a Pt atom in the center of the square (see Figure 2). The distance of 2.88 Å between Pt atoms along the stacking axis is only slightly larger than the Pt-Pt distance in a metal (2.77 Å). Consequently, as sketched in Figure 2, the $5d_z^2$ orbitals of the Pt atoms overlap strongly in c-direction and form an energy band, whereas the interchain overlapping is very small. Principally, such a band should be completely filled by two electrons given by each Pt atom resulting in an insulating state. The presence of 0.3 bromine atoms per formula unit, however, partially oxidizes the Pt by transferring 0.3 electrons per Pt atom to the bromine atoms giving rise to Br^- ions. Hence, the energy band is only 85% filled up to the wavevector $k_F = 1.7 \cdot \pi/c \simeq 0.92$ Å^{-1} leading to a one-dimensional conducting state. It is noteworthy how important the Br^- content is for the conduction properties of KCP forming a hole-like conduction band of width

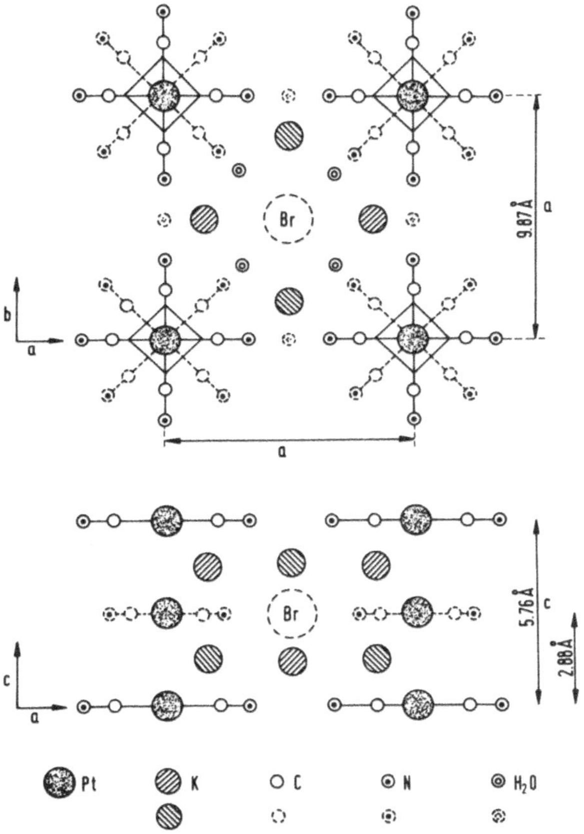

Fig. 1 Crystal structure of KCP (Br) projected on the crystallographic planes *ab* (upper part) and *ac* (lower part).

$0.3 \cdot 2\pi/c$ in the top Pt band. Because of the random occupancy of Br sites the conduction electrons are exposed to a random potential along the c-axis.

In KCP three different non-equivalent sites of the water molecules exist in the crystal structure, usually referred to as $H_2O(I)$, $H_2O(II)$, and $H_2O(III)$ [19]. The characteristics of these three different types of water molecules in the crystal are

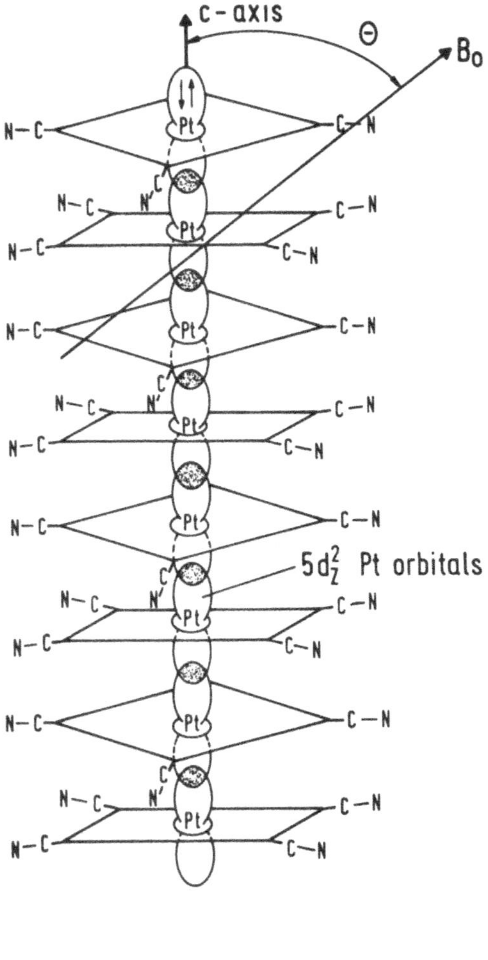

Fig. 2 Stacking of square planar $Pt(CN)_4$ units along the *c*-axis showing the strong overlap of the $5d_{z^2}$ orbitals of the Pt atoms.

the following:

- the $H_2O(I)$ molecules are perfectly ordered with their H-H vector parallel to the crystallographic *a*-axis,

- the $H_2O(II)$ molecules present hydrogen bonds with the bromine and the C-N groups and have a large mobility, *i.e.*, dynamical disorder above 150 K,
- the $H_2O(III)$ molecules occupy the bromine vacant sites distributed randomly along the chains.

Further NMR experiments show the existence of two distinct bromine sites, Br(I) and Br(II), in KCP with a ratio of population of about three [20].

2.2. PEIERLS TRANSITION AND CDW

Peierls [7] has shown that a purely 1D metallic chain, formed by conduction electrons localized on an equidistant array of ions would be unstable compared with a 1D semiconductor with an insulating Fröhlich ground state. Due to electron-phonon interaction a periodic lattice distortion (PLD; Peierls distortion or Kohn anomaly) with amplitude x_0 arises whose periodicity results in the opening of a small gap, 2Δ, in the electronic energy spectrum at the Fermi surface. The relation between Δ and the strength x_0 of the PLD is given by

$$\Delta = g \cdot 2x_0/a_0 \tag{2.1}$$

where g denotes the electron-phonon coupling constant which describes the change in electronic energy due to relative compression of the ions in the conducting chain and a_0 is the separation of these ions.

For KCP the PLD was shown to be incommensurate with the lattice structure [11]; the periodicity can be expressed as

$$\Delta x = x_0 \cos\left(2k_F \cdot x + \Phi\left(x\right)\right) \tag{2.2}$$

where Δx denotes the shift of a Pt atom at site x along the c-axis, $x_0 = 5 \cdot 10^{-3} a$, and $k_F = 1.7\pi/c \simeq 0.92$ Å^{-1} (see section 2.1) corresponding to a Fermi energy $E_F = \hbar^2/(2m) \cdot k_F$ of 3.3 eV. The PLD is accompanied by a rearrangement of the charge density of the conduction electrons leading to a charge density wave (CDW) of the form (plane wave approximation)

$$\rho(x) = \rho_0 + \rho_1 \cos(2k_F \cdot x + \Phi(x)) \tag{2.3}$$

The ground state of the CDW is pinned by a periodic pinning potential

$$V(x) = V_0 + V_1 \cos(2k_F \cdot x + \Phi(x)) \tag{2.4}$$

In the mean-field picture, the Peierls instability leads to a second-order phase transition at the critical temperature T_p. Using the mean-field relation $\Delta = 1.76kT_p$ [3] the transition temperature of KCP turns out to be of the order of 10^3 K. Experimental data suggest $T_p = 600$ K. The ordinary concept of a critical temperature, however, breaks down in one dimension, since critical fluctuations, which are not considered in a mean-field theory, prevent any phase transition from occurring at finite temperatures. What reestablishes the phase transition in real compounds, is the existence of finite transverse chains (quasi 1D system). The 1D fluctuations at low temperatures will create correlated domains over fairly long parts of uncoupled chains. This introduces 3D properties into the system which leads to a phase transition at T_{3D} often referred to as the transition temperature at which the Peierls instability manifests itself. Introducing a relative electronic transverse coupling constant η the dependence of T_{3D} on η has been computed by Horowitz *et al.* [21]. For KCP, T_{3D} is about 120 K [11]. As mentioned in the Introduction, however, complete 3D ordering is never achieved for the Peierls distortion of KCP. At 120 K the interchain correlation reaches one interchain distance a, increasing rapidly with decreasing temperature to about $3a$, at about 100 K, and then remains constant down to 6 K. Probably the random occupation of Br sites by bromine and H_2O(III) in the crystal lattice and by other impurities will tend to prevent longe-range 3D order in KCP.

The Peierls gap, 2Δ, gives rise to an activated electrical conductivity along the chains of the form

$$\sigma_{\parallel}(T) = \sigma_0 \exp(-\Delta/kT) \tag{2.5}$$

where σ_0 is characterized by details of the mechanism governing the mobility of the conduction electrons. According to Soda *et al.* [22] the ratio of the conductivity perpendicular and parallel to the c-axis, $\sigma_{\perp}/\sigma_{\parallel}$, is determined by the transverse electronic coupling constant η as follows

$$\sigma_{\perp}/\sigma_{\parallel} = \eta^2 \cdot (2a/c)^2 \tag{2.6}$$

The experimental value for $\sigma_{\parallel}/\sigma_{\perp}$ is of the order of 10^5 [6], but decreases to $3 \cdot 10^3$ at 35 K leading to a very small value for η of about $3 \cdot 10^{-3}$. The slope of the $\ln(\sigma_{\parallel}) - vs - 1/T$ curve yields a gap energy 2Δ of the Peierls instability of about 0.16 eV while optical data suggest a value of 0.2 eV [15]. It should be noted that one does not see any clear evidence of a phase transition around $T_{3D} = 120$ K from the $\sigma - vs - 1/T$ curves.

2.3. SOLITARY EXCITATION MODES OF CDW

As discussed in the last section the usual mechanism for the electrical conductivity is the independent particle excitation of electron-hole pairs across the Peierls gap 2Δ. If the forces which pin the CDW's to the equilibrium lattice are sufficiently small, it is possible to visualize excited states where the phase of the PLD moves through the crystal carrying the CDW. The existence of low-lying excitations of the CDW condensate was first suggested by Rice *et al.* [23]. These non-linear modes describe localized charge excitations ("phase solitons") and may account for the unusual temperature dependence of ^{195}Pt nuclear spin relaxation observed at low temperatures in KCP. These non-linear charged excitations arise from "solitary wave" solutions of the non-linear wave equation

$$\Phi - c_0^2 \nabla_x^2 \Phi + \omega_F^2 dV/d\Phi = 0 \tag{2.7}$$

for the local phase $\Phi(x,t)$ of the CDW. Here, $V(\Phi)$ denotes the periodic pinning potential of the CDW (see Eq. (2.4)), $c_0 = V_F \cdot (m/m^*)$, where $V_F = \hbar k_F/m$, and m and m^* denote, respectively, the effective electronic mass in the uncondensed and the condensed state, and ω_F is the pinning frequency describing the strength of the pinning potential. Typically, ω_F is in the range of 0.1 - 1 cm^{-1}. If a cosine potential of the form of Eq. (2.4) is used, Eq. (2.7) is the Sine-Gordon equation which has a class of exact non-linear solitary wave solutions given by

$$\Phi_{\pm}(x - vt) = 4 \tan^{-1} \exp\left[\pm \frac{1}{d}(x - v \cdot t)\right] . \tag{2.8}$$

The solution corresponds to phase kinks $\Delta\Phi = \pm 2\pi$ localized over a length $d = (1 - v^2/c_0^2)^{1/2} \cdot c_0/\omega_F$ propagating with a velocity v.

The phase kinks illustrated in Figure 3 cause an additional modulation of the CDW $\rho(x)$ according to

$$\tilde{\rho}(x) = \rho(x) + \frac{n_s \cdot e}{2k_F} \cdot \frac{d\Phi}{dx} = \rho(x) + \rho_\Phi(x) \tag{2.9}$$

where n_s is the 1D density of the condensed electrons. Thus, the phase kinks can be thought of as charged quasiparticles (solitons) with a charge density

$$\rho_\Phi(x) = \frac{n_s \cdot e}{2k_F} \cdot \frac{d\Phi(x)}{dx} \tag{2.10}$$

carrying a net excess charge of $\Delta e \simeq \pm 2e$. As the pinning potential $V(x)$ is of the type $1 - \cos\Phi(x)$ the spatial dependence of $\Phi(x)$ can be obtained explicitly from a solution of the corresponding Sine-Gordon equation (2.7) as

$$\Phi(x) = \frac{4}{p} \arctan \exp(\frac{x}{d}) \tag{2.11}$$

with

$$\frac{d\Phi}{dx} = \frac{2}{p} \cdot d \cdot \text{sech}(\frac{x}{d}) \tag{2.11a}$$

when p describes the amount of the phase kink $\Delta\Phi = 2n/p$. Typically, $p = 1$.

The mass M_Φ of the quasiparticles is given by

$$M_\Phi = 8n_s \cdot M_F^*/q^{*2} \cdot d \tag{2.12}$$

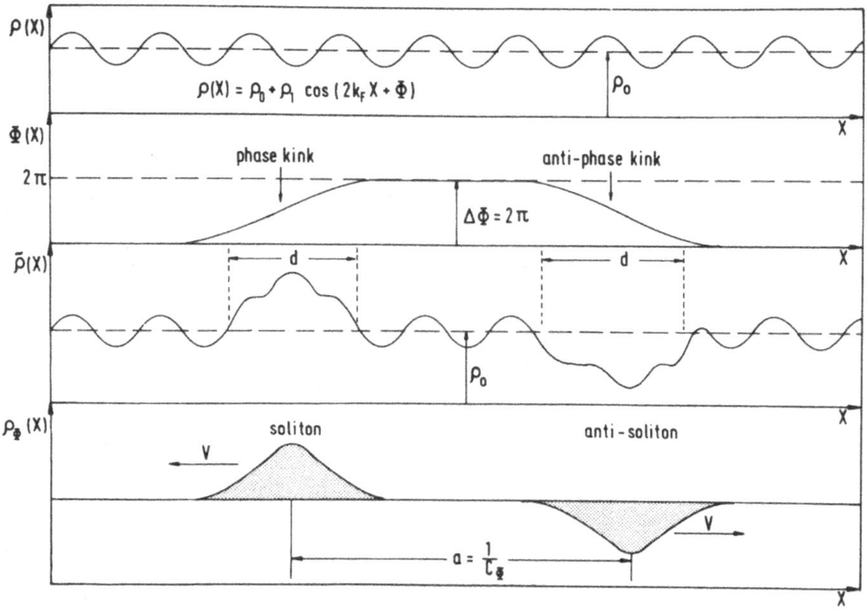

Fig. 3 Rigid charge density wave (CDW) $\varrho(x)$, phase kinks $\Phi(x)$ of the CDW, charge density $\bar{\varrho}(x)$ in the presence of phase kinks, and charge density $\varrho_\Phi(x)$ of phase solitons and antisolitons.

where $M_F^* \simeq 10^4 m^*$ denotes the Fröhlich effective mass and $q^* = 2n/\lambda_s$ with λ_s being the wave length of the incommensurate CDW. For KCP, $\lambda_s = 6.7a$. Assuming the solitons are thermally created, then the concentration c_Φ of a soliton-antisoliton pair can be expressed as [23]

$$c_\Phi = N_s \cdot \frac{d}{L} = \left(\frac{E_\Phi}{kT}\right)^{1/2} \cdot \exp\left(-\frac{E_\Phi}{kT}\right) \tag{2.13}$$

where $E_\Phi = M_\Phi \cdot c^2$ and L is the total length of the conducting chain, *i.e.*, L/d represents the magnitude of the available particle "sites". The analogy to the thermal formation of classical defects like Schottky defects is apparent. For KCP, the formation energy E_Φ of solitons was estimated to be of the order of 200 K while the extent of the solitons, d, is roughly 30 times the Pt-Pt separation in the chain [24].

Cohen and Heeger [25] have assumed that the solitons are nearly free and perform a continuous thermal diffusion. Then, their velocity can be expressed as

$$v = v_{th} \simeq (2kT/M_{\Phi})^{1/2} \tag{2.14}$$

Alternatively, the solitons may be pinned by impurities and/or distortions of the chain. Supposing a uniform strength of the pinning potential E_p, the velocity of the solitons is given by a thermally activated process according to

$$v = b/\tau(T) = b/\tau_0 \cdot \exp(-E_p/kT) \tag{2.15}$$

where $\tau(T) = \tau_0 \cdot \exp(E_p/kT)$ denotes the mean waiting time of a soliton at an impurity between two consecutive jumps with jump time $t_j \ll \tau(T)$, $1/\tau_0$ is the attempt frequency determined by the shape of the potential E_p, and b is the mean jump width of the moving soliton.

If the pinning potential E_p is sufficiently large, the solitons are fixed on the chain in the entire temperature range forming a one-dimensional "soliton lattice" in the simplest case.

It is obvious that ^{195}Pt NMR measurements in KCP should be able to give insights in the details of solitary excitations. In particular, ^{195}Pt nuclear spin relaxation should be influenced remarkably by the dynamics of the solitons.

3. NMR Investigations in KCP

3.1. Introductory remarks

While information about structural properties of solids can be obtained from the position, width, shape, and intensity of the different lines in the NMR spectrum, dynamical effects originating from time-dependent spin-spin and spin-lattice interactions due to fluctuations in the solid are studied through the related strength of the resulting nuclear spin relaxation process, namely $1/T_2$, the spin-spin relaxation rate, and $1/T_1$, the spin-lattice relaxation rate (see for instance [26,27]).

The NMR properties observed with a given nuclear probe are governed by the actual spin Hamiltonian $\hat{H}_{int}$ which describes the interaction of the nuclear spin with its surroundings.

According to Mehring [26] the interaction Hamiltonian can be written generally as

$$\hat{H}_{int} = \sum_k \sum_{q=-k}^{+k} (-1)^q A_{kq}(\vec{r}) \cdot \hat{T}_{k,-q}(\hat{I}) \tag{3.1}$$

where A_{kq} refers to the spatial coordinates $(\vec{r})$ and $\hat{T}_{k-q}$ to the spin variables $(\hat{I})$. The summation index k is 0,1 for shielding interactions, while k=0,1,2 for operators bilinear in the spin variable. $k = q = 0$ yields the secular part of the Hamiltonian responsible for the structure of the NMR spectrum, while $|q| = 1$ and $|q| = 2$ describe the transition $|\Delta m| = 1$ and $|\Delta m| = 2$, respectively, in the Hamiltonian. In particular, a nuclear spin relaxation process is due to time-dependent fluctuations of $\hat{H}_{int}$, $\hat{H}_{int}(t)$, which are assumed to be weak compared to nuclear Zeeman energies between the strongly coupled spins. The time dependence of $\hat{H}_{int}$ may be caused by phonons, phonon-induced fluctuations of conduction electrons or thermally induced hopping motion of defects like vacancies or solitons. The strength of the relaxation process is described by the nuclear spin relaxation rate. In the weak-collision approximation [28] the fluctuations characterized by the correlation time τ_c are much more rapid than the variation $\delta\omega$ in the spin-spin interaction, *i.e.* $\tau_c^{-1} \gg \delta\omega$. Then, the nuclear spin-lattice relaxation rate can be written as

$$\frac{1}{T_1} = \Delta\omega^2 \cdot \sum_{q>0} J^{(q)}(q \cdot \omega_0) \tag{3.2}$$

where $J^{(q)}(\omega)$ denotes the Fourier transform of the correlation function $G(t)$ of the fluctuation and $\Delta\omega^2$ describes the strength of the spin-lattice coupling determined by the actual Hamiltonian $\hat{H}_{int}$. In the usual semi-classical approximation the spin-operator part $\hat{T}$ (see Eq. (3.1)) is treated quantum-mechanically, while the spatial part $A(\vec{r}(t))$ in Eq. (3.1) is assumed to form a classical pair correlation function between spin pairs

$$G(t) = < A(t') \cdot A^*(t - t') >_{t'} \tag{3.3}$$

where the brackets $<>_{t'}$ denote the time average. As discussed above, depending on the type of the Hamiltonian Eq. (3.2) consists of only one ($q = 1$) or two ($q = 1, 2$) terms. Similarly, in the weak collision limit the spin-spin relaxation rate describing the strength of the energy transfer inside the spin system is related to the Fourier transform $J^{(q=0)}(\omega)$ of $G(t)$ as

$$\frac{1}{T_2} = \Delta\omega^2 \left[J^{(q=0)}(0) + \sum_{q\neq 0} \gamma_q J^{(q)}(q \cdot \omega_0) \right] \tag{3.4}$$

where the coefficients γ_q are of the order of one depending on the type of interaction Hamiltonian $\hat{H}_{int}$ responsible for the relaxation process.

Often, the condition $J^{(q)}(0) \gg \sum \gamma_q J^{(q)}(q \cdot \omega_0)$ is fulfilled leading to the approximation

$$\frac{1}{T_2} = \Delta\omega^2 \cdot J^{(q=0)}(\omega = 0). \tag{3.5}$$

Obviously, the leading problem for calculating spin relaxation rates is to derive the correlation function $G^{(q)}(t)$ of the actual fluctuation based on an assumed microscopic model of the motional process. A large number of review articles and monographs have been published in the past dealing with the calculation of correlation functions $G(t)$ and their corresponding spectral densities $J(\omega)$ based on different microscopic models (see for instance [28]). Comparison of measured rates with theoretically calculated spectral densities yields the characteristic time, the correlation time τ_c, of the fluctuation process as a function of thermodynamic properties such as temperature or pressure and distributions of τ_c due to a random distribution of motional parameters like a distribution of potential barriers or of jump widths. Further, the frequency dependence of the spectral density $J(\omega)$ reveals information about the dimensionality of the motional mechanism [29]. For instance, in the limit of rapid motion ($\omega\tau_c \ll 1$) unlimited continuous diffusion without correlation effects leads to $J(\omega) \propto \omega^{-1/2}$ and $J(\omega) \propto \ln\omega$ in the 1D case and the 2D case, respectively, while $J(\omega)$ is independent of ω in the 3D case [30]. On the contrary, if the fluctuation consists of discrete single jumps or excitations the resulting spectral density exhibits the same frequency dependence in the 1, 2, and 3D case [31]. Spectral density functions $J(\omega)$ applicable to dynamic properties of KCP will be presented in the corresponding section of the next chapter.

3.2. PROPERTIES OF THE CONDUCTING CHAINS BY ^{195}Pt NMR

3.2.1. *NMR spectrum and chemical shift below 120 K* Figure 4 shows the NMR spectrum of naturally abundant (33.8%) ^{195}Pt nuclei (spin $I = \frac{1}{2}$) in single-crystalline KCP with the Pt-Pt chain axis along the external magnetic field B_0 ($\Theta = 0°$; see Figure 2) at a temperature of 53 K and a field strength of 6.34 T. In order to appreciate the resolution of the NMR spectrum depicted in Figure 4 one should note that ^{195}Pt Mössbauer experiments as performed by Rüegg *et al.* in KCP at 4 K [32] revealed only a broad Lorentzian line with a width of about 1.3 GHz. No charge localization could be observed from the single-line spectrum. The observed splitting $\Delta\nu$ of the NMR line can be explained in terms of CDW's inducing a shielding $\sigma(x)$ of the ^{195}Pt nuclear probe at site x which is proportional to the CDW $\rho(x)$ (see Eq. (2.3))

$$\sigma(x) = \sigma_0 + \sigma_1 \cdot \cos(2k_F \cdot x + \Phi). \tag{3.6}$$

Taking into account an additional site-independent chemical shielding σ_2 due to the chemical structure of $Pt(CN)_4$ complexes, the resonance frequency can be written as

$$\omega(x) = \omega_c + \omega_1 \cdot \cos(2k_F \cdot x + \Phi) \tag{3.7}$$

with $\omega_c = \omega_0(1 - \sigma_0 - \sigma_2) = \omega_0(1 - \sigma')$ being the center of gravity and $\omega_1 = \sigma_1 \cdot \omega_0$.

Assuming a continuous distribution of Pt nuclei at the chain the resulting line shape $I_\sigma(\omega)$ is given by the density of spin frequencies $dn(\omega)/d\omega$. Using Eq. (3.7) one obtains

$$I_\sigma(\omega) = \text{const}\left[1 - \left(\frac{\omega - \omega_c}{\omega_1}\right)^2\right]^{-1/2}. \tag{3.8}$$

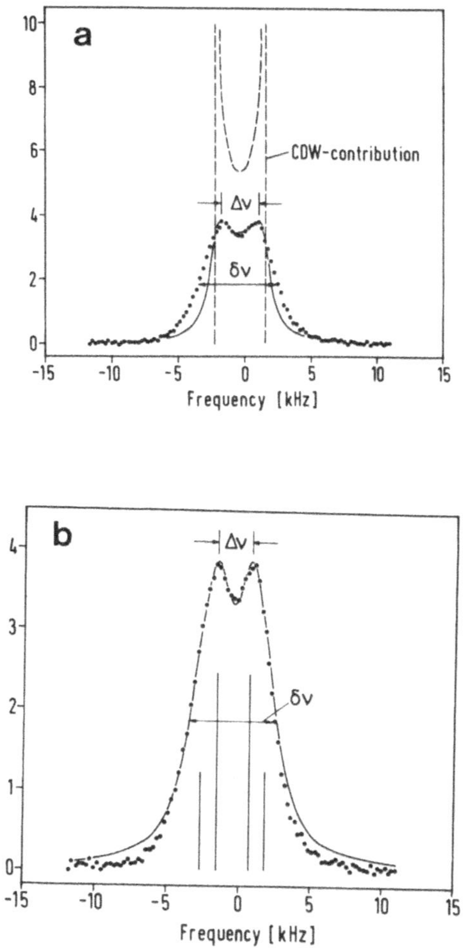

Fig. 4 ^{195}Pt NMR spectrum at $T = 53$ K and magnetic field B_0 along the crystallographic c-axis ($\Theta = 0°$). Resonance frequency: 58.3 MHz [33].
a) Solid line is a fit to the data using the continuum model (Eq. (3.8) and (3.9)).
b) Solid line is a fit to the data using the discrete model (Eq. (3.11)) [33].

The line shape $I_\sigma(\omega)$ has a symmetric form with two singularities at $\omega_c \pm \omega_1$ as shown in Figure 4a. In order to compare the experimentally observed line with Eq. (3.8) one has to take into account that besides the CDW interaction other interactions such as dipole-dipole interactions do exist leading to an additional broadening of the NMR line which may be described by a broadening function $f(\omega)$. Then, the total line shape $I(\omega)$ is a convolution of $I_\sigma(\omega)$ and $f(\omega)$:

$$I(\omega) = \int_{-\infty}^{+\infty} I_\sigma(\omega') f(\omega - \omega') d\omega'. \tag{3.9}$$

The solid line in Figure 4a is a least-squares fit to the data by means of Eq. (3.9) using a Lorentzian shape for $f(\omega)$ with a half width $\delta\nu_f$. The values of the fit parameters $2\nu_1$ and $\delta\nu_f$ are listed in Table 1 [33].

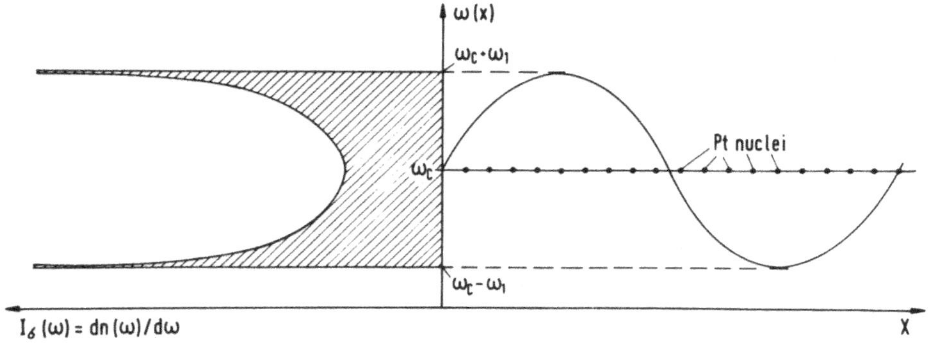

Fig. 5 Density of states $I_\sigma(\omega)$ of the shift frequencies $\omega(x)$ caused by an incommensurate CDW $\varrho(x)$.

model	$2 \cdot \nu_1$ (kHz)	$\delta\nu_f$ (kHz)
continuum	3.78	1.37
discrete	2.24	2.50

Table 1 Splitting frequency $2 \cdot \nu_1$ and half width $\delta\nu_f$ of the broadening function $f(\omega)$ as obtained from the fit procedure (solid lines in Figures 4a and 4b).

As the wave length λ_s of the incommensurate CDW is of the order of the Pt-Pt distance ($c/2 = 2.88$Å, see Figure 1) in the chain ($\lambda_s = 6.7c/2$) the number of Pt nuclei per wave length is rather limited.

Then, instead of treating the line shape by a continuum model one may introduce a discrete model. Approximating the incommensurate wave length $\lambda_s = 6.7c/2$ by a commensurate wave length $\lambda_s' = 6c/2$ one obtains according to Eq. (3.7) four discrete shift frequencies

$$\omega(x_i) = \omega_c + \omega^{(i)}, \qquad i = 1...4 \tag{3.10}$$

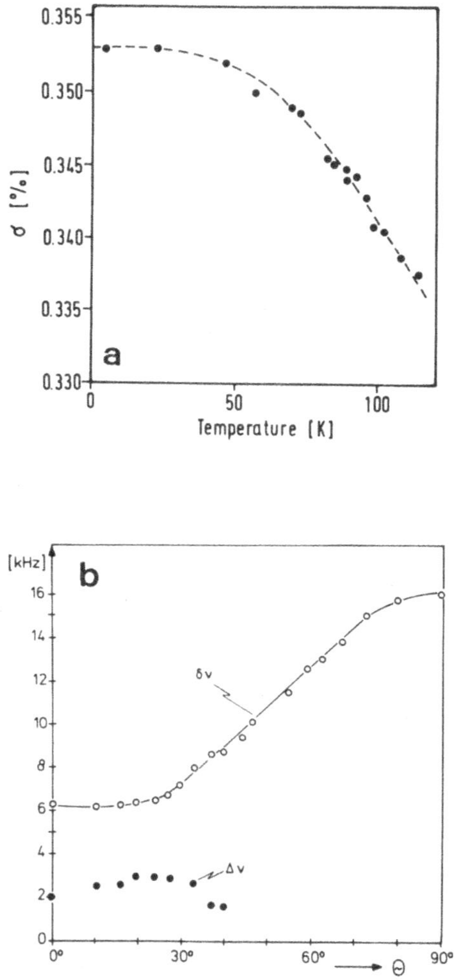

Fig. 6 a) Temperature dependence of ^{195}Pt shielding constant σ with respect to an aqueous solution of H_2PtCl_6 for $\Theta = 20°$ which is the angle between the chain axis c and the magnetic field [34].
b) Orientational dependence of the line width $\delta\nu$ and the line splitting $\Delta\nu$ of the ^{195}Pt NMR line shown in Figure 4 at $T = 53$ K [34].

with $\omega^{(1)} = -\omega^{(2)} = 0.5\omega_1$ and $\omega^{(3)} = -\omega^{(4)} = \omega_1$. The intensity of the four different lines is proportional to the number of the corresponding Pt nuclei per wave length, *i.e.*, $I(\omega^{(1)}) = I(\omega^{(2)}) = 2 \cdot I(\omega^{(3)}) = 2 \cdot I(\omega^{(4)})$ (see also Figure 4b). Analogous to the continuum approximation discussed above one has to assume additional broadening mechanisms of each line. Supposing a Lorentzian broadening function $f(\omega)$ with a uniform half width $\delta\omega_f$ for the four lines, the resulting line shape can be written as

$$I(\omega) = \text{const} \sum_{i=1}^{4} \alpha_i \left[1 + \left(\frac{\omega - \omega^{(i)}}{\delta\omega_f} \right)^2 \right]^{-1} \tag{3.11}$$

with $\alpha_i = 1$ for $i = 1, 2$ and $\alpha_i = 1/2$ for $i = 3, 4$. The corresponding fit of the data by means of Eq. (3.11) is shown in Figure 4b (solid line) using the best-fit parameters listed in Table 1 [33]. Comparing the quality of the fit procedure presented in Figure 4 one has to conclude that both models (continuous and discrete distribution of Pt atoms in the chain) are able to explain the observed line shape with practically the same degree of accuracy. Taking into account, however, the incommensurate CDW structure which spreads the frequencies of a discrete number of nuclei belonging to an infinite chain over the frequency range $(\omega_c - \omega_1, \omega_c + \omega_1)$ one has to favor the continuum model. Finally, as can be seen from Table 1, the magnitude of the two best-fit parameters, namely the CDW-induced splitting frequency $2\nu_1$ and the half width $\delta\nu_f$ of the broadening function, respectively, differs considerably for the two models. It should be noted here, that one did not observe a soliton-induced central peak in the split ^{195}Pt NMR line. Hence, one has to conclude that the concentration of solitons is fairly low ($\leq 20\%$).

Measurements for other orientations Θ of the magnetic field B_0 (see Figure 2) showed the following variation of the resonance frequency ω_c of the line's center of gravity:

$$\omega_c(\Theta) = \omega_0(1 - \sigma'(\Theta)) \tag{3.12a}$$

with

$$\sigma'(\Theta) = \sigma_i + \frac{1}{3}\Delta\sigma(3\cos^2\Theta - 1) \tag{3.12b}$$

as is expected for an axially symmetric shielding tensor with the symmetry axis parallel to the Pt chain ($\Theta = 0°$). Referring to the ^{195}Pt line of an aqueous solution of H_2PtCl_6 one obtains at a temperature of 50 K the following values for the shielding tensor [34]:

Isotropic part $\sigma_i = 0.465\%$

Total anisotropy $\Delta\sigma = \sigma_{\parallel} - \sigma_{\perp} = -0.212\%$.

Since the reference compound itself exhibits a chemical shift estimated to be -0.7% [35] with respect to the "naked " Pt nucleus, the σ values for KCP are actually negative. The negative value of the anisotropy $\Delta\sigma$ shows that the Pt nuclei are less shielded when the magnetic field is parallel to the Pt chains. The observed shift parameters in KCP are clearly in the range of tetracyanoplatinates chemical shifts as measured by Rupp [36]. What is even more surprising, they are nearly exactly equal to the chemical shift values found by Rupp on the unoxidized compound $K_2Pt(CN)_4$. In other words, the change in the Pt band structure due to oxidation by the bromines, which is supposed to create holes in a d subband, does not affect considerably the chemical shift. Otherwise one would expect an increase of the shift due to orbital paramagnetism.

Further one has to conclude from the data that the CDW-induced part of the shielding, $\sigma(x)$, is fairly small compared to σ_c. However, due to the better resolution at high fields the CDW-induced part can also be detected resulting in the line splitting $\Delta\nu$ shown in Figure 4.

As depicted in Figure 6a the chemical shift σ_c decreases slowly with rising temperature above about 50 K. As the shielding tensor was found to be axially symmetric, the observed temperature dependence can not be explained by a temperature dependent libration of the Pt chains around the c-axis. Probably the anisotropic part $\Delta\sigma$ increases slightly with increasing temperature caused by a thermal expansion of the planar $Pt(CN)_4$-complexes. The Knight shift of the resonance frequency which sets in at about 120 K will be discussed in the following section (3.2.2).

The line splitting, $\Delta\nu$, and the total line width, $\delta\nu$, at half height as obtained from the data like those in Figure 4a were investigated for different magnetic field orientations and strengths. The orientational dependence of both parameters at 50 K is plotted in Figure 6b. The total broadening increases with the angle Θ whereas the line splitting $\Delta\nu$ is approximately constant up to $\Theta = 40°$. Beyond $\Theta = 40°$ the splitting can no longer be resolved due to the broadening of the resonance line. The total line width $\delta\nu$ is caused by inhomogeneous broadening, as determined from spin-echo measurements to be discussed in the following section. The dipole-dipole interaction among the dilute ^{195}Pt nuclei which can be calculated by means of van Vleck's second-moment relations [27], contributes to the linewidth $\delta\nu$ an amount of 986 Hz for $\Theta = 0°$ and 243 Hz for $\Theta = 45°$,respectively. Also heteronuclear dipole interaction which is of the order of 1 kHz, cannot explain the observed large width. The width $\delta\nu$ (for $\Theta = 45°$ at T=50 K) is found to be directly proportional to the field strength with $\delta\nu = 8.9$ kHz at 6.3 T. Hence, one has to conclude that the total line width $\delta\nu$ is mainly caused by inhomogeneous broadening effects due to a distribution of chemical shifts along the chain. This is supported by the results of spin-echo experiments. As discussed in the next chapter the spin echo decays with a time constant T_2, which is much longer than the free induction decay time $T_2^* = 1/\delta\nu$.

Such a distribution of chemical shifts could arise from lattice imperfections and/or from the incommensurate CDW. The effect of the statistical distribution of defect water and bromine is expected to give a small contribution to $\delta\nu$ since the more pronounced effect of the nonequivalence of the Pt sites shows up only in a small splitting $\Delta\nu$ of the line width.

3.2.2. *Knight shift and relaxation above 120 K* As discussed in the previous section (3.2.1) the ^{195}Pt resonance frequency is only weakly temperature dependent below 120 K (see Figure 6a). However, above this temperature the frequency exhibits a pronounced temperature dependence [37,38]. A summary of measurements carried out between 1.2 and 330 K in various samples using again H_2PtCl_6 as the reference substance is shown in Figure 7 taken from Ref. [38]. (Note that these data were taken at constant *frequency*.) The shifts measured below 120 K [37,38,39] agree with the more precise data discussed above in section 3.2.1..

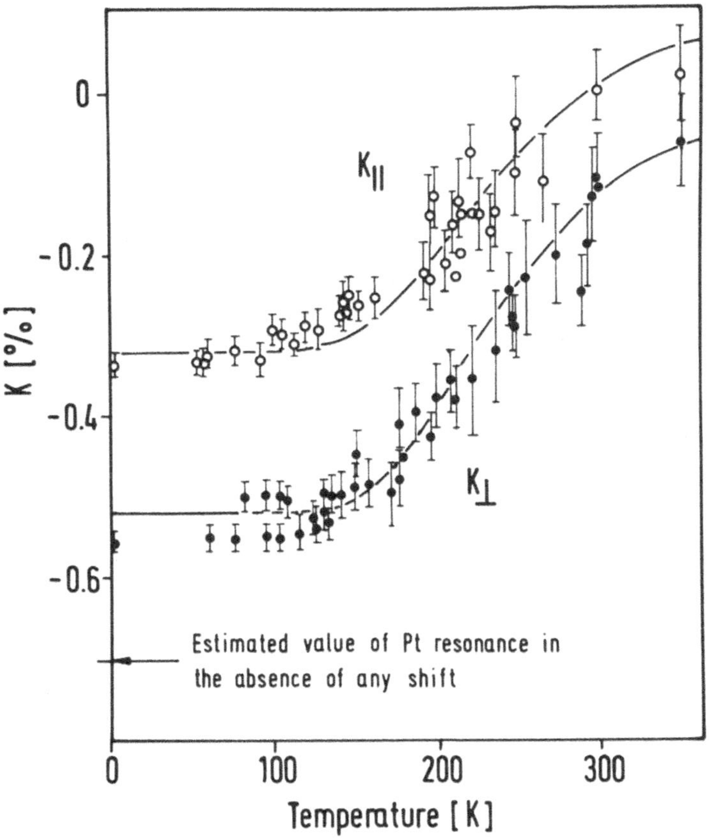

Fig. 7 ^{195}Pt Knight shift of 9 different samples versus temperature with H_2PtCl_6 as reference. The solid curves represent the Pauli susceptibility calculated by Lee *et al.* [40] normalized to a mean-field Peierls temperature of 625 K [38].

Above 120 K the Pt shift increases monotonously towards positive values. However, at room temperature where the values agree with those of Rupp [36], the shift not yet levels off; the measurements were limited by the decomposition of the samples. The anisotropy $K_{\parallel} - K_{\perp}$ seems to decrease slightly with temperature. Parallel to the increase of the shift the Pt line broadens excessively due to life time shortening.

The increase of the frequency shift above 120 K where three-dimensional ordering disappears [11] is interpreted as the onset of a Knight shift due to excitations across the Peierls gap. The variation of the Knight shift (with respect to the shift at 0 K, K(0)) due to a Peierls transition of the d_{z^2} electrons may be written as

$$K(T) - K(0) = (H_{hf})_{VV}\chi_{VV} + (H_{hf})_d\chi_d \ . \tag{3.13}$$

Here, $(H_{hf})_{VV}$ and $(H_{hf})_d$ are the van Vleck and d-hyperfine fields, respectively, and χ_{VV} and χ_d are the van Vleck and the d-component of the static susceptibility, respectively. Since the first contribution is negligible due to the small value of $(H_{hf})_{VV}$ the temperature variation of the Knight shift is then proportional to the static susceptibility or to the actual density of electron states at the Fermi level, $n(E_F)$. The d-hyperfine fields are usually negative. However, the strong covalency of the $Pt(CN)_4^{--}$ complex perturbs the d-orbitals and some s-character of the d-functions is sufficient to dominate the hyperfine field and to make the Knight shift positive. Obviously $n(E_F)$ increases steadily up to 330 K without saturation, thus the metallic state is not yet completely established at this temperature.

The strong variation of the Knight shift with temperature is accompanied by a similar behavior of the spin-lattice relaxation time T_1 [38] as shown in Figure 8 (together with linewidth data). Up to 150 K the relaxation is non-exponential presumably due to a distribution of T_1 values; thus the data points represent a "mean" relaxation rate. Above 150 K the poor signal-to-noise ratio prevented any check of the exponential character of the relaxation.

Since the rate is very large for a spin 1/2 nucleus it must arise from interactions with conduction electrons. For the contact interaction of s-electrons the Korringa relation connects relaxation and Knight shift:

$$(1/T_1)_{Korr} = c\,[K(T) - K(0)]^2\, T \ . \tag{3.14}$$

If d-electrons are present they contribute a similar term related to the d-electron Knight shift. However, the Korringa rate as deduced from the experimental Knight shift data is much smaller than the observed relaxation rate; at room temperature $(1/T_1)_{Korr}$ is about 60 times smaller (see Figure 8). Since an orbital relaxation process is ruled out in the absence of any orbital shift, the constant c in Eq.(3.14) could be enhanced by the presence of $s-d$ mixing. However, Figure 8 reveals that the enhanced curve $60(1/T_1)_{Korr}$ adjusted to the room temperature experimental value strongly deviates from the experimental values at temperatures below 300 K. It had therefore been suggested [38] that the fast relaxation arises from critical fluctuations of $n(E_F)$ which one would expect to be present in 1D systems. These

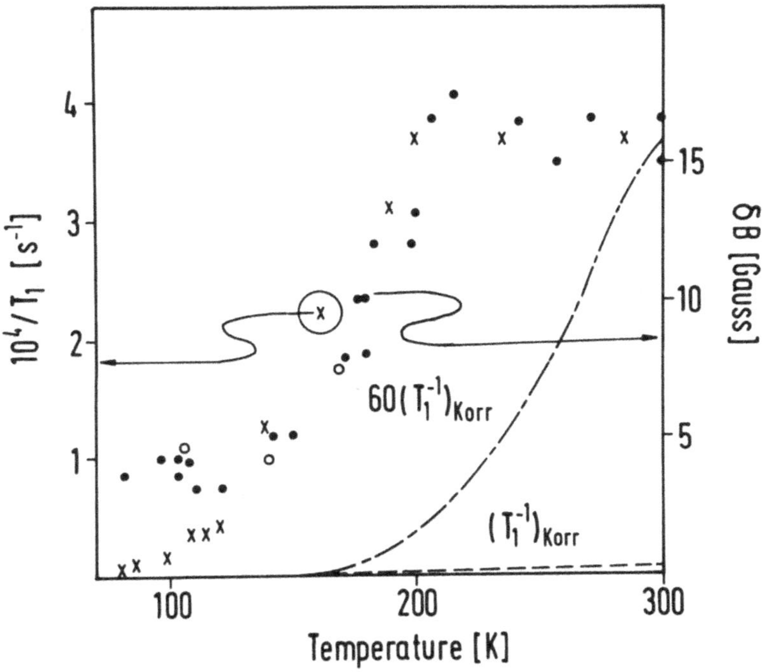

Fig. 8 Temperature dependence of the line width $\delta B_{\|}$ (open circles) and $\delta B_{\perp}$ (full circles) and the mean [see Eq. (3.17)] spin-lattice relaxation rate $T_{1\|}^{-1}$ (crosses) of ^{195}Pt. The dotted curves represent the Korringa relaxation rate deduced from the experimental shift, and an enhanced Korringa rate where the constant factor has been adjusted to the experimental room temperature value [38].

fluctuations would shorten the electron correlation time and would cause a very efficient relaxation process.

The effect of fluctuations on the Peierls transition in a 1D system has been calculated by Lee, Rice and Anderson [40] taking a functional average over variations of the order parameter. The authors found that the transition is suppressed to a temperature T_{3D} of approximately one quarter of the mean-field Peierls transition temperature T_c but remains fairly sharp. In addition, Lee *et al.* have calculated the Pauli spin susceptibility χ normalized to the susceptibility in the metallic state; their result is plotted in [40] as a function of T/T_c. Thus, χ can be fitted to the experimental Knight shift data of Figure 7 as this shift is proportional to χ by using only two adjustable parameters: a value for T_c and the scale of the vertical axis. The solid curves in Figure 7 represent a fit assuming $T_c = 625$ K and arbitrary but different scaling for $K_{\parallel}$ and $K_{\perp}$ which implies a small anisotropy of the d contribution [38]. In view of the large scatter of the experimental data the fit appears satisfactory. It is thus concluded that the pronounced temperature dependence of both the spin-lattice relaxation time and the Knight shift at temperatures above 120 K provide evidence for the presence of critical fluctuations of the electronic density of states. In the temperature range 30 to 120 K, the relaxation is governed by the non-linear excitations of the CDW as will be shown in 3.2.3..

The influence of hydrostatic pressures up to 20 kbar on K and T_1 has been investigated by Müller *et al.* [41]. Because of a poor signal-to-noise ratio the error bars are much larger than those of the normal pressure data shown in Figure 7; measurements were performed with the c-axis parallel to the magnetic field. The K data at 8 kbar are nearly indistinguishable from the 1 bar data. At 20 kbar, however, the variation of the shift with temperature seems to become larger between 120 and 250 K; at 300 K the data agree with the 1 bar data. Since the Knight shift is proportional to the density of electron states at the Fermi level the larger temperature variation of the shift reflects a reduction of the metal-insulator transition by the application of pressure, *i.e.* an enhancement of the 3D coupling due to pressure.

As to the influence of pressure on T_1, there is a different response at temperatures below and above about 100 K. At 77 K, application of 20 kbar increases the relaxation rate by nearly one order of magnitude. If the relaxation is induced by free carriers as Müller *et al.* [41] propose, the increase of the rate will be due to a reduction of the low temperature band gap. If the relaxation in this region is governed by non-linear excitations of the CDW to be discussed in the following section, it remains to be shown how this mechanism is affected by pressure. Above 120 K, pressure *reduces* the relaxation rate; at about 220 K the reduction factor is about 30. In other words, the enhancement of the experimental relaxation rate with respect to the Korringa rate as discussed above is nearly cancelled by the application of pressure. This again seems to be a manifestation of increased 3D character.

3.2.3. *Relaxation beween 30 and 120 K* As shown in Figure 9a the time evolution of the ^{195}Pt nuclear magnetization in KCP following a $\frac{\pi}{2} - \tau - \frac{\pi}{2}$ (T_1-) pulse sequence turns out to be non-exponential. On the other hand, the magnetization recovers **homogeneously** indicating that the fluctuations which induce the nuclear transitions are not caused by phase fluctuations but by dynamics of solitons. As discussed by Mali *et al.* [34] and by Masin *et al.* [42] the non-exponential behavior of the magnetization recovery arises from a distribution of spin-lattice relaxation times due to the spatial distribution of solitons along the chain axis. Using Eq.(3.2) the "local relaxation rate" $1/T_1(x)$ can be written as

$$1/T_1(x) = \Delta\omega^2(x) \cdot J^{(1)}(\omega_0). \tag{3.15}$$

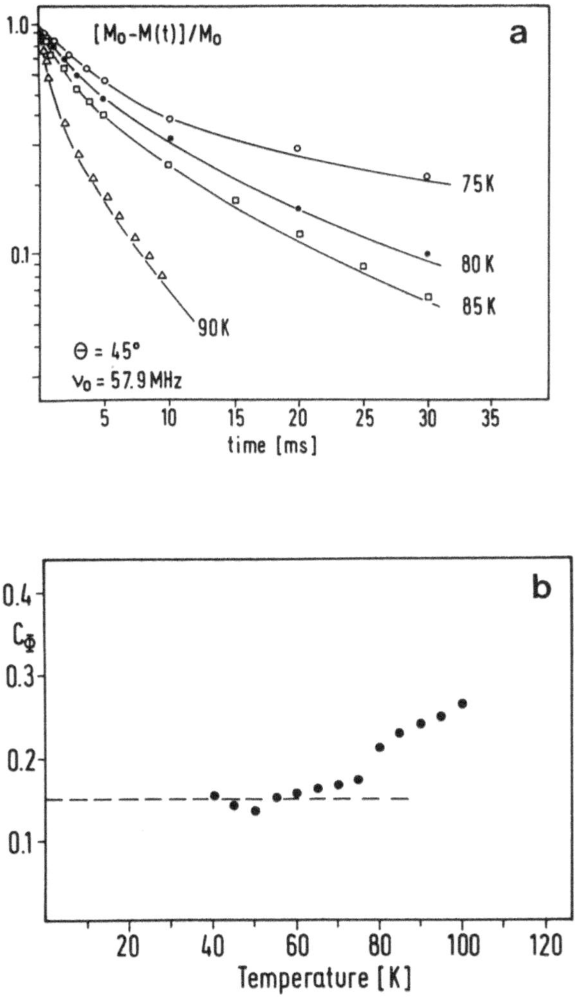

Fig. 9 a) Non-exponential spin-lattice relaxation of ^{195}Pt at 58.3 MHz and $\Theta = 45°$ for different temperatures. The solid lines correspond to Eq. (3.17) based on a "soliton lattice" model [33].
b) Soliton concentration c_Φ vs. temperature obtained from a tentative analysis of magnetization decay curves like those shown in Figure 9a [33].

Assuming that the coupling strength $\Delta\omega^2$ is given by the interaction of a soliton charge with the Pt nuclei (see Eq. (2.10, 2.11)), $1/T_1(x)$ can be expressed as

$$1/T_1(x) = 1/T_{1s} \cdot \mathrm{sech}^2(x/d) \tag{3.16}$$

where $1/T_{1s}$ denotes the strength of the relaxation at the center of the soliton. Since the total magnetization is an integral over local magnetizations, a non-exponential relaxation function results:

$$M(t) = \frac{1}{a}\int_0^a dx \exp\left[-\frac{t}{T_{1s}}\left(\mathrm{sech}^2\left(\frac{x}{d}\right) + \mathrm{sech}^2\left(\frac{x-a}{d}\right)\right)\right] \tag{3.17}$$

where a is the mean distance between two solitons (see Figure 3). It has to be noted that relation (3.17) is valid only for fixed or slowly moving solitons, *i.e.* the motional correlation time $\tau = b/v$ of the soliton (Eq. (2.15)) has to be large compared to the inverse of the Larmor frequency, ω_0 ($\omega_0\tau \gg 1$). For $\omega_0\tau \gg 1$ the magnetization decays exponentially independent of the coupling mechanism. The solid lines in Figure 9a are fits of the data by means of Eq. (3.17). Tentatively one can extract a soliton concentration $c_\Phi = 1/a$ for different temperatures (see Figure 9b). The soliton density seems to increase above 80 K and may eventually lead to a "melting of the soliton structure" above 100 K. Neutron scattering shows a loss in the transverse correlation at this temperature [43] which may be caused possibly by the same effect.

As shown by Nixdorf [33] and exhibited in Figure 10a for T=80 K, the nuclear magnetization decays slightly faster with increasing Larmor frequency ω_0 between 30 K and 100 K, *i.e.* the spin-lattice relaxation rate increases with rising ω_0. The findings can be explained by the two assumptions:

(i) As suggested by the analysis of the ^{195}Pt spectrum discussed in the last section the soliton-nucleus coupling $\Delta\omega$ in Eq. (3.15) is caused by a shielding interaction, *i.e.* $\Delta\omega \propto \omega_0$.

(ii) The solitons perform a slow ($\omega_0\tau \gg 1$) continuous diffusional motion along the Pt chains. For such a 1D-motion as shown by Kimmich and Voigt [29], the frequency dependence of the spectral density function $J(\omega_0)$ is given by $J(\omega_0) \propto \omega_0^{-3/2} \cdot \tau^{-1/2}$ where τ denotes the correlation time of the soliton motion. Then, the relaxation rate $1/T_{1s}$ in Eq. (3.17) is found to be proportional to $\omega_0^{1/2}$ which is in rough agreement with the experimental data presented in Figure 10a.

Empirically, the non-exponential magnetization decay can be described in a good approximation by a stretched exponential function

$$M(t) = M_0 \exp\left[-\left(t/T_1\right)^{\eta}\right] \tag{3.18}$$

where the exponent η is about 0.6 and varies slightly with the direction of the magnetic field relative to the crystallographic c-axis as shown in Figure 10b. For

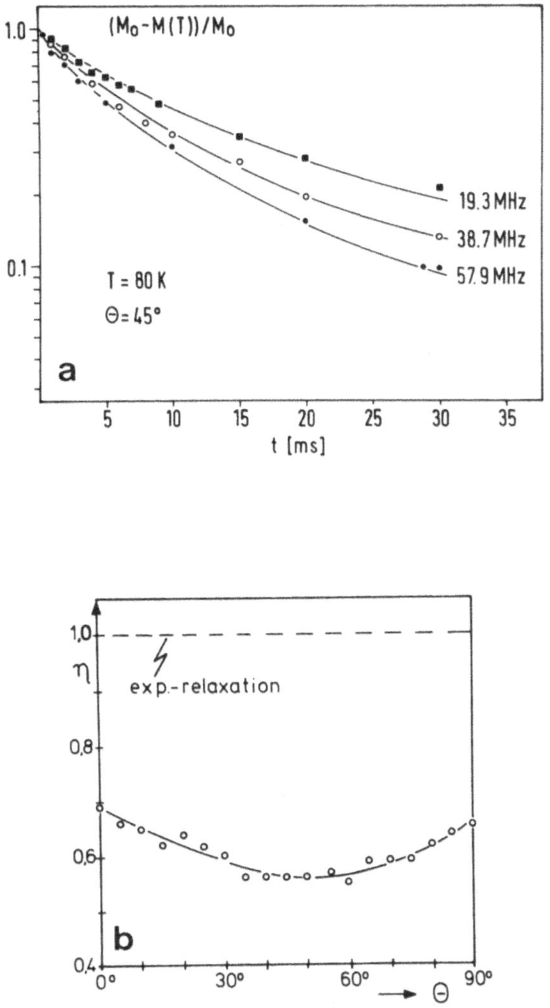

Fig. 10 a) Frequency dependence of the non-exponential ^{195}Pt magnetization decay at T=90 K for $\Theta = 45°$. Note that the corresponding relaxation rates increase slightly with rising frequency [see also Eq. (3.17)].
b) Orientational dependence of the exponent η of the "stretched" exponential function $\exp(-t/T_1)^{\eta}$ describing empirically the observed non-exponential decay (T=53 K) [34].

data evaluation a mean spin-lattice relaxation time T_1 is defined according to Eq. (3.18) as the time where $M(T_1) = M_0 \cdot e^{-1}$.

The temperature dependence of T_1 such defined is shown in Figure 11 for three different orientations of the Pt chains with respect to the magnetic field. Three different regions can be distinguished:

(i) Low-temperature region $T < 30$ K, where the relaxation time T_1 is longer than 10^2 s and shows a weak temperature dependence. The relaxation in this regime is most likely due to paramagnetic impurities and/or by interruptions in the Pt strands.

(ii) Medium temperature region $30 \leq T \leq 100$ K, which is of interest here and where the relaxation times range from $2 \cdot 10^{-3}$ s up to 10^2 s. The relaxation in this region is governed by the non-linear excitations (solitons) of CDW's.

(iii) High-temperature region $T > 100$ K with relaxation times $T_1 < 2 \cdot 10^{-3}$ s and only slight orientational dependence. Excitation across the Peierls gap leads to electrons in the conduction band which produce a Knight shift and electronic relaxation above 100 K as discussed in section 3.2.2..

In the following we shall focus on the soliton-region (ii).

Figure 12 displays the observed strong orientational dependence of the spin-lattice relaxation rate $1/T_1$ at 53K which can be fairly well accounted for by a $\sin^2 2\Theta$ dependence. Such an orientational dependence of $1/T_1$ is expected from an axially symmetric shielding tensor $\tilde{\sigma}$ with one principal axis along the c-direction caused by the charge distribution of a soliton. Labelling the corresponding shift frequencies as $\omega_{\parallel} = \omega_0 \cdot \sigma_{\parallel}$ and $\omega_{\perp} = \omega_0 \cdot \sigma_{\perp}$ the orientation dependence of the coupling constant $\Delta\omega^2$ in Eq. (3.15) can be written as [44]:

$$\Delta\omega^2 = (\sigma_{\parallel} - \sigma_{\perp})^2 \cdot \omega_0^2 \cdot \sin^2 2\Theta \, . \tag{3.19}$$

It is evident from Figure 12 and Eq. (3.19) that for $\Theta = 45°$ the relaxation time T_1 is most strongly influenced by the shielding coupling of the solitons whereas for $\Theta = 0°$ ($\vec{c} \| \vec{B}_0$) and $\Theta = 90°$ ($\vec{c} \perp \vec{B}_0$) other relaxation mechanisms may play a dominant role. The discussion of the soliton-induced relaxation mechanism presented in the next section is therefore based on the T_1-data for $\Theta = 45°$.

The temperature dependence of T_1 for $\Theta = 0°$ and $\Theta = 90°$ as displayed in Figure 11 shows evidence for an additional relaxation mechanism setting in above about 60 K and causing a bent to shorter T_1 values in the plot of $\log T_1$ vs $1/T$ in Figure 11. This relaxation mechanism may be explained by an activation across the Peierls gap with $\Delta E/2k \approx 1200$ K corresponding to $\Delta E \approx 0.2$ eV.

The ideas of a soliton-induced relaxation mechanism are supported by the temperature and the orientational dependence of the spin-spin relaxation time T_2 measured by a $\pi/2 - \tau - \pi$ spin echo pulse sequence. The observed spin echo decays exponentially with the relaxation time T_2, which is much longer than the free-induction decay time, *i.e.* the line width is essentially due to inhomogeneous broadening (see also section 3.2.1). On the other hand, T_2 is much shorter than expected from

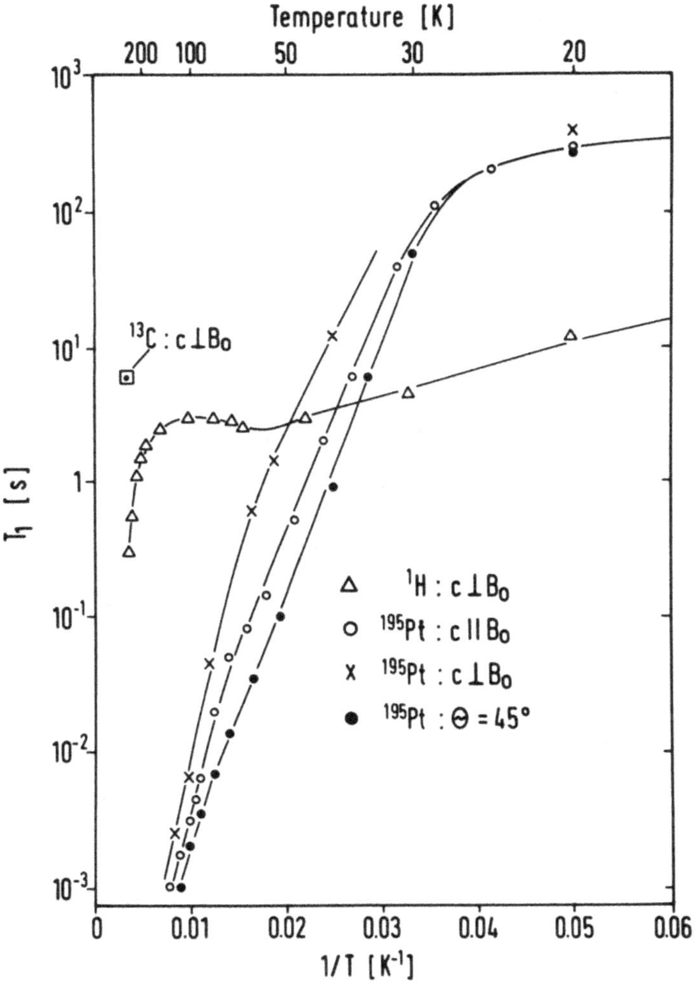

Fig. 11 Spin-lattice relaxation time T_1 vs. inverse temperature of ^{195}Pt for different orientations of the platinum chains with respect to the magnetic field B_0. T_1 values of the water protons ^{1}H and one value for natural abundant ^{13}C are included for comparison [34].

dipolar interactions alone and exhibits a pronounced temperature and orientational dependence as shown in Figure 13. With decreasing temperature T_2 increases con-

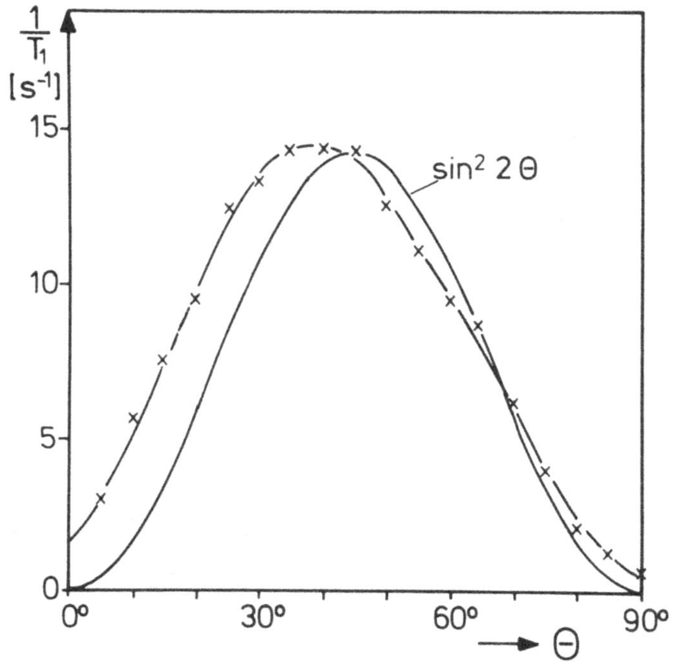

Fig. 12 Orientational dependence of the spin-lattice relaxation rate $1/T_1$ of ^{195}Pt at T=53 K, where Θ is the angle between the chain axis c and the magnetic field B_0 [34].

tinuously reaching a finite value of about 100-150 μsec at 4 K thus exhibiting the presence of appreciable fluctuations also at lower temperatures. The orientational dependence of $1/T_2$ parallels the one of the line width $\delta\nu$ (see Figure 6b) in the respect that both reach their maximum value for $\Theta = 90°$ thus pointing to a common mechanism responsible for both parameters.

In order to be sure that the ^{195}Pt relaxation in region (ii) is predominantly caused by the Pt chain dynamics and not by fluctuations of off-chain atoms or by statistically distributed paramagnetic centers one has to look also at the relaxation of the protons in KCP (see also section 3.5.). Results of such measurements are also presented in Figure 11. By comparing the relaxation behavior of ^{195}Pt and ^1H a common relaxation mechanism can be excluded. Obviously, the ^{195}Pt relaxation mechanism is highly restricted to the Pt chains. It should be noted that the sharp decrease of the proton relaxation time around 200 K has to be attributed to localized motions of water molecules (see section 3.5.).

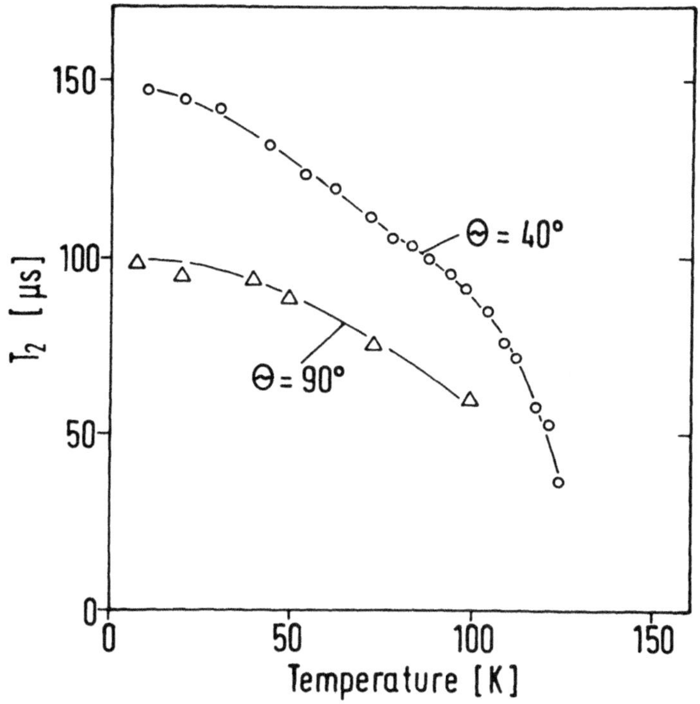

Fig. 13 Temperature dependence of the spin echo decay time T_2 for two different orientations Θ of the Pt chain with respect to the magnetic field B_0 [34]. Note that the relaxation times $T_2^* = 1/\delta\nu$ [see Figure 6b] are remarkably smaller than T_2 ($T = 50$ K): $T_2^*(\Theta = 90°) = 60\mu s$; $T_2^*(\Theta = 40°) = 120\mu s$.

Summarizing the experimental findings in the temperature region (ii) one has to conclude that fluctuating local fields at the ^{195}Pt nuclei in the metal chains of KCP are caused by non-linear excitations of the CDW's. Other possible interactions like fluctuating homo- and heteronuclear dipolar interactions can be excluded.

Non-linear excitations of the CDW, *i.e.* solitons can be shown to produce strong temperature dependent relaxation rates as observed experimentally. Within the model for solitary excitations presented above the following expressions for the mean relaxation rate $1/T_1$ and for the line shift $\omega - \omega_0$ result for $\omega \cdot \tau \gg 1$:

$$\frac{1}{T_1} \simeq e^{-E_{\Phi}/kT} \cdot (\sigma_{\|} - \sigma_{\perp}) \cdot \omega^{1/2} \cdot \sin^2 2\Theta \cdot e^{-E_p/2kT} \tag{3.20}$$

$$\omega - \omega_0 \simeq e^{-E_{\Phi}/kT} \cdot \left[\cos^2\Theta \left(\sigma_{\|} - \sigma_{\perp}\right) + \sigma_{\perp}\right] \cdot \omega_0 = \sigma \cdot \omega_0 \quad . \tag{3.21}$$

Here E_Φ denotes the formation energy of a soliton-antisoliton pair (see Eq. (2.13)) and τ is the correlation time of the soliton motion which is assumed to be thermally activated, *i.e.* $\tau = \tau_0 \exp(E_p/kT)$ with E_p being the migration energy of the soliton (see Eq. (2.15)).

Taking into account the finite shielding observed at low temperatures Eq. (3.21) can be rewritten as

$$\sigma(T) = \sigma_0 - \sigma = \sigma_0 - e^{-E_\Phi/kT}\left[\cos^2\Theta \cdot (\sigma_\parallel - \sigma_\perp) + \sigma_\perp\right] \tag{3.22}$$

A fit of T_1 and σ according to Eqs. (3.20, 3.22) to the experimental data presented in Figures 6a and 11 is shown in Figure 14. The fit is consistent with the following parameters: $E_\Phi/k = 295$ K (25 meV), $E_p/k = 240$ K (21 meV), $\tau_0 = 2 \cdot 10^{-10}$ s, and $\sigma_0 = 0.353\%$. The interaction strength $\sigma_\perp$ was estimated to be about 0.8% while the orientation dependence of T_1 leads to $\sigma_\parallel \simeq 0.6\sigma_\perp$.

The formation energy $E_\Phi/k = 295$ K obtained from the analysis is in the range of what has been estimated from soliton particle theory ($E_\Phi/k \sim 200$ K; [23]) and is considerably less than the Peierls gap of 2300 K. Furthermore, it should be noted that one cannot unambiguously distinguish from nuclear spin relaxation experiments if the soliton is migrating with a correlation time τ determined by the "migration energy" $E_p/k = 240$ K or if the solitary excitation is localized and relaxing with the correlation time τ which itself is thermally activated with energy E_p. The temperature dependence of the NMR relaxation and line shift of the ^{195}Pt nuclei in the linear chain of KCP, however, is consistent with the model of solitary excitations of the CDW.

3.3. LATTICE FLUCTUATIONS NEAR THE 3D ORDERING TRANSITION BY ^{14}N NMR

The observed absence of truly established long range order above the 3D ordering temperature of about 120 K may be attributed to low-frequency critical Peierls fluctuations. As pointed out by Soda *et al.* [12] the study of the quadrupolar relaxation of ^{14}N nuclei belonging to the $(CN)_4$ square planes (see Figure 2) is a good way to observe these fluctuations. No itinerant electron density exists on nitrogen atoms, ^{14}N spins should relax mainly via fluctuations of the EFG at the ^{14}N sites due to lattice vibrations. Since the $(CN)_4$ ligands are rigidly coupled to the Pt of the conducting chains the measurement of ^{14}N spin-lattice relaxation provides a very direct probe for the existence of critical lattice fluctuations [45]. It should be noted here, that the influence of critical Peierls fluctuations on the quadrupolar relaxation has been reported also in an other Q1D conductor, namely $^{87}Rb_{0.3}$ MoO_3 showing a Peierls transition at 181 K [46].

Figure 15 shows the temperature dependence of the ^{14}N spin-lattice relaxation rate in KCP as measured by Soda *et al.* [12]. The relaxation rate minimum occurring around 220 K indicates a change of the mechanism of the nuclear quadrupole relaxation. The observed increase of the relaxation rate above 220 K is due to

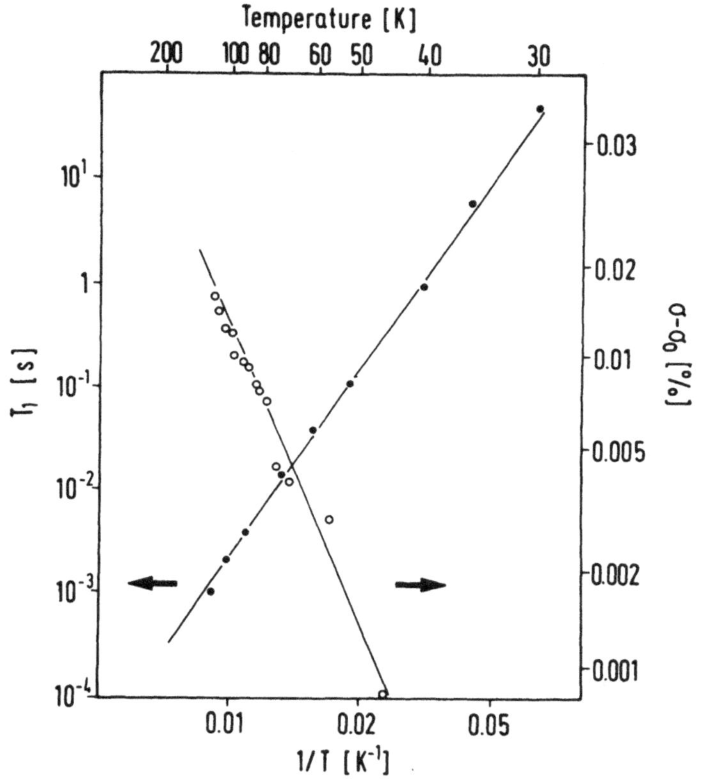

Fig. 14 Arrhenius representation of the soliton induced spin-lattice relaxation time T_1 and the line shift $\sigma - \sigma_0$ of ^{195}Pt. The slope of the curves yields the formation and migration energy of the solitons [see Eq. ((3.20),(3.21))].

phonon-induced quadrupole fluctuations. Since the temperature domain is higher than the Debye temperature of KCP the second order Raman process leading to $1/T_1 \propto T^2$ [27] should dominate. The observed data are in qualitative agreement with the existence of such a Raman process.

Below 220 K the $1/T_1$ temperature dependence cannot be explained by ordinary phonon fluctuations in a regular lattice. The divergence of $1/T_1$ observed when approaching the Peierls transition has to be ascribed to a critical slowing down of the lattice fluctuations [45]. The log-log presentation of the $1/T_1$ data versus $T - T_c$

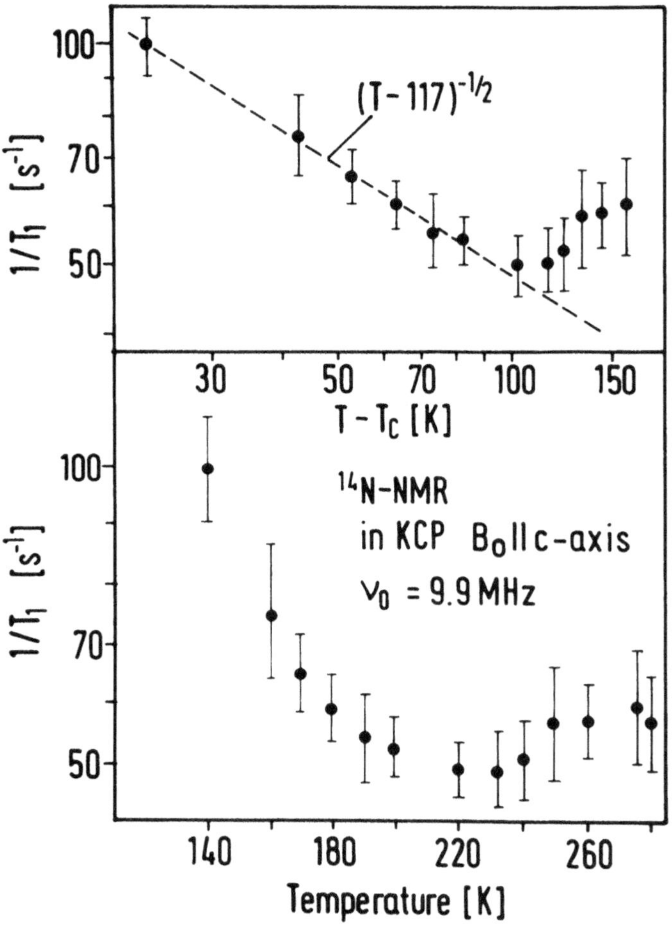

Fig. 15 Temperature dependence of the ^{14}N spin-lattice relaxation rate $1/T_1$ (bottom). Log-log plot of $1/T_1$ vs. $T - T_c$ with $T_c = 117$ K (top). The dashed line is a power law with a mean-field exponent $\beta = 1/2$ [12].

in Figure 15 clearly supports a power law of the form $1/T_1 \propto [(T - T_c)/T_c]^{-\beta}$ with the fit parameters $\beta = 1/2 \pm 0.05$ and $T_c = 117$ K. It is well established from several experiments that the temperature of the Peierls "ordering" in KCP occurs around

110-120 K, showing the possible relevance of the 3D "ordering" on the enhancement of $1/T_1$. The observed exponent $\beta = 1/2$ is mean-field like and is well known to result from strongly damped 3D fluctuations which effect the spin relaxation via a direct process.

Summarizing, while ^{195}Pt NMR probes mainly electronic effects the ^{14}N relaxation in KCP observes lattice fluctuations. In particular, the square root divergence of $1/T_1$ of ^{14}N which occurs below 220 K can be fairly well understood in terms of a critical slowing down of lattice fluctuations which couple to the spins by a direct spin-phonon mechanism.

3.4. ^{13}C CHEMICAL SHIFT AND RELAXATION

Another probe to study the electronic behavior of the $(CN)_4$ ligand of the Pt chains is ^{13}C NMR. As we will see the results are partly similar to the findings of the ^{14}N studies. We will discuss chemical shift and relaxation time investigations. The technique of measuring carbon chemical shift tensors in solids allowed to determine the ^{13}C chemical shift and to furnish information on the anisotropy in the electronic structure of the CN group [47]. Single crystals grown from cyanide 60% enriched in ^{13}C were used. In addition to the chemical shift the carbon spectra exhibited the effect of dipolar interactions with other moments especially with ^{14}N and ^{195}Pt. Since the ^{14}N quadrupole and Zeeman interactions are of similar size in KCP large effects of the ^{14}N quadrupole interaction were observed showing up as asymmetries in the ^{13}C spectra.

The principal values of the chemical shift tensor relative to the carboxyl carbon in CH_3COOH are: $\sigma_{11} = 261 \pm 10$ ppm (parallel to C-N bond), $\sigma_{22} = -48 \pm 10$ ppm (perpendicular to C-N bond and Pt chain) and $\sigma_{33} = -10 \pm 5$ ppm (parallel to Pt chain). This yields an isotropic average of $\bar{\sigma} = 68$ ppm and an anisotropy $\Delta\sigma = \sigma_{11} - (1/2)(\sigma_{22} + \sigma_{33})$ of 290 ppm both values being in agreement with HCN values. Thus, the effects of the cyanide bonding to Pt and effects due to the solid state structure affect the ^{13}C chemical shift tensor by producing an asymmetry of $(\sigma_{33} - \sigma_{22})/(\sigma_{11} - \bar{\sigma}) = 0.2$. On the other side, the ^{13}C nuclei although being closest to the Pt chains do not seem to "feel" the CDW. One thus concludes that the CDW is restricted to the Pt chains.

Brenni *et al.* [48] found no dependence of the ^{13}C chemical shift on temperature between 25 K and room temperature. Some of the lines broaden at lower temperature probably due to the freezing of water rotations (see section 3.5.). The authors have studied the ^{13}C spin-lattice relaxation in the same temperature range. Similar to Pt the relaxation is non-exponential and therefore T_1 was defined as the time at which the signal has decayed to $1/e$ of its initial value. The results are shown in Figure 16. No pronounced frequency dependence could be detected. However, an unusual thermal hysteresis denoted by the arrows in Figure 16 has been discovered between 140 and 70 K. The "bump" in the rate is only observed on *lowering* the temperature after waiting at room temperature for several hours.

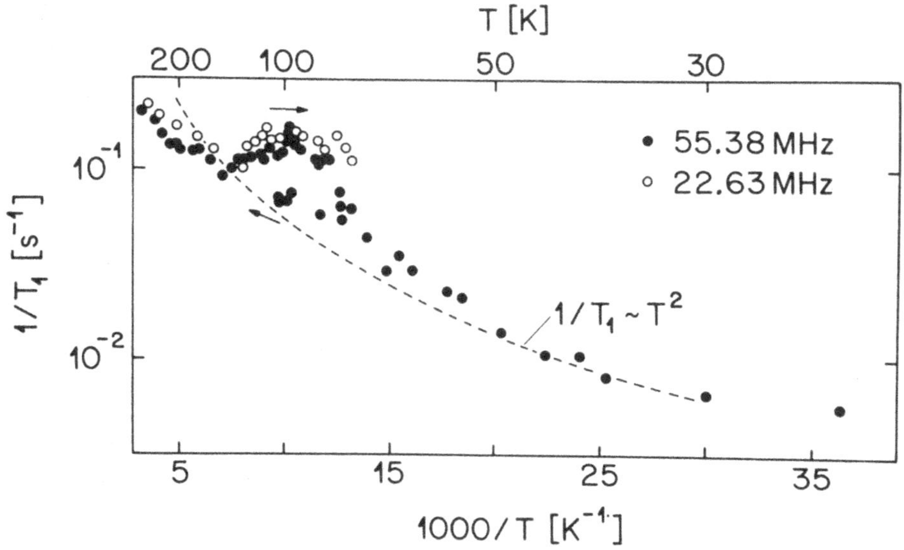

Fig. 16 Arrhenius representation of the ^{13}C spin-lattice relaxation rate $1/T_1$. The slope of the curve above 30 K yields E = 12.5 meV. Note the hysteresis behavior of the data around 120 K (3D ordering temperature). An alternative fit using a T^2 law is also indicated.

Relaxation below 30 K arises from interactions with paramagnetic impurities. Taking into account the different gyromagnetic ratios the data agree with those for Pt. However, above 30 K the ^{13}C relaxation rate increases much slower than the Pt rate does, the apparent activation energy is only 0.0125 eV. If both ^{195}Pt and ^{13}C nuclei would relax via a coupling to the CDW one would assume contact interaction for Pt and dipolar interaction for C, but the C relaxation disagrees with such an assumption. In other words: the ^{13}C relaxation supports the conclusion that the CDW is restricted to the Pt chains.

Thus, the major contribution to the ^{13}C relaxation must arise from other mechanisms. In view of the similar behavior of the ^{14}N relaxation and the absence of a pronounced frequency dependence of the ^{13}C rate, it is tempting to attribute the rate's increase above 50 K to a two-phonon (Raman) process which induces fluctuations of the ^{13}C anisotropic chemical shift tensor. As indicated by the dotted line in Figure 16, a T^2 law is in rough agreement with the data. As in the case of

^{14}N relaxation, the ^{13}C relaxation rate increases when approaching the 3D ordering temperature T_{3D} (about 120 K) from the high-temperature region; this increase is ascribed to a critical slowing down of lattice fluctuations.

In conclusion, both "nearby off-chain" nuclei ^{13}C and ^{14}N probe only the phonon fluctuations in the KCP lattice; "on-chain" electronic properties can be observed only via the ^{195}Pt nuclei.

3.5. SITES AND MOTION OF WATER MOLECULES BY ^{1}H NMR

We now turn to NMR studies of more distant "off-chain" nuclei namely the ^{1}H nuclei in the water molecules and the Br ions. The NMR of these nuclei can provide detailed information on the structure (and relations to *e.g.* neutron studies) but also give additional clues to the understanding of the overall behavior of KCP. We will start with proton NMR investigations.

KCP contains two types of water molecules: stoichiometric and defect molecules. According to X-ray and neutron diffraction studies the stoichiometric molecules occupy two inequivalent sites: the $H_2O(1)$ molecules with the oxygen positional parameters (0, 0.5, 0.4044) and the $H_2O(2)$ molecules with oxygen parameters (0.3390, 0.3390, 0.0696). The defect water molecules $H_2O(3)$ are assigned to an additional atomic site (1/2, 1/2, ≈0.34). Originally, this site had been tentatively identified as a second Br-site by Williams *et al.* [49,50] employing neutron diffraction techniques. Subsequently, Peters and Eagen [51], Heger *et al.* [52,53] and Deiseroth and Schulz [54] using X-ray and neutron investigations found this site to be occupied by a third type of water molecule in such cases when the nearby Br site (1/2, 1/2, ≈0.5) is empty.

The question how the water molecules are ordered has been addressed via c.w. NMR by measuring the orientational dependence of the splitting in the well-known Pake doublet. The intra-molecule dipole-dipole interaction of the protons of one water molecule splits the NMR line into two lines with a separation given by

$$\Delta B = (3\mu/r^3)(3\cos^2\Phi - 1) \tag{3.23}$$

Here, μ is the proton magnetic moment, r is the length of the vector connecting the two H-atoms, and Φ is the angle between this vector and the external magnetic field.

The first studies of this kind performed at *room temperature* [55,56,19] revealed three types of Pake doublets. Figure 17 shows the dependence of ΔB for a rotation of a single crystal around the [110] axis [19]. The doublets denoted by a are assigned to the stoichiometric water molecules $H_2O(1)$ which are ordered with a H-H vector being parallel to the crystallographic a axis, as shown in Figure 18a. By c we denote doublets whose *maximum* splitting is only one quarter of the maximum splitting observed for regular and static molecules. The c doublets are produced by those $H_2O(2)$ molecules at site 2 which are dynamically disordered. The motion of these molecules can be described phenomenologically by two independent rotations of the H-H vector: one around the R_1 axis, the other around R_2 with both axes perpendicular to each other. R_1 is fixed with respect to the crystal being parallel to the c-axis, while R_2 is rotating around R_1. The result is an "averaged" H-H vector parallel to the c-axis (see Figure 18a). Physically these rotations probably correspond to tunneling of hydrogen atoms between equivalent potential wells. More detailed information requires knowledge of the second moment and its temperature dependence; because of the complex proton spectrum these measurements had not been performed.

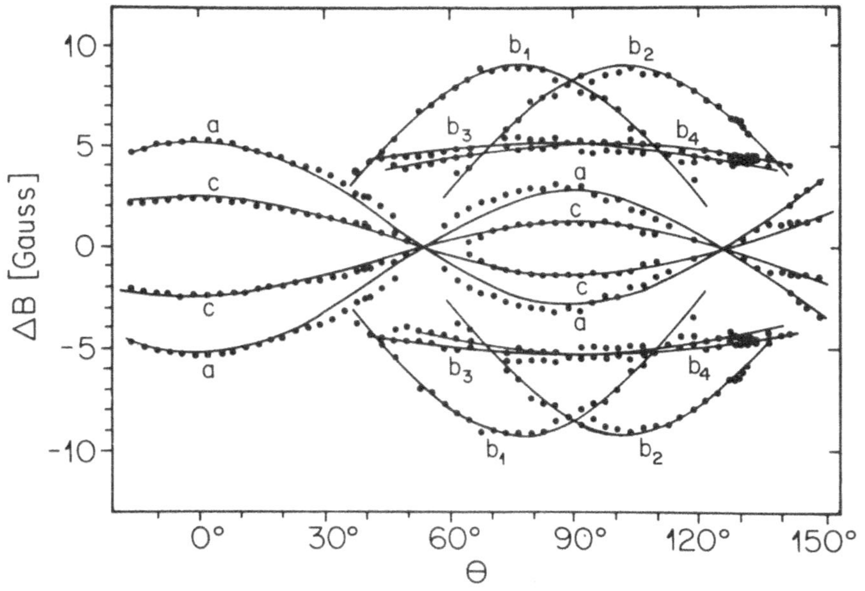

Fig. 17 Angular dependence of the splitting ΔB of the proton doublets. Solid lines whose labelling is explained in the text are least-square fits to the data using Eq. (3.23). [110] is perpendicular to the magnetic field B_0, Θ is the angle between B_0 and the c-axis [19].

The doublets b_1 through b_4 were originally attributed to defect water molecules since these signals taken at *room temperature* are about 5-10 times smaller than the c signals and thus accidentally matched the occupation probability of about 0.4 for the third water position [51,52]. However, using the results of temperature dependent measurements of Niki *et al.* [16] to be discussed below, we now assign

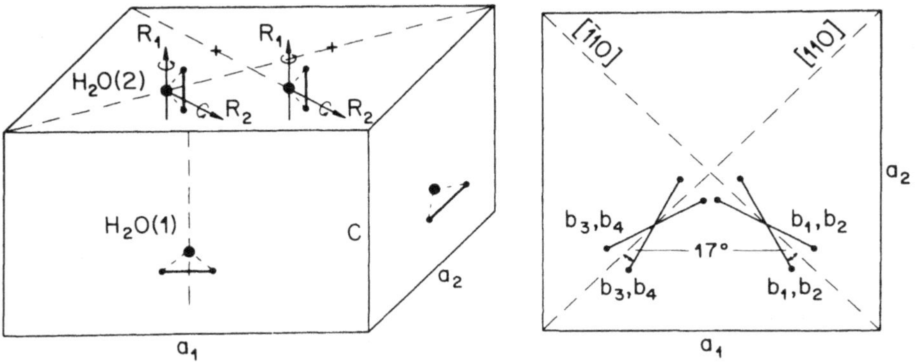

Fig. 18 Sites and direction of the H-H distance vector with respect to the crystal axes for the stoichiometric water molecules $H_2O(1)$ and $H_2O(2)$ [19].
a) Static $H_2O(1)$ and rotating $H_2O(2)$ molecules. The crosses mark further $H_2O(2)$ sites. Above about 250 K $H_2O(2)$ molecules perform a rotational motion about the axes R_1 and R_2.
b) Projection of the H-H vectors (solid lines with full circles) of *static* $H_2O(2)$ molecules onto the (001) plane. Each projection refers to two vectors inclined against (001) by $+12°$ and $-12°$, respectively.

the *b* signals to non-rotating $H_2O(2)$ molecules. These molecules exhibit a high degree of positional disorder which is most likely of static character resulting from a space distribution of their H-H vectors over at least 16 different orientations as indicated in Figure 18b [19]. Each projection refers to two H-H vectors inclined against the (001) plane by $\alpha = \pm 12°$. The projections themselves are not parallel to [110] and [$\bar{1}$10], respectively; instead they make an angle $\beta = 17°$ with the face diagonals. These "misalignments" have been carefully checked by measuring the Pake doublet with the external field oriented parallel to the respective H-H vector. Measurements at 163 K [57] showed a small change of the H-H vector orientations: α slightly increases to 14° while β decreases to 11°. These results support the conclusion that the positional disorder is not due to dynamic effects such as hindered internal motions of the molecules.

The room temperature positional disorder of the static $H_2O(2)$ molecules has been verified by Heger *et al.* [58] using neutron diffraction. Introducing the disorder into their analysis improved the R-values significantly. A value of 16° was obtained for β which is in excellent agreement with the NMR result.

The measurements of Niki *et al.* [16] between 50 and 300 K in a KCP single crystal aimed at the librational vibrations of the water molecules. In terms of the orientation of the H-H vectors Niki *et al.* distinguished three sites of water molecules without relating them to specific crystallographic sites in the KCP unit cell. The H-H vector at site I is parallel to [100] or [010]. At site II the vector is inclined by $\pm 13°$ against the (001) plane with the projection on that plane being *parallel* to [110] or [$\bar{1}$10]. At site III the H-H vector is parallel to [001]. These results presumably refer to 77 K. The ratio of water content at these sites I, II and III (presumably measured also at 77 K) is 1 : 1.8 : 0.2. Given the chemical formula of KCP we identify the sites I and II with the crystallographic sites 1 and 2 for stoichiometric water molecules and site III with site 3 for defect water.

Following the signal intensities as a function of temperature Niki *et al.* [16] determined the number of static water molecules as shown in Figure 19. Raising the temperature decreased the signal intensity of site II molecules while at the same time a narrowed Pake doublet appeared (corresponding to the *c* signals in Figure 17) produced by rotating molecules. Therefore the rotating water molecules were assigned to site II which we identify with the crystallographic site 2. At 300 K the ratio of non-rotating molecules at sites 2 and 1 is roughly 0.28 (Figure 19) which agrees with the value 0.23 obtained in [19] for the intensity ratio of the *b* and *a* signals. This proves that the *b* signals in Figure 17 do not arise from defect molecules, instead they must be associated with non-rotating $H_2O(2)$ molecules.

We thus conclude that in terms of assignments the different investigations can consistently be interpreted. However, we like to emphasize the different results for the orientation of the static $H_2O(2)$ molecules: while Brinkmann *et al.* [57,19] obtained $\pm 11°$ and $\pm 17°$ for the angle β at 163 and 300 K, respectively, Niki *et al.* [16] measured $\beta = 0°$ at 77 K. Whether the zero value results from a low signal resolution or signifies a strong temperature dependence of β below 163 K remains an open question.

Figure 19 demonstrates how at a given temperature the number of mobile molecules is different for various sites. The ordered $H_2O(1)$ and the disordered $H_2O(2)$ molecules start to become mobile at about the same temperature; however, the ordered molecules are more "resistant" to rotations than the others. On the other hand, defect molecules are dynamic already at 150 K and above 205 K all of them are rotating although Ref. [56] reports a strong signal at 203 K which must arise from *static* defect molecules. These differences are probably due to variations in the water content of samples. Because of low intensity and masking by strong signals the temperature dependence of the signals of rotating water molecules could not be studied in detail [16].

Niki *et al.* [16] have measured the temperature dependence of ΔB of Eq. (3.23) for *constant* orientations of the crystal with respect to the external field B_0. For example, the $H_2O(2)$ molecules were studied with B_0 either parallel or perpendicular to [110]. The decrease of ΔB with rising temperature is explained in terms

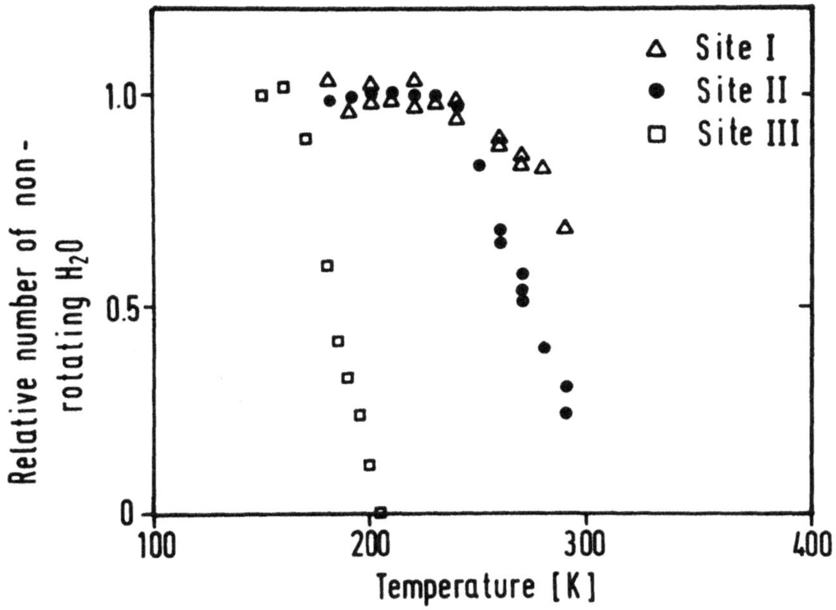

Fig. 19 Temperature dependence of the relative number of static H_2O molecules at sites I, II and III [16].

of librational motions of the molecules. The potentials of the rocking and twisting oscillations are well explained by the model of a simple harmonic oscillator and the oscillations in a three-well potential, respectively. The analysis is valid if the averaged orientation of the H-H vector does not change with temperature. However, as mentioned above, at 163 and 300 K the projection of the H-H vector of $H_2O(2)$ molecules was found to deviate from [110]. These temperature dependent "misalignments" can also account quantitatively for the decrease of ΔB Niki *et al.* have observed.

Butler and Guggenheim [59] investigating the temperature dependence of the proton *powder* spectrum found that some of the water molecules are free to rotate in such a manner as to narrow the resonance and that this motion freezes out at temperatures below about 200 K. The narrowing is interpreted to arise from a distribution of tumbling times, *i.e.* of activation energies, for some of the water molecules. The activation energy results from an interaction of the water electric dipole moment with the local electric field. Because of the irregular occupancy of the K^+ and Br^- (or Cl^-) sites there is a distribution in electric fields. If the electric field at the site of the water molecule is strong enough it might align the dipole moment thus making this water molecule "rigid". It is worthwhile to note that the slowing down of rotation occurs in the same temperature range where the width of the ^{195}Pt line narrows and the spin-lattice relaxation time becomes longer [37,38].

Butler and Guggenheim [59] propose that the importance of the water molecules stems from the shielding of platinum chains from the random potentials generated by the K^+ and Br^- (or Cl^-) sites. The spatially varying potential provides a mechanism for electron scattering which lowers the conductivity and a removal of water would decrease the conductivity which is an experimental fact.

Masin and Leo [17] have investigated the motion of the dynamically disordered $H_2O(2)$ molecules by measuring the temperature dependence of the spin-lattice relaxation time T_1 between 50 and 300 K employing pulse techniques (see Figure 20). At 230 K a T_1 minimum of about 0.13 s was observed. The data were fitted to a model assuming thermally activated 180° flips of the H-H vectors, *i.e.* an interchange of the protons, and a dipole-dipole interaction between different pairs of spins including an additional background relaxation rate (possibly due to paramagnetic ions). A value of 1.71 G for the fluctuating magnetic field was obtained which is in good agreement with a calculation based on the crystallographic parameters and first neighbors only, *i.e.* the bromine and the two closest protons outside the molecule itself. The activation energy of 5.8 kcal/mole (0.25 eV) is typical of a hydrogen bond thus providing further evidence for the hydrogen bonding of $H_2O(2)$ with bromine. At temperatures above 280 K the interactions with $H_2O(1)$ and $H_2O(3)$ cannot be neglected anymore.

3.6. SITES AND DYNAMICS OF BROMINE BY ^{81}Br AND ^{39}K NMR

The bromine NMR played a role in the discussion of alternatives to the picture of the Peierls transition. For instance, Inkson [60] and Gusman and Deltour [61] postulated fairly mobile bromine atoms jumping or tunneling along the Pt strands from one cell to an empty neighboring cell; transverse movements are forbidden because of the strong anisotropy of the structure. In this way a maximum in the conductivity-vs-pressure function could be explained [61]. How can NMR contribute to these questions?

Nagasawa and Miyauchi [62] reported at room temperature a narrow ^{81}Br signal in KCP powder with a large chemical shift of -0.2%. The linewidth was smaller than a quarter of the dipolar field at the Br site calculated by using crystal structure data. This was taken as the result of motional narrowing provided the correlation time of the fluctuations of the local field at the Br nucleus is much shorter than the Larmor period in the local field. It was suggested that this result supports the Inkson model [60].

When repeating these measurements Brinkmann *et al.* [57] could not detect a Br signal with a chemical shift of -0.2% although their spectrometer was sensitive enough to record the neighboring weak resonance of the ***naturally*** abundant isotope

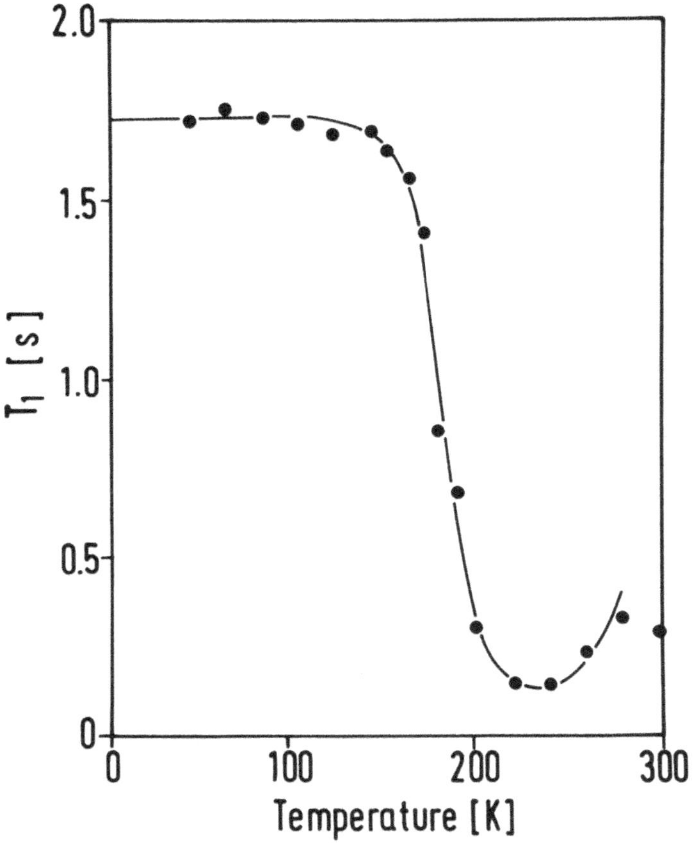

Fig. 20 ^{1}H spin-lattice relaxation time of $H_2O(2)$ molecules as a function of temperature [17]. The solid line is a fit to the data using the Bloembergen-Purcell-Pound relation [27].

^{13}C. However, a Br signal with a linewidth of 1 G could be observed in samples containing traces of mother liquid. The signals were not shifted with respect to the Br resonance in an aqueous solution of NaBr. In addition to these Br signals a strong "liquid water" proton signal was detected superimposed on the proton spectrum of the stoichiometric water molecules. From subsequent measurements it was concluded that this Br signal arises from bromine in aqueous solution which are liquid inclusions (of about 6% of the crystal volume) in single crystals. Thus there was no evidence for fast Br motions like jumping or tunneling and hence no

NMR support of the alternative models mentioned above.

We return now to the question concerning the controversy whether there is a "second Br site" or not. Employing a high magnetic field of 5.17 T Brenni *et al.* [20] provided experimental evidence for the existence of a second Br site and a defect water site in KCP. Figure 21a shows two ^{81}Br signals which are interpreted as the central lines of two Br spectra split by interactions between the ^{81}Br quadrupole moment eQ and the electric field gradient (EFG) at the nuclear site; the satellite signals are "wiped out". These signals exist at least between 24 and 300 K; their quadrupole coupling constants eQV_{zz} [V_{zz} is the maximum principal component of the EFG] were measured in this interval, they are of the order of 17 and 22 MHz. The linewidth (Figure 21b) depends strongly on orientation suggesting a distribution of EFG's which are believed to arise to an appreciable amount from the charge density waves (CDW) on the Pt chains. The line narrowing starting around 60 K and being complete around 120 K is attributed to an uncorrelated motion of CDW's at different Pt chains.

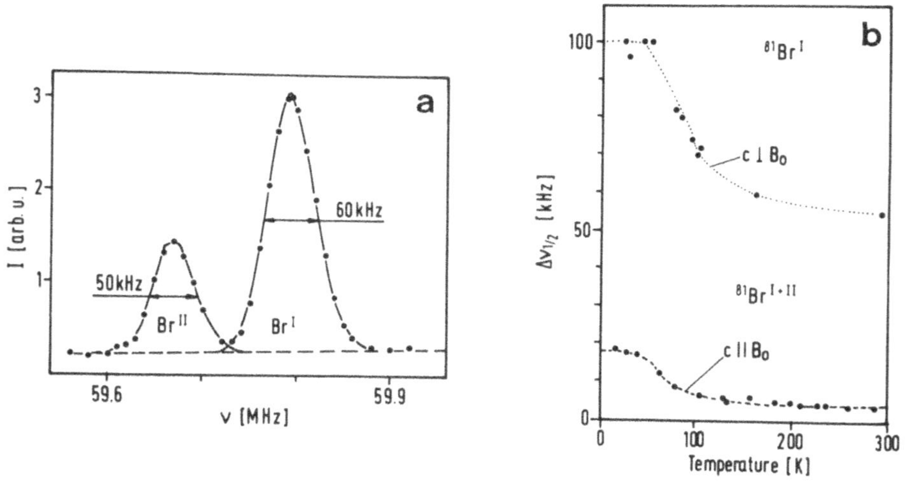

Fig. 21 a) The two ^{81}Br NMR signals (central transitions) Br^{I} and Br^{II} at 163 K for the c-axis perpendicular to the magnetic field B_0 [20].
b) Temperature dependence of the $^{81}Br^{I}$ line width $\Delta\nu_{1/2}$ for $c \perp B_0$ (top) and $c || B_0$ (where the two lines Br^{I} and Br^{II} coincide) [20].

From the fact that two Br signals were found with no indication for a back and forth jumping between two sites in the same unit cell, it was concluded that two types of unit cells exist with Br either in site I or in site II. These sites are probably identical with those Williams *et al.* [49] have suggested and which are located at (1/2, 1/2, 1/2) and (1/2, 1/2, 0.36). Williams *et al.* estimated the ratio of the populations of these sites to be 3 which comes close to the value 2.5 ± 0.3 of the NMR intensities (see Figure 21a).

Further support for the suggestion that two Br sites exist comes from potassium NMR [20]. Figure 22 displays the ^{39}K spectrum at 73 K when the c-axis is parallel to the external field and hence the four K positions of each unit cell are equivalent. The spectrum can be resolved into 3 signals with Gaussian lineshapes (dotted lines) designated by K^0, K^I and K^{II}. The signals are interpreted as the central signals of three quadrupolar split spectra whose satellite lines are wiped out. It is suggested that the light H_2O molecule creates a smaller EFG at the K sites than the heavy Br ion does and that therefore the ^{39}K nuclei of bromine free cells exhibit a smaller quadrupolar shift than the other K nuclei. Thus the K^0 signal is attributed to unit cells free of Br and the K^I and K^{II} signals to cells containing Br^I or Br^{II}, respectively. Taking into account the NMR intensities, one concludes that about 43% of all unit cells contain Br^I, 17% contain Br^{II} and the remaining 40% are occupied by defect water.

An interesting question arises whether the two Br ions and the defect water could be ordered along the c-axis. An array of 11 Br^I, 4 Br^{II} and 10 defect water cells is in accord with the Br-defect water ratio and the ratio of the Br signal intensities and would represent a very large superstructure. Fukuyama [63] has proposed that a quasi-periodic distribution of Br ions acts like a periodic potential on the CDW leading to the commensurate pinning of the CDW at low temperatures; this explains the long correlation length at this temperature.

Mali *et al.* [64] have performed some measurements of the spin-lattice relaxation time T_1 of ^{81}Br at 59.4 MHz with the magnetic field B_0 parallel to the c-axis. From 18 to 50 K the rate increases exponentially and attains a broad maximum of 10^3 s^{-1} around 67 K. It then decreases by a factor of 10 and is at a minimum at 188 K. Above this temperature the rate shows a steep increase. It is interesting to note that in the same temperature range where the ^{81}Br relaxation rate reaches a maximum, Mali *et al.* [64] have also observed a maximum of the ^{14}N relaxation rate (for $c \| B_0$). Together with the ^{1}H rate maximum (see Figure 11) at about the same temperature the occurrence of these maxima suggests a common relaxation mechanism for all three isotopes which probably is associated with the water molecule dynamics. It should be added that the ^{14}N rate maximum at 50 K is only 8 s^{-1} reflecting the steep decrease of the rate on the low temperature side of its peak value around 120 K [12]. That divergence had been associated with the slowing down of lattice fluctuations.

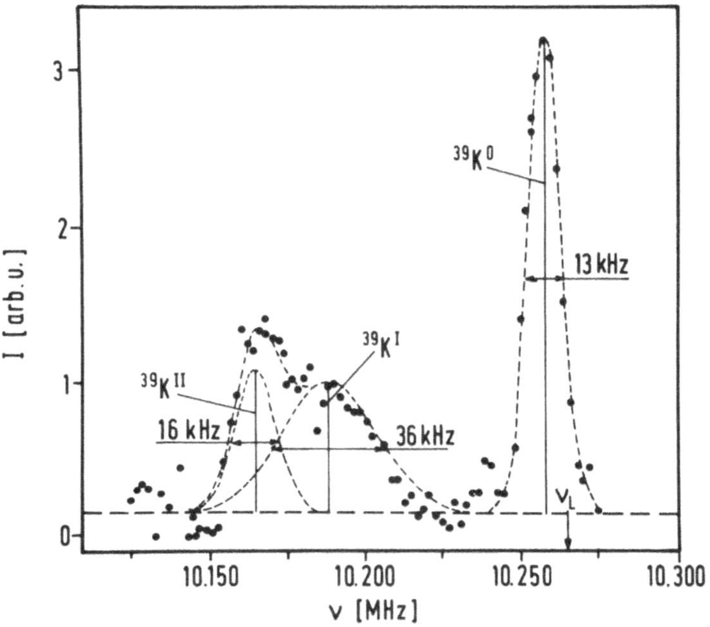

Fig. 22 ^{39}K NMR spectrum at 73 K for $c||B_0$. The spectrum indicates three different sites (0, I, II) for potassium [20].

4. Conclusions

The NMR work on KCP has employed several NMR isotopes to deal with various aspects of this quasi one-dimensional (Q1D) conductor and a wealth of information on a variety of properties has been accumulated. Roughly speaking, three kinds of properties have been studied: "chain properties" by ^{195}Pt NMR, "nearby off-chain" properties by ^{14}N and ^{13}C NMR and finally characteristics of the compound only indirectly related to the Pt chains by combined ^{81}Br, ^{39}K and ^{1}H NMR.

^{195}Pt NMR has supported and confirmed the conception that KCP is described well by the Q1D model of the Peierls instability provided fluctuation effects are included. Evidence for the presence of critical fluctuations of the electron density of states has been provided by measurements above 120 K, where the Knight shift and the spin-lattice relaxation are governed by electron excitations over the Peierls gap. These experiments also show that the complete metallic state is not yet established at room temperature.

Very detailed investigations of line splittings, chemical shifts and relaxation have convincingly demonstrated the existence of the charge density wave (CDW) which is a consequence of the Peierls distortion. In particular, it has been shown that relaxation and line shifts in the medium temperature range of 30 to 100 K are governed by non-linear excitations (solitons) of the CDW. Though it could not be decided whether the solitons are actually migrating or, in the case of localization, are relaxing with a certain correlation time, the relevant parameters such as formation and migration energies and prefactors could be determined.

NMR studies of the nearby off-chain isotopes ^{14}N and ^{13}C seem to indicate that the CDW is restricted to the Pt chains. Thus the relaxation of these isotopes provides insight into the lattice dynamics. In particular the slowing down of lattice fluctuations at the onset of 3D ordering below 140 K appears to be responsible for features of the ^{14}N and ^{13}C spin-lattice relaxation.

Finally, ^{81}Br, ^{39}K and ^{1}H nuclei have been used to investigate several other questions pertinent to the understanding of KCP. For instance, the ^{81}Br studies did not provide support for models which were designed as alternatives to the Peierls picture. On the other side, bromine NMR together with ^{39}K NMR settled a controversy about the second Br site in KCP. Such a site exists in those cells which do not contain defect water.

Proton NMR has been employed to answer several questions concerning for instance the ordering and bonding of the water molecules, the type of their localized motions, the role these molecules play with respect to the conducting chains. In terms of assignment of the different proton NMR signals to the various crystallographic sites, the results of different investigations can consistently be interpreted. However, there are still some discrepancies left concerning details of the proton-proton distance vector orientation which seem to have relevance for the interpretation of the temperature dependence of localized water motions.

In conclusion, we believe that despite the remarkable results obtained so far KCP is still worthy of a fresh look by NMR. Especially further studies of the relaxation of ^{14}N, ^{13}C and ^{1}H are expected to yield valuable answers to questions still open in this fascinating compound.

References

1 K. Krogmann, *Angew. Chem. Int. Ed. Engl.* **8**, 35 (1969).

2 A.G. Toombs, *Physics Reports* **40**, 181 (1978).

3 S. Kagoshima, H. Nagasawa, and T. Sambongi: *One-dimensional Conductors*, ed. P. Fulde, Springer Series in Solid State Sciences, Vol. 72, Berlin 1982.

4 H. R. Zeller and P. Brüesch, *Phys. Stat. Sol. (b)* **65**, 537 (1974).

5 D. Kuse and H. R. Zeller, *Sol. State Comm.* **11**, 355 (1972).

6 H. R. Zeller and A. Beck, *J. Phys. Chem. Sol.* **35**, 77 (1974).

7 R. E. Peierls, *Quantum Theory of Solids*, Oxford, Univ. Press, London 1955, p. 108.

[8] W. Kohn, *Phys. Rev. Lett.* **2**, 393 (1959).

[9] R. Comès, M. Lambert, H. Launois, and H. R. Zeller, *Phys. Rev.* **B8**, 571 (1973).

[10] B. Renker, H. Pietschel, L. Pinschovius, W. Gläser, P. Brüesch, D. Kuse, and M.J. Rice, *Phys. Rev. Lett.* **30**, 1144 (1973).

[11] B. Renker, L. Pintschovius, W. Gläser, H. Rietschel, R. Comès, L. Liebert, and W. Drexel, *Phys. Rev. Lett.* **32**, 836 (1974).

[12] G. Soda, C. Bourbonnais, and D. Jérome, *J. Phys. Soc. Japan* **56**, 3951 (1987).

[13] C.F. Eagan, S.A. Werner, and R.B. Saillant, *Phys. Rev.* **B12**, 2036 (1975).

[14] H.R. Zeller, *Festkörperprobleme XIII - Advances in Solid State Physics*, ed. H.J. Queisser, Pergamon-Vieweg, 1973.

[15] P. Brüesch, S. Strässler, and H.R. Zeller, *Phys. Rev.* **B12**, 219 (1975).

[16] H. Niki, H. Doi, and H. Nagasawa, *J. Phys. Soc. Japan* **51**, 2470 (1982).

[17] F. Masin and V. Leo, *Sol. State Comm.* **43**, 61 (1982).

[18] P.L. Johnson, A.J. Schulz, A.E. Underhill, D.M. Watkins, D.J. Wood, and J.M. Williams, *Inorg. Chem.* **17**, 839 (1978).

[19] D. Brinkmann and K. Meier, *Sol. State Comm.* **19**, 1099 (1976).

[20] P. Brenni, D. Brinkmann, H. Huber, M. Mali, J. Roos, and H. Arend, *Sol. State Comm.* **47**, 415 (1983).

[21] B. Horowitz, M. Weger, and H. Gutfreund, *Phys. Rev.* **B9**, 3174 (1975).

[22] G. Soda, D. Jérome, M. Weger, J. Alizon, J. Gallice, H. Robert, J.M. Fabre, and L. Giral, *J. Phys.* **38**, 931 (1977).

[23] M.J. Rice, A.R. Bishop, J.A. Krummhansl, and S.E. Trullinger, *Phys. Rev. Lett.* **36**, 432 (1976).

[24] L. Gunther and Y. Imry, *Phys. Rev. Lett.* **44**, 1225 (1980).

[25] M. J. Cohen and A.J. Heeger, *Phys. Rev.* **B16**, 688 (1976).

[26] M. Mehring: *Principles of High Resolution NMR in Solids*, Springer-Verlag, Berlin 1983.

[27] A. Abragam: *The Principles of Nuclear Magnetism*, Clarendon Press, Oxford 1961.

[28] O. Kanert, *Phys. Reports* **91**, 184 (1982).

[29] R. Kimmich and G. Voigt, *Z. Naturforsch.* **33a**, 1294 (1978).

[30] J.L. Bjorkstam and M. Villa, *Magn. Resonance Rev.* **6**, 1 (1980).

[31] M. Mehring, unpublished results.

[32] W. Rüegg, D. Kuse, and H.R. Zeller, *Phys. Rev.* **B8**, 952 (1973).

[33] R. Nixdorf, Diploma Thesis, Dortmund 1982.

[34] M. Mali, O. Kanert, M. Mehring, and D. Brinkmann, *Phys. Rev.* **B20**, 4442 (1979).

[35] R.E. Walstedt, J.H. Wernick, and V. Jaccarino, *Phys. Rev.* 162, 301 (1967).

[36] H.H. Rupp, *Z. Naturforsch.* **26a**, 1937 (1971).

[37] D. Brinkmann and H.U. Keller, *Helv. Phys. Acta* **47**, 387 (1974).

[38] H. Niedoba, H. Launois, D. Brinkmann, and H.U. Keller, *J. de Physique* **35**, L-251 (1974).

[39] H. Niedoba, H. Launois, D. Brinkmann, R. Brugger, and H.R. Zeller, *Phys. Status Solidi (b)* **58**, 309 (1973).

[40] P.A. Lee, T.M. Rice, and P.W. Anderson, *Phys. Rev. Lett.* **31**, 462 (1973).

[41] W.H.-G. Müller, H. Niedoba, and D. Jérome, *Sol. State Comm.* **16**, 655 (1975).

[42] F. Masin, G. Gusman, and R. Deltour, *Sol. State Comm.* **40**, 291 (1981).

[43] J.W. Lynn, M. Iizumi, G. Shirane, S.A. Werner, and R.B. Saillant, *Phys. Rev.* **B12**, 1154 (1975).

[44] M. Mehring, O. Kanert, M. Mali, and D. Brinkmann, *Sol. State Comm.* **33**, 225 (1980).

[45] A. Rigamonti, *Adv. Magn. Res.* **33**, 115 (1984).

[46] C. Berthier and P. Ségransan: *Low-Dimensional Conductors and Superconductors*, ed. L. Caron and D. Jérome, Plenum Press, New York 1987.

[47] M.E. Stoll, R.W. Vaughan, R.B. Saillant, and Terry Cole, *J. Chem. Phys.* **61**, 2896 (1974).

[48] P. Brenni, D. Brinkmann, M. Mali, J. Roos, O. Kanert, and M. Mehring, *Bull. Magnetic Resonance* **2**, 269 (1981).

[49] J.M. Williams, J.L. Petersen, H.M. Gerdes, and S.W. Peterson, *Phys. Rev. Lett.* **33**, 1079 (1974).

[50] J.M. Williams, M. Iwata, F.K. Ross, J.L. Petersen, and S.W. Peterson, *Mat. Res. Bull.* **10**, 411 (1975).

[51] C. Peters and C.F. Eagen, *Phys. Rev. Lett.* **34**, 1132 (1975).

[52] G. Heger, B. Renker, H.J. Deiseroth, H. Schulz, and G. Scheiber, *Mat. Res. Bull.* **10** (1975) 217.

[53] G. Heger, B. Renker, H.J. Deiseroth, and H. Schulz, *Sol. State Comm.* **18**, 517 (1976).

[54] H.J. Deiseroth and H. Schulz, *Mat. Res. Bull.* **10**, 225 (1975).

[55] R. Brugger and D. Brinkmann, *Sol. State Comm.* **13**, 889 (1973).

[56] D. Brinkmann and H.U. Keller, *Phys. Status Solidi (b)* **67**, K43 (1975).

[57] D. Brinkmann, G. Finger, and H. Arend, *Phys. Status Solidi (b)* **83**, K45 (1977).

[58] G. Heger, H.J. Deiseroth, and H. Schulz, *Acta Cryst.* **B 34**, 752 (1978).

[59] M.A. Butler and H.J. Guggenheim, *Phys. Rev.* **B 10**, 1778 (1974).

[60] J.C. Inkson, *Sol. State Comm.* **14**, 1229 (1974).

[61] G. Gusman and R. Deltour, *Sol. State Comm.* **15**, 1597 (1974).

[62] H. Nagasawa and N. Miyauchi, *Phys. Status Solidi (b)* **71**, 671 (1975).

[63] H. Fukuyama, *J. Phys. Soc. Japan* **45**, 1266 (1978).

[64] M. Mali, J. Roos, and D. Brinkmann, unpublished results.

HIGH RESOLUTION NMR ON ORGANIC RADICAL SALTS

MICHAEL HELMLE AND MICHAEL MEHRING
2. Physikalisches Institut, Universität Stuttgart
Pfaffenwaldring 57
W-7000 Stuttgart 80
Federal Republic of Germany

1. Introduction

NMR (Nuclear Magnetic Resonance) is one of the most versatile spectroscopic techniques for the investigation of solids because it can provide information about local magnetic and electric fields at the atomic level as well as fluctuations of these fields caused *e.g.* by conduction electrons in organic conductors. In ordinary organic materials proton and carbon NMR spectroscopy in solution has become a routine method to study the diamagnetic interaction of these nuclei with electrons in closed shells (chemical shift). The NMR investigation of organic radical salts, however, is aimed at gathering information about the interaction of delocalized unpaired electron spins (hyperfine interaction) thereby probing basically the electronic structure of the conduction band which can be viewed as a linear combination of the singly occupied molecular orbitals (SOMO). In isolated organic radicals the hyperfine coupling is usually very large, whereas in organic conductors it is rather small due to electron delocalization (organic conductors) or due to a strong exchange coupling among the electron spins (Mott Hubbard insulators). In organic conductors high resolution NMR spectroscopy is therefore mandatory in order to observe the hyperfine interaction between conduction electrons and the nuclei resulting in the Knight shift.

Some early solid state NMR investigations were performed on Pyrene [1] and DPPH [2]. These and other investigations in organic conductors were mostly based on broadline proton NMR which has too limited resolution in order to determine the hyperfine interaction accurately enough. Although it was possible in some radical systems to resolve parts of the strongly coupled protons by going to extremely low temperatures, in order to utilize the strong increase in the Curie susceptibility, this method is inappropriate, however, for organic metals. Some attempts have been made to separate the protons in the spectrum by a moment analysis and by selective deuteration [3,4]. However, in none of these experiments it was possible

T. Butz (ed.), Nuclear Spectroscopy on Charge Density Wave Systems, 275–312.

to map out the complete spin density distribution. The development of multiple pulse experiments to remove the dipolar couplings among protons opened the way to resolve the different protons under favourable circumstances, but only very few experiments carried out at room temperature have been reported [5,6]. It was demonstrated that the application of special solid state high resolution techniques can result in a locally resolved picture of the SOMO. Even the application of these highly sophisticated techniques leaves some blind spots in the spin density map at the sites of the non-protonated carbons. The advent of magic angle spinning (MAS) ^{13}C-NMR, however, made high resolution NMR spectroscopy of naturally abundant carbons possible. Although this technique just began to be applied to organic radical salts, the results are highly promising. All magnetically non-equivalent carbons can in principle and most often in praxi be resolved. In most cases the temperature range was limited to 150K, however recently CP/MAS experiments down to 10 K and below have been reported [7,8,9].

In this article we will restrict ourselves to ^{1}H and ^{13}C NMR investigations. Some representative examples will be discussed. All equations will be given in SI units. The abbreviations for substances mentioned in this article are summarized at the end of the text.

2. Solid State Theory of Radical Salts

In this section simple models will be developed to describe the properties of organic radical salts. Both the metallic and the insulating (semi-conducting) phase will be covered and special emphasis will be placed on the magnetic susceptibility. Thereby a link will be established between the hyperfine interaction detected by high resolution NMR and solid state properties which are represented by typical parameters like diffusion constant, localization length, dimensionality of the transport parameters, bandwidth, on-site electron-electron correlation, amount of charge transfer and density of states.

2.1. SIMPLE BAND MODELS

The description of a classical metal is usually based on the free electron gas model. In the case of organic metals, however the bandwidth is much smaller and because of the low dimensionality the effects of electron-phonon coupling and electron-electron interaction will become much more pronounced. Therefore a description in the framework of tight binding theory seems to be more appropriate.

The one-dimensional tight-binding model considers an infinite linear chain of molecular units, which are uniformly spaced at a distance a. These units are weakly coupled due to orbital overlap, which is characterized by the transfer integral t.

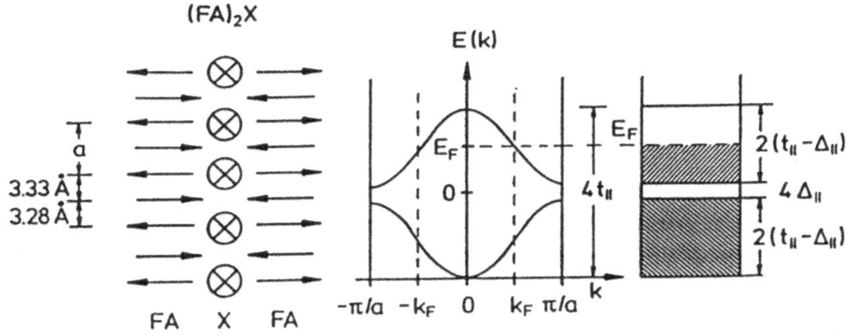

Fig. 1 Left: Stacking arrangement of fluoranthene (FA) molecules and counterions (X) in $(FA)_2X$. Middle and right: 1D bandstructure including dimerization gap. The Fermi energy is sketched for a threequarter filled band [35].

In Fig. 1 we have sketched the development of the one-dimensional tight-binding bandstructure. In the limit of nearest neighbour interaction the corresponding Hamiltonian reads

$$\hat{H}_{tb} = \sum_j \hat{H}_0(j) + t \sum_j c^+_{j,\sigma} c_{j+1,\sigma} + c^+_{j+1,\sigma} c_{j,\sigma} \tag{2.1}$$

where H_0 represents the Hamiltonian of each molecular unit and c^+, c are Fermion creation and annihilation operators, respectively, for electrons located at site j with spin orientation σ. With the Bloch-Ansatz for the total wave function

$$\Psi = \frac{1}{\sqrt{N}} \sum_j \varphi_j \cdot e^{ikna} \tag{2.2}$$

and zero differential overlap approximation one obtains the band structure as a function of the wave vector $\mathbf{k}$ as follows

$$E(k) = E_0 + 2t \cos(ka) \quad . \tag{2.3}$$

It has tacitly been assumed above that each molecular unit contributes only one electronic state, which can be occupied with two electrons having opposite spin orientation. Therefore N states belong to the first Brillouin zone. In consequence the density of states in k-space is given by

$$N(k) = \frac{Na}{2\pi} \quad . \tag{2.4}$$

The density of states as a function of energy is related to the density of states in **k**-space by

$$dN = N(E)\,dE = N(k)\,dk \tag{2.5}$$

which leads to

$$N(E) = \frac{N}{4\pi\mid t\mid\sqrt{1-\left(\frac{E-E_0}{2|t|}\right)^2}} \quad . \tag{2.6}$$

An important parameter for the characterization of organic radical salts is the band filling ρ, which is defined as the average charge per unit. Only a partially filled band can provide a metallic conductivity. A metallic state can be established via an incomplete charge transfer from donor to acceptor (*e.g.* TTF-TCNQ, NMP-TCNQ) or by choosing a complex stoichiometry. Especially a 1:2 stoichiometry is often found ($(FA)_2X$, $(TMTSF)_2X$, $(TMTTF)_2X$). The degree of charge tranfer is intimately related with the electron affinity of the acceptor and the ionization potential of the donor molecule. The crystal structure may play some role in refining these quantities.

Up to now we have ignored any correlations between the electrons. It will turn out in the experimental section, however, that in almost every low-dimensional organic system electron-electron correlations are in the same order of magnitude as the tranfer integral t leading to substantial effects in the physical properties. A simple model which includes electron-electron correlation was proposed by J. Hubbard [10,11] in which the electronic system is represented by

$$\hat{H} = \hat{H}_{tb} + U\sum_j n_{j,\sigma}\,n_{j,-\sigma} \qquad n_{j,\sigma} = c^+_{j,\sigma}\,c_{j,\sigma} \tag{2.7}$$

where the short-range ("on-site") correlation is expressed by the parameter U. One of the basic results of this model is that double occupancy of a certain site costs the energy U, whereas in the tight-binding model each k-state could be occupied by two electrons without such an energetic penalty. For $U > 4t$ the former tight-binding band splits into two sub-bands, i.e. a metal with a half-filled tight-binding band will turn into a Mott-Hubbard insulator. In the simple model above only short-range correlations have been taken into account. However, as has been pointed out by Mazumdar *et al.* [12], the long-range correlations provide a considerable amount of dielectric screening which results in a much smaller effective U. In consequence the effective U depends strongly on the bandstructure.

One of the most important quantities which can be derived from NMR hyperfine investigations is the local spin-susceptibility χ_e. Other solid state parameters are then derived from χ_e invoking some model. In metallic systems the susceptibility can be related to the density of states at the Fermi energy.

$$\chi^{mol} = \frac{1}{2}\mu_0 N_L\,(g\mu_B)^2\,N(E_F) \quad . \tag{2.8}$$

However, the calculation for the Hubbard model gets rather involved, therefore only the results for $T = 0$ will be presented as given *e.g.* in references [13,14,15] and the references quoted there. The magnetic susceptibility for the tight binding model can be expressed as

$$\chi_0 = \frac{\mu_0 \mu_B^2 N_L}{\pi t \sin\left(\frac{1}{2}\pi\rho\right)} \tag{2.9}$$

where μ_0 is the permeability of free space, μ_B is the Bohr's magneton and N_L is the Avogadro number. The bandfilling ρ equals one for half filling. In case of the Hubbard model the susceptibility is enhanced and is given in the limit of small U by

$$\chi = \frac{\chi_0}{1 - U/(4\pi t)} \quad . \tag{2.10}$$

If U is large, however, the following expression for the susceptibility applies

$$\chi = \frac{2\mu_0 \mu_B^2 N_L}{\pi^2 J}\rho \qquad J = \frac{2t^2}{U}\rho\left(1 - \frac{\sin 2\pi\rho}{2\pi\rho}\right) \quad . \tag{2.11}$$

2.2. INSTABILITIES IN ONE-DIMENSIONAL SYSTEMS

Peierls showed in his textbook [16] that a one-dimensional metal is inherently unstable with respect to a lattice dimerization. An analogue phenomenon was proposed by Beni and Pincus [17] for a one-dimensional chain of spins, which is therefore often referred to as spin-Peierls transition. In order to explain superconductivity Fröhlich [18] set up a model which demonstrated in a self-consistent treatment that within a certain parameter range the ground state of a one-dimensional free electron system exhibits a modulation with wave vector $\mathbf{q}$ equal to $2k_F$, which creates an energy gap at the Fermi level. Therefore the instability proposed by Peierls should be regarded as a special case for half-filled bands of the Fröhlich charge density wave instability.

The charge modulation must include both the conduction electrons and the lattice in order to avoid large Coulomb energies. A similiar extension for the spin chain leads to a spin density wave with the same wave number as the CDW [19]. The underlying reason for both kinds of instabilities is the highly degenerate Fermi surface of a one-dimensional metal, which can be split by an infinitesimal perturbation resulting in a periodic lattice distortion with the so-called nesting wave vector $2k_F$. Therefore these instabilities are referred to as nesting instabilities. The nesting instability manifests itself in the electronic dielectric susceptibility and is accompanied by a softening of the corresponding phonon mode (Kohn-anomaly). For two- and even more for three-dimensional systems the possibilities for nesting of the Fermi surface are extensively reduced. For details the reader will be referred to [20]. Since these instabilities are most pronounced in one-dimensional systems much of the research on organic radical salts was focussed on these topics. The simple one-electron theory however has to be extended in order to include electron-electron correlation to describe the experimentally observed coexistence of $2k_F$- and $4k_F$-fluctuations.

3. High Resolution Techniques

3.1. THE NEED FOR HIGH RESOLUTION TECHNIQUES

The Knight shift discussed in the preceding section is certainly one of the most pronounced features in the NMR of organic conductors. Since the conduction band orbitals in organic conductors is predominantly of π character the hyperfine interaction with nuclei like ^{13}C and ^{1}H is usually weak, leading to rather small values for the Knight shift of these nuclei. It amounts to a few ppm (10^{-6}) for protons and from about 10 ppm up to several 100 ppm for ^{13}C. The experimentally observed ^{13}C and proton NMR linewidth in solids is usually some 10 kHz wide. This masks completely the Knight shift. There are two main causes for line broadening in solids, namely

- nuclear dipolar interaction and
- shift anisotropies.

The nuclear dipolar interaction can be of homonuclear (*e.g.* ^{1}H - ^{1}H) or heteronuclear (^{13}C - ^{1}H) type. The homonuclear case is much more difficult to deal with than the heteronuclear case. The shift anisotropy, however, appears to be a line broadening mechanism only in powders, where all orientations of the crystallites with respect to the magnetic field are present. The overlap of many different anisotropy patterns obscures completely the shift of a particular site. In single crystals, however, the anisotropy can be put to advantage for investigations of the local site symmetry and the local Knight shift tensor.

Different techniques have been invented to obtain high resolution NMR spectra in the solid state. For a review of these techniques consult references [21,22]. We will discuss two different approaches, which have been proven usefull for organic conductors, in the next two sections.

3.2. DOUBLE RESONANCE AND MAGIC ANGLE SPINNING

In order to eliminate the heteronuclear broadening mechanism (*e.g.* ^{13}C - ^{1}H) one has to decouple the non-observed spin (*e.g.* ^{1}H). This can simply be achieved by strong irradiation exactly on-resonance for the non-observed spin. A double resonance setup is required for this. Moreover, the same double resonance setup can advantageously be utilized in order to cross-polarize the observed spins (*e.g.* ^{13}C) by transfering spin order from the abundant non-observed spins to the rare observed spins. Different pulse schemes have been devised in order to perform this experiment efficiently.

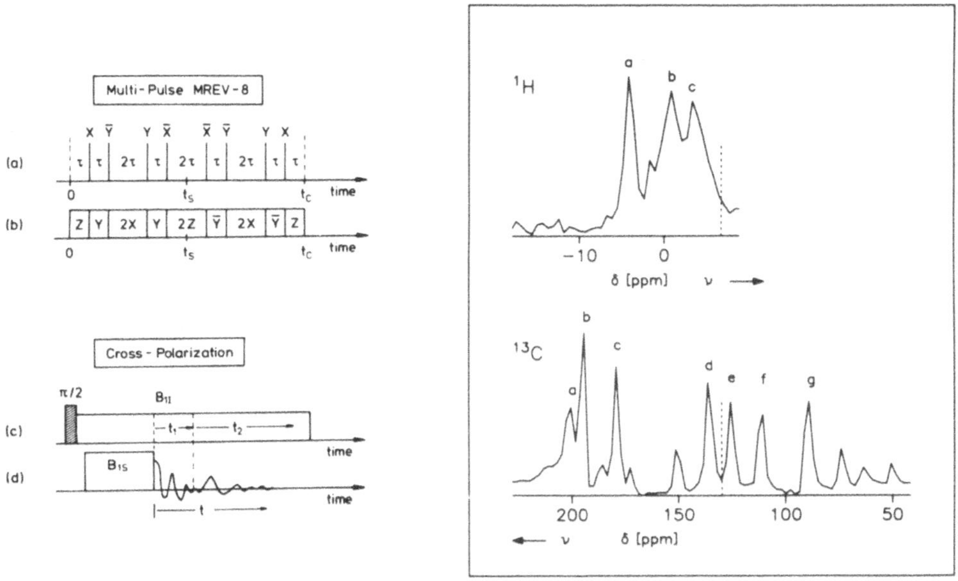

Fig. 2 Left: (a), (b) Timing of the MREV-8 multiple pulse sequence for line narrowing of dipolar coupled spins like protons in radical salts. (c), (d) Cross-polarization sequence including decoupling for rare spin NMR like ^{13}C in radical salts. Right: High resolution proton NMR spectrum of $(FA)_2X$ obtained with an MREV-8 sequence (top). Bottom: High resolution ^{13}C NMR spectrum of $(FA)_2X$ from cross-polarization including magic angle spinning (MAS) [35].

The standard sequence is shown in Fig. 2. In single crystals a highly resolved NMR spectrum is readily obtained by this technique. In powders, however, the spectrum is still broad and usually unresolved because of the shift anisotropies. Since the shift anisotropies are caused by second rank tensors, a rotation of the sample with the rotation axis tilted by 54.7° (the "magic angle") with respect to the magnetic field leaves the trace of this tensor, i.e. the isotropic part invariant. If the rotation is performed rapidly enough compared with the frequency spread of the anisotropy, only the isotropic part is retained in the spectrum. This "magic angle spinning" (MAS) technique provides highly resolved NMR spectra even in an unoriented powder.

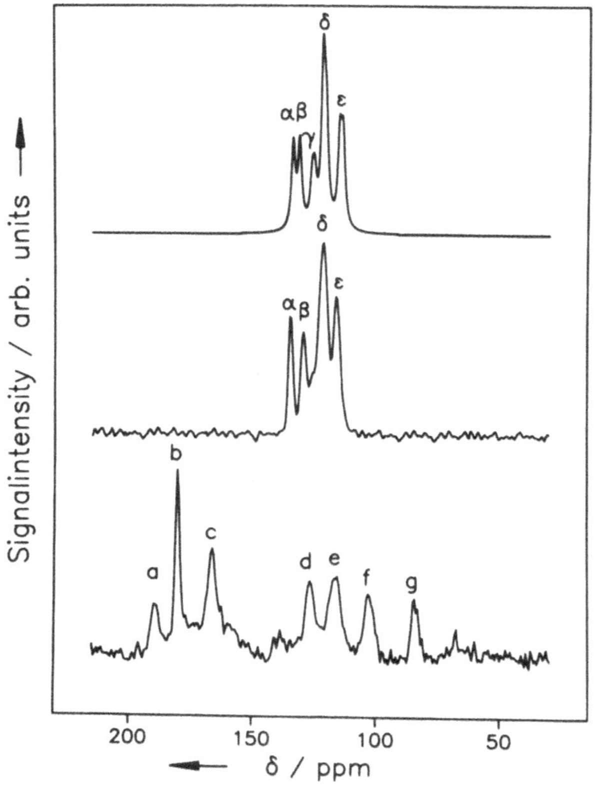

Fig. 3 High resolution ^{13}C NMR spectra [37]. Top: Liquid solution of FA molecules. Middle: MAS spectrum in FA powder with resolution of individual ^{13}C lines. Bottom: MAS spectrum in $(FA)_2X$ powder. This spectrum demonstrates the influence of the hyperfine coupling to the conduction electrons resulting in Knight shifts of the individual ^{13}C lines.

One of the early applications of this technique to organic conductors is shown in Fig. 3 where the ^{13}C spectrum of the fluoranthene molecule in solution (top) and the ^{13}C MAS NMR spectrum of fluoranthene powder (middle) are compared with the ^{13}C spectrum in the organic conductor $(Fluoranthene)_2SbF_6$. Note that the spread in shifts in the organic conductor is much larger than in the neutral and non-conducting compounds. The resolution of the Knight shifts led to the molecular map of the spin density, i.e. the conduction band molecular orbital in this organic conductor.

3.3. MULTIPLE PULSE TECHNIQUES

The technique described in the preceding section is only feasible when homonuclear dipolar interaction is not limiting the resolution of the observed spins. If abundant spins are observed like ^1H the homonuclear dipolar interaction is dominant. In this situation WAHUHA type multiple pulse techniques must be applied in order to reduce the homonuclear dipolar interaction considerably [21,22]. The basic pulse sequence consists of a cycle of four phase shifted pulses which are repeated a few hundred times within several milliseconds. The technique is very demanding and needs some expertise before reliable results can be obtained. Extremely high resolution has been achieved with this technique in non-conducting solids. In organic conductors the skin depth problem can severely deteriorate the resolution, however. The multiple pulse technique has been combined with MAS in order to achieve high resolution in powder samples. This was used in one instance also for observing the proton Knight shift in the organic conductor $(\text{Fluoranthene})_2SbF_6$.

4. Magnetic Electron Nuclear Interaction

Since we are dealing in this article with spin 1/2 nuclei we will discuss only magnetic interactions. There are basically two different kinds which have to be distinguished, namely those caused by orbital currents and those due to the electron spin. We will treat them in the following separately.

4.1. CLOSED SHELL INTERACTIONS

As the investigation of organic radical salts is focused on the hyperfine interaction due to the conduction electrons the closed shell interactions will be discussed only very briefly. The closed shell interactions manifest themselves basically in two different observable parameters:

a) Chemical shift

The magnetic field at the position of a nucleus is shielded due to dia- magnetic and paramagnetic (second order) orbital currents. The amount of shielding is very sensitive to the bond geometry. Therefore the chemical shift tensor can serve as a fingerprint of a certain molecular fragment. In addition the chemical shift slightly depends on the partial charge at each site. The chemical shift tensor is usually represented by three principle values and the corresponding axes.

b) J-Coupling

The magnetic moment of a nucleus can also induce orbital currents in the electronic shell. Another nucleus will couple to this induced magnetic field which results in weak nuclear-nuclear interaction. Due to the second order character and corresponding weakness of this coupling compared to dipolar couplings among nuclear spins the J coupling can usually be neglected in solid state samples.

4.2. OPEN SHELL INTERACTIONS

In partially filled bands there are similar electron nucleus interactions as discussed in the preceding section, namely orbital shifts (van Vleck type interaction) as well as J-couplings (RKKY interactions). In the following we will focus, however, on the spin dependent interactions of conduction electrons. The hyperfine interaction between a nuclear spin and an unpaired electron spin consists of a dipolar part and the Fermi contact contribution and can be written as [23,24,25]:

$$\hat{H}_{hfi} = -\frac{\mu_o}{4\pi}\left\{\hat{\mu}_e\tilde{\mathbf{D}}\hat{\mu}_n + \frac{8\pi}{3}\delta(r)\hat{\mu}_e\cdot\hat{\mu}_n\right\} \quad . \tag{4.1}$$

Hereby μ_0 is the permeability of free space, μ_e and μ_n are the the magnetic moments of the electron and the nucleus, respectively, and δ is Dirac's delta function. The dipolar coupling tensor $\tilde{\mathbf{D}}$ is written explicitly

$$\tilde{\mathbf{D}} = \frac{1}{r^3}(3\frac{\hat{r}\hat{r}}{r^2} - \mathbf{1}) \tag{4.2}$$

where $\mathbf{r}$ is a vector connecting electron and nuclear moment and $\mathbf{1}$ represents the unity matrix. The interaction with the electron orbital momentum has been neglected. This is justified in the case of hydro-carbons due to the fact that unless there is an orbitally degenerate ground state the orbital momentum will be quenched [26]. The orbital momentum can become partially dequenched through the action of spin-orbit coupling, but in pure hydrocarbons this effect is too small to cause any substantial effect. For the same reason the electron g-factor is essentially scalar and any pseudo-contact shifts are weak. The situation will change, however, in the case of heavy hetero-atoms or organo-metallic compounds.

4.3. TIME AVERAGED HYPERFINE INTERACTION

For a non-fluctuating electron spin the hyperfine interaction would produce a line splitting in the NMR spectrum in the order of 10-100 MHz which would lead to undetectable NMR lines in powder samples. The NMR investigation of paramagnetic molecules is limited to cases in which the magnetic moment of the electron undergoes spin flips which are rapid compared to the strength of the electron-nucleus interaction. This leads to a small shift of the NMR line instead of a large splitting. In this case it is permissible to replace the electron magnetic moment by its time average:

$$< \hat{\mu}_e >= \frac{1}{\mu_0} \frac{\chi^{mol}}{N_L} \mathbf{B} \quad \text{where} \quad \chi^{mol} = \frac{N_L}{N/V} \chi \tag{4.3}$$

is the molar spin-susceptibility. Replacing the magnetic moment in Eq. (4.1) by the expression above results in

$$\hat{H}_{hfi} = -\frac{1}{4\pi} \hbar \gamma_n I_Z \left\{ \left\langle \frac{3\cos^2\Theta - 1}{r^3} \right\rangle + \frac{8\pi}{3} \langle \delta(r) \rangle \right\} \frac{\chi^{mol}}{N_L} B_0 \quad . \tag{4.4}$$

It can be seen that the resulting frequency shift is proportional to the strength of the external field B_0, therefore it is convenient to relate the frequency shift to the nuclear Larmor frequency. This results in the field independent Knight shift K

$$K = \frac{1}{4\pi} \left\{ \left\langle \frac{3\cos^2\Theta - 1}{r^3} \right\rangle + \frac{8\pi}{3} \langle \delta(r) \rangle \right\} \frac{\chi^{mol}}{N_L} \quad . \tag{4.5}$$

The first expression in the bracket (dipolar contribution) depends on the orientation between the vector connecting electron and nuclear spin and the external field and has zero isotropic value, whereas the second part, the so-called Fermi contact part gives only an isotropic shift contribution if the electronic g-factor is isotropic. It should be noted that all core polarization contributions which are in fact dominant in organic conductors must be included in the contact term.

4.4. HYPERFINE INTERACTION IN ORGANIC MOLECULES

Up to now no attention has been paid to the orbital structure of the molecular units. This structure, however, will influence the local hyperfine coupling parameter a_i and therefore its knowledge is required in order to extract the spin-susceptibility from Knight shifts and relaxation rates. On the other hand for a known susceptibility the study of the local hyperfine interaction will reveal the spin density distribution. The reliable theoretical calculation of the spin density distribution in organic molecules is still a formidable quantum chemical problem due to the many electron character and the large molecules involved. Usually a simple semi-empirical approach based on the Hückel theory can serve as a good starting point. For more advanced theories the reader is referred to the books of Salem [27], Pople and Beveridge [28], Sadlej [29] and the classical review on paramagnetic shifts by La Mar *et al.* [30]. The Hückel molecular orbital formalism is based on the assumption that the SOMO is built up from a linear combination of p_z-orbitals centered at the carbon atoms which do not interact with the σ-shell.

Furthermore the overlap of the p_z-orbitals which each other will be neglected. The molecular orbital which contributes to the conduction band is a linear combination of the molecular p_z-orbitals. Therefore the complete information is contained in the amplitude c_i of each p_z-orbital. The dipolar part of the hyperfine interaction can be calculated approximately in the following way: each carbon site is replaced by a magnetic dipole carrying an amount of spin density equal to the square of the amplitude c_i. The on-site hyperfine interaction is taken as the one experimentally observed in the glycin radical [31] scaled by the on-site spin density according to c_i^2. This approach has been used earlier for the interpretation of ESR spectra of radicals. The dipolar interaction with the spin density on neighbouring sites can be included by calculating the electron-nuclear dipolar interaction with these sites and by using the spin density at the neighbouring sites as a weighting factor. The calculation of the Fermi contact interaction, however, is clearly beyond Hückel theory. As mentioned before the existence of a contact term is the result of a core polarization of the σ-shell by the unpaired electron spin in the π-orbital, which itself has a node at the site of the carbon nucleus and does not contribute directly to the isotropic hyperfine interaction. The core polarization is basically a consequence of exchange interaction between π- and σ- electrons and must be included somewhat artificially in the Hückel framework. This problem has been tackled sucessfully in connection with proton hyperfine interactions by H. McConnell ([32] and references there) who showed that the observed hyperfine interaction of the proton is directly related to the spin density of the adjacent carbon atom. The proportionality factor has been calibrated empirically and led to the McConnell-equation:

$$a_H = -2\pi \cdot 65 \text{ MHz} \cdot \rho \tag{4.6}$$

For isotropic carbon hyperfine interactions the situation is more complicated because also neighbouring carbons will contribute some amount of core polarization through the σ-bonds. Karplus and Fraenkel [33] set up the following empirical equation for carbon fragments

$$a_j = Q \cdot \rho_j + \sum_k Q_k \cdot \rho_k \tag{4.7}$$

where $Q_k = -2\pi{\cdot}38.9$ MHz and $Q = 2\pi{\cdot}99.7$ MHz if two carbon atoms (k=1,2) and one proton are connected to the carbon atom i or $Q = 2\pi{\cdot}85.4$ MHz if three carbon atoms (k=1,2,3) are connected. This simple equation might be also useful as an empirical rule with other Q-parameters for heteroatoms. The Hückel theory predicts only positive spin densities whereas experimentally also negative values are observed. A refined model which takes exchange interactions between π-electrons in the form of 1st order perturbation theory into account has been worked out by McLachlan [34]. The exchange interaction leads to a spin polarization of the paired π-electrons in the underlying orbitals which results in a small negative spin density at sites where simple Hückel theory predicts none.

4.5. HYPERFINE FLUCTUATIONS

As mentioned above only the part of the hyperfine coupling that commutes with the Zeeman interaction contributes in the high field limit to the NMR spectrum. The non-secular parts represent processes in which one spin is flipped or two spins are flipped simultaneously. These processes do not conserve energy and are therefore forbidden. The situation changes, however, if the hyperfine coupling is modulated through lattice vibrations which can balance the energy. Obviously the spin flip terms establish exchange of energy between the spin system and the surrounding (lattice) thus leading to spin lattice relaxation phenomena. In organic conductors the conduction electron spins undergo spin flips due to scattering processes with phonons, other elctrons and impurities. The scattering is usually very strong leading to a diffusive motion for small scattering wavevectors. Soda *et al.* therefore developed a theory of nuclear spin relaxation in organic conductors which considers diffusive motion for small scattering wavevectors and an additional $2k_F$ contribution which may become dominant if the organic conductor is highly one-dimensional. Variants of this approach have been applied to the analysis of $(FA)_2X$ relaxation by [35] and to the proton spin lattice relaxation due to soliton motion in polyacetylene by [36]. We summarize here the basic formalism.

It was pointed out in [35] that the classical Korringa relation for metals is conveniently expressed in organic conductors in the form of a sum rule which can be rewritten for the nuclear spin lattice relaxation rate as

$$1/T_1 = K^2 T C_0 S_K \qquad \text{with} \qquad C_0 = \frac{4\pi k_B}{\hbar}\left(\frac{\gamma_n}{\gamma_e}\right)^2 \tag{4.8}$$

and where K is the isotropic Knight shift, T is the temperature, and S_K is a dimensionless parameter with $S_K = 1.0$ for a three-dimensional free electron gas, i.e. ordinary metals. Indeed S_K is observed to be in the order of 0.5 in alkali metals, aluminum and copper. However, S_K contains all the subtleties of the electron spin dynamics. In organic conductors it can be expressed as [35]

$$S_K = \frac{1}{2}\left(\frac{\tau_\perp}{\tau_s}\right)^{1/2}\left[\frac{3}{5}\epsilon j(\omega_n) + (1+\frac{7}{5}\epsilon) j(\omega_e)\right]\kappa_0(\alpha) + \frac{1}{2}(1+2\epsilon)\kappa_{2k_F}(\alpha) \quad . \tag{4.9}$$

In this expression the dipolar contribution to the hyperfine interaction is represented by d and the isotropic contribution by a, respectively. Both contributions enter the expression for the S_K parameter via the ratio $\epsilon = d^2/a^2$. The spectral density of the fluctuating conduction electrons $j(\omega)$ due to diffusive quasi one-dimensional motion can be expressed as

$$j(\omega) = \left[\frac{(1+\omega^2\tau_\perp^2)^{\frac{1}{2}} + 1}{2(1+\omega^2\tau_\perp^2)}\right]^{\frac{1}{2}} \tag{4.10}$$

where $\tau_\perp$ is the correlation time for interchain hopping. The intrachain electron scattering time is related to the intrachain diffusion constant $D_\parallel$ and the corresponding intrachain hopping time $\tau_\parallel$ by the following relations

$$D_\parallel = \frac{a^2}{\tau_\parallel} \quad \text{and} \quad D_\parallel = v_F^2 \tau_s \tag{4.11}$$

where v_F is the Fermi velocity and τ_s is the scattering time that characterizes the diffusive motion along the stack. Coulomb correlations enter through the quantities

$$\kappa_0(\alpha) = (1-\alpha)^{\frac{1}{2}} \quad \text{and} \quad \kappa_{2k_F}(\alpha) = \frac{(1-\alpha)^2}{[1-\alpha F(2k_F)]^2} \tag{4.12}$$

which contain the Stoner enhancement factor $\alpha = UN(E_F)$ and the Lindhard function $F(2k_F) = 1/2\ln(4.56\,T_F/T)$ with the Fermi Temperature T_F. In [35] it was shown that in highly one-dimensional radical salts like $(FA)_2X$ the $2k_F$ fluctuations can become dominant leading to a rather large value of S_K.

5. Experimental Results

5.1. Local Hyperfine Interaction and Spin Density Distribution

The intermolecular overlap integral t is usually considerably smaller than the intramolecular one. Consequently the spin density distribution is in most organic conductors in first order determined by the molecular orbitals. The molecular spin density distribution basically corresponds to that of an isolated radical scaled by the total value of the spin density per molecule in the radical salt, which is closely related to the amount of charge tranfer. Consequently Eq. (4.5) can be rewritten in the form

$$K_j = \frac{a_j}{\mu_0 \hbar \gamma_e \gamma_n} \frac{\chi^{mole}}{N_L} \tag{5.1}$$

for the Knight shift at the nuclear site j on a specific molecule and where a_j is the z-component of the hyperfine interaction tensor at the particular site, which can be related to the spin density ρ_j by using Eq. (4.6) in the case of protons or by Eq. (4.7) for carbon sites. Note that the susceptibility is normalized to one formula unit.

In order to calculate the amount of charge transfer the hyperfine interactions determined from the Knight shifts must be compared with those of the radical devided by the number of molecules in the formula unit. In order to establish a complete spin density map ultimate resolution is needed to resolve all atomic sites

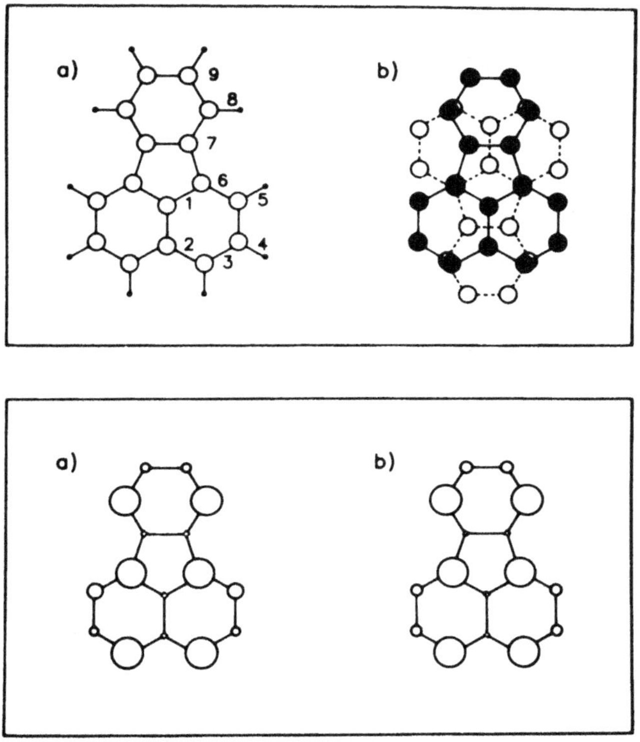

Fig. 4 Top: Molecular structure and carbon position numbering of the FA molecule (a). Alternate stacking of FA molecules in an FA dimer (b). Bottom: Spin density distribution of the conduction electrons on the FA molecule. The diameter of the circles is proportional to the spin density. (a) Experimental results from proton and ^{13}C NMR data. (b) Calculated spin density based on Hückel MO calculations.

carrying spin densities. Therefore the application of high resolution NMR methods is required as described in section 3. (Fluoranthenyl)$_2$X, abbreviated as (FA)$_2$X was the first organic radical salt which was investigated by applying these methods [37,6]. The corresponding ^{13}C spectrum is shown in Fig. 3.

In Fig. 4 we have summarized the results from ^{13}C as well as ^{1}H high resolution NMR spectroscopy together with a sketch of the molecular overlap and the spin density distribution.

In $(FA)_2SbF_6$ the pure hydrocarbon fluoranthene forms one-dimensional stacks in face-to-face configuration separated by SbF_6 groups. The crystal structure was determined by Enkelmann *et al.* [38]. The interest in $(FA)_2X$ results from its specific features like the extremely narrow ESR line and the fairly high conductivity which shows an extremely one-dimensional character and the appearance of an electronic phase transition near 180 K. Upon comparison of the locally resolved isotropic Knight shifts [37] with the ^{13}C spectrum of pure FA in solution and in a powdered sample (Fig. 3, see also Fig. 4) one immediately recognizes large positive and negative shifts. Knight shifts were extracted from the experimental spectra by subtracting the corresponding chemical shifts of the neutral compound. The attempted correlation with the spin densities from a Hückel calculation for a FA radical [39] by using a McConnell type relation failed because the orbital for the radical anion was chosen. A later analysis using the Karplus-Fraenkel equation [33] resulted in the spin density shown in Fig. 4. An assignment of the lines to certain carbon positions was achieved by means of gated-decoupling experiments and by comparison with the theoretical calculations. Clearly the spin density is concentrated at the positions where the molecular overlap between adjacent molecules is a maximum. The spin density distribution is not considerably altered when going from the radical to the one-dimensional stack pointing at a rather weak intermolecular overlap leading to narrow bands.

In an NMR/ESR double resonance experiment it was shown by Wind *et al.* [40] that the isotropic Knight shift can be suppressed by saturating the ESR line. Fig. 5 displays the lineshift of the ^{13}C MAS spectra in $(FA)_2AsF_6$ under two different ESR saturation conditions. This technique allows in principle to determine the Knight shift independendly from the knowledge of the chemical shift. This is important since the usual procedure where the chemical shift of the neutral compound is taken as the reference for vanishing Knight shift is strictly speaking incorrect. It ignores the fact that the ionic character of the molecular ion should have a chemical shift which is different from the neutral molecule. The shift difference can be considered small, because of the delocalization of the charge. Nevertheless it should be observable. The isotropic proton hyperfine interaction was derived from proton Knight shifts [6] and gave comparable values for the spin density distribution by using the McConnell equation. For comparison all results are summarized in Fig. 4.

A similar study has been reported for $(Perylene)_2(AsF_6)_{0.75}(PF_6)_{0.35}\cdot$ 0.85 CH_2Cl_2 [5]. Both proton and carbon Knight shifts were determined by application of CRAMPS and CP/MAS high resolution techniques, respectively. However, substantially less resolution was obtained in the ^{13}C spectra in comparison to $(FA)_2X$, regardless of the high symmetry of the perylene molecule. This is explained by the incomplete charge transfer q=0.55 which leads in combination with the mixed counter ions to electronic disorder. An incomplete charge tranfer has also been demonstrated for $(Perylene)_2PF_6\cdot 2/3THF$ in an Overhauser shift experiment, which showed isotropic hyperfine coupling of the fluorine nuclei [41]. Again the molecular overlap between adjacent molecules is found to be optimal at sites

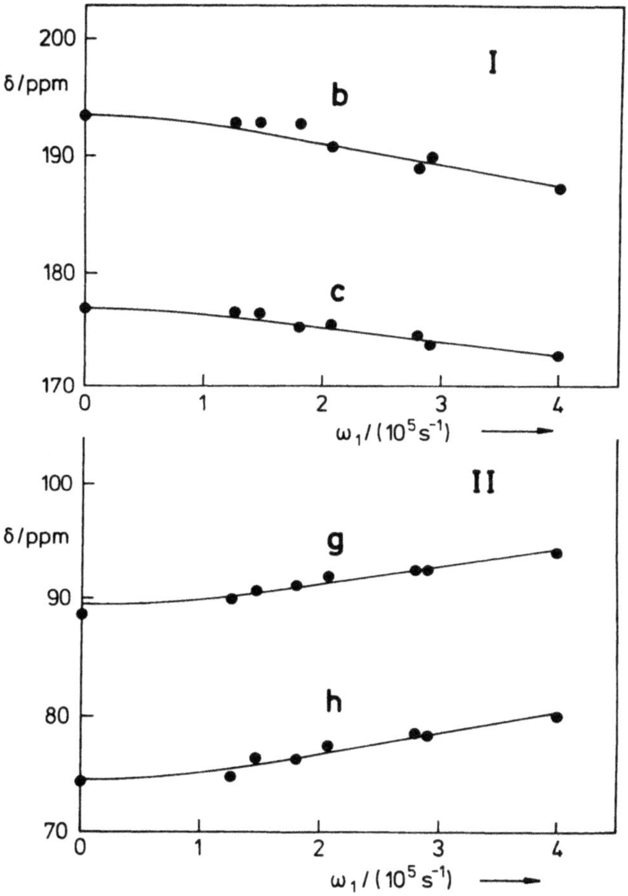

Fig. 5 Variation of the ^{13}C Knight shift in $(FA)_2X$ with the strength ω_1 of the saturating microwave field applied at the electron spin resonance frequency of the conduction electrons [40]. The strongly Knight shifted lines b, c and g, h approach the center of the spectrum (vanishing Knight shift) when the electron spin polarization is reduced by saturation.

which carry a high spin density. However, it is questionable whether it can be concluded that there is really an accumulation of spin density at the maximum overlap positions in the radical salts by comparing the experimental values with MO calculations as was done by Wieland *et al.* [5]. Instead one should rather compare the ^1H hyperfine couplings in the radical salt with the ones determined for free (Perylene)$^+$-radicals in solution [42,43]. However, the hyperfine values of the radical must be scaled by the factor 0.275 which takes the incomplete charge transfer in the radical salt into account. It turns out that the hyperfine coupling constants and accordingly the spin densities are in fact diminished in the radical salt at the sites where large overlap in the molecular stack occurs.

Well resolved ^{13}C Knight shifts were also observed in the Bechgaard salt $(TMTSF)_2$ ClO_4 [44]. A qualitative analysis of the shifts shows that the spin density distribution is compatible with the theoretical calculation of Metzger [45] and the findings of Devreux [3] from ^{1}H relaxation measurements. However, the observed line splitting is much more pronounced compared with the theory. This is indicative of a reduced symmetry in the radical salt compared to the neutral TMTSF which shows no such splittings. This reduced symmetry might be explained by the fact that in the radical salt the TMTSF molecule is no longer planar [46]. Therefore one can expect chemical shift differences to account at least for part of the line splittings. In later publications Knight shifts for $(TMTSF)_2ReO_4$ [47,48,49,50] and $(TMTSF)_2PF_6$ [49,50] were reported. No significant differences for samples with different counter ions were observed. In the case of the PF_6 system, however, an additional line with a dramatic temperature dependence was observed. No explanation has been given for this line so far.

An alternative way to extract the spin density distribution is the exploitation of the anisotropic hyperfine interaction. This was demonstrated for $(FA)_2X$ by Köngeter and Mehring [51] in a single crystal study and for β-$(BEDT\text{-}TTF)_2I_3$ by Vainrub *et al.* [52], by analysing the side band intensities of the MAS spectrum (Fig. 6) according to the formalism of Herzfeld and Berger [53].

The ratio of side band intensities to center band intensity can be used to recover the anisotropy of the sum of chemical shift and Knight shift. The spin density distribution is then determined via a point dipole model, where the hyperfine tensor for the on-site electron spin is taken as the one found in the glycin radical [31] scaled by the on-site spin density. A more detailed study of the Knight shift tensor was published by Hennig *et al.* [54] and more recently by Klutz *et al.* [55] which resulted in a complete spin density map.

5.2. MAGNETIC SUSCEPTIBILITY

Locally resolved Knight shifts can be used to determine the individual susceptibilities in two-chain conductors. This was demonstrated by Takahashi *et al.* [56] for TTF-TCNQ. The decomposition of the total susceptibility in two parts belonging to the TTF and TCNQ stack, respectively, was achieved by monitoring the temperature dependence of the ^{13}C Knight shift of a labelled carbon position on both TTF and TCNQ [57] and by assuming that the hyperfine coupling constant does not change with temperature. The different susceptibilities determined in this way

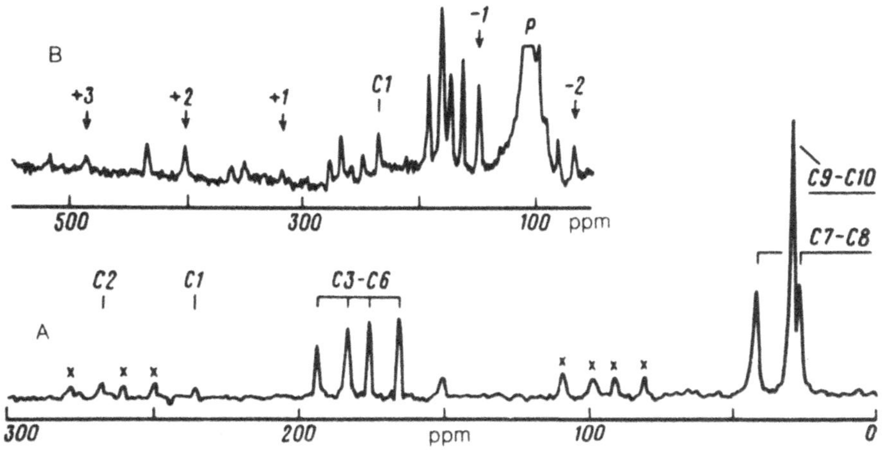

Fig. 6 ^{13}C MAS spectrum of $(BEDT\text{-}TTF)_2I_3$ according to Vainrub *et al.* [52]. Note the appearance of spinning sidebands labelled by 1,2,3. Assignment of the lines to different carbon positions is shown.

are presented in Fig. 7. This procedure turned out to be more precise than the g-factor analysis performed by Tomkiewicz *et al.* [58,59].

From a series of temperature dependent ^{13}C MAS NMR spectra from 300 K down to 48 K for $(DMe\text{-}DCNQI)_2Cu$, shown in Fig. 8, it was possible to determine the temperature dependence of the spin-susceptibility not only of the Cu salts but also for other metal ion salts of DMe-DCNQI [60] as is shown in Fig. 9. In comparison with the alkali DMe-DCNQI salts the copper salt was shown to possess a smaller susceptibility. It was concluded that part of the total spin density must be located on the Cu ion resulting from an incomplete charge transfer. Indeed a large copper Knight shift has been observed in these salts [61,62]. The oxidation state of copper in $(DMe\text{-}DCNQI)_2$-Cu must be of the form $Cu^{+(1-\delta)}$, where δ is estimated to be 0.3 [60] in contrast to $Cu^{+(2-\delta)}$ with $\delta = 0.67$ proposed before [63].

5.3. Localized States and Disorder

In many organic radical ion salts disorder seems to play an important role for a detailed understanding of the electronic properties. There are two different approaches for a theoretical treatment of disordered systems depending on the starting point: choosing the delocalized state as the unperturbed one leads to a description of disorder in terms of impurities which scatter the electrons between different delocalized states. On the other hand starting with the localized state treats the

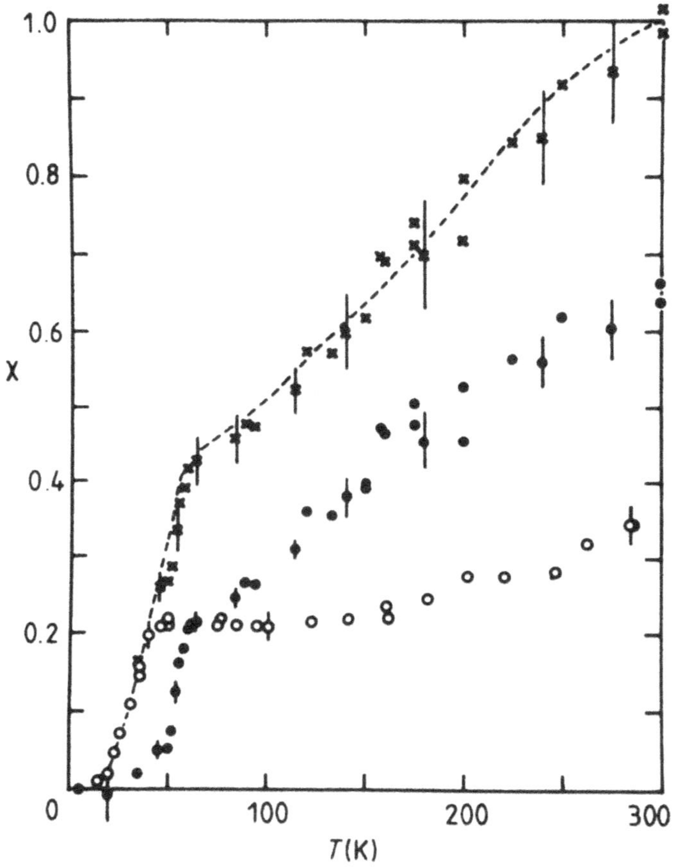

Fig. 7 Normalized magnetic susceptibility versus temperature for TTF-TCNQ according to [56]. Total susceptibility (×), TCNQ stack (•) and TTF stack (∘) susceptibility were determined from combining ^{13}C NMR and ESR data.

problem as an impurity pinning of the CDW state [64]. The conductivity in those samples is regarded as a thermally activated hopping process between localized states.

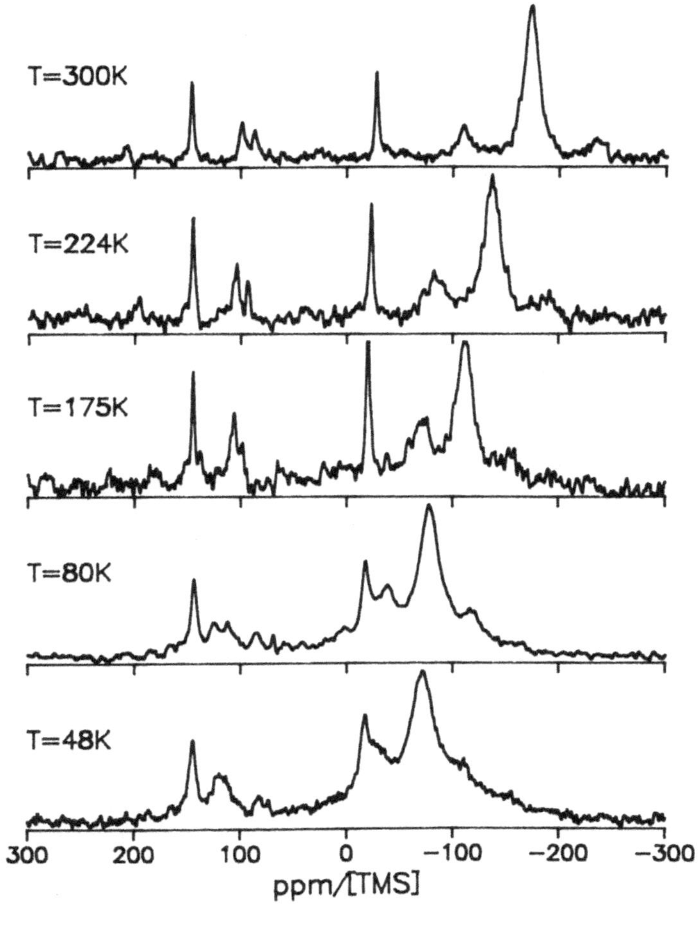

Fig. 8 A series of ^{13}C MAS spectra of $(DMe\text{-}DCNQI)_2Cu$ at different temperatures between 300 K and 48 K. The strong line at the right hand side is due to ^{13}C enriched CN groups. Note the temperature dependent Knight shift of this line.

Charge localization in organic conductors has been observed in $Qn(TCNQ)_2$. The completeness of the charge transfer in $Qn(TCNQ)_2$ was proven by X-ray scattering [65] and by ^1H NMR [3]. Early investigations of $Qn(TCNQ)_2$ were focussed on the study of the one-dimensional diffusion processes. Since relaxation studies are probing the spin dynamics, additional information is required in order to distinguish

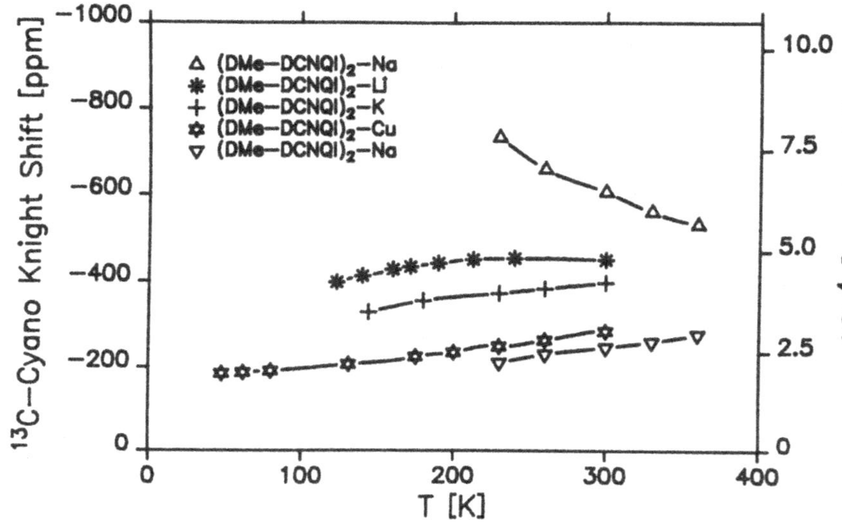

Fig. 9 Temperature dependence of the ^{13}C Knight shift and the corresponding susceptibility of different DMe-DCNQI salts. Note the small susceptibility of (DMe-DCNQI)$_2$Cu in comparison with the other salts. In (DMe-DCNQI)$_2$Na two different stacks with different spin (charge)-susceptibility are observed [60].

whether the fluctuations originate from the diffusion of charge carriers or spins. Devreux *et al.* [66] tried to answer this question by comparing the relaxation rate in a neutron irradiated sample to a non-irradiated one. In the irradiated sample the relaxation rate decreases according to the observed decrease in the conductivity whereas the spin-susceptibility determined from ESR measurements remained unchanged. In addition the determined diffusion constant displayed the same temperature dependence as the conductivity. A more recent ^{13}C CP/MAS investigation of Qn(TCNQ)$_2$ by Vainrub *et al.* [67] gave direct evidence for the disordered state. The carbon resonance from the ^{13}C labeled cyano groups of the TCNQ molecule is inhomogeneously broadened, and at the same time the relaxation rate measured across this line obeys a Korringa-type law, which indicates that hyperfine relaxation is the dominant relaxation mechanism (Fig. 10). Following Avilov *et al.*[68], the line broadening can be quantitatively analysed in order to determine the localization length. It turned out that the electrons are strongly localized in the sense of Anderson localization with a localization length in the order of 15 intermolecular spacings corresponding to about 50 Å.

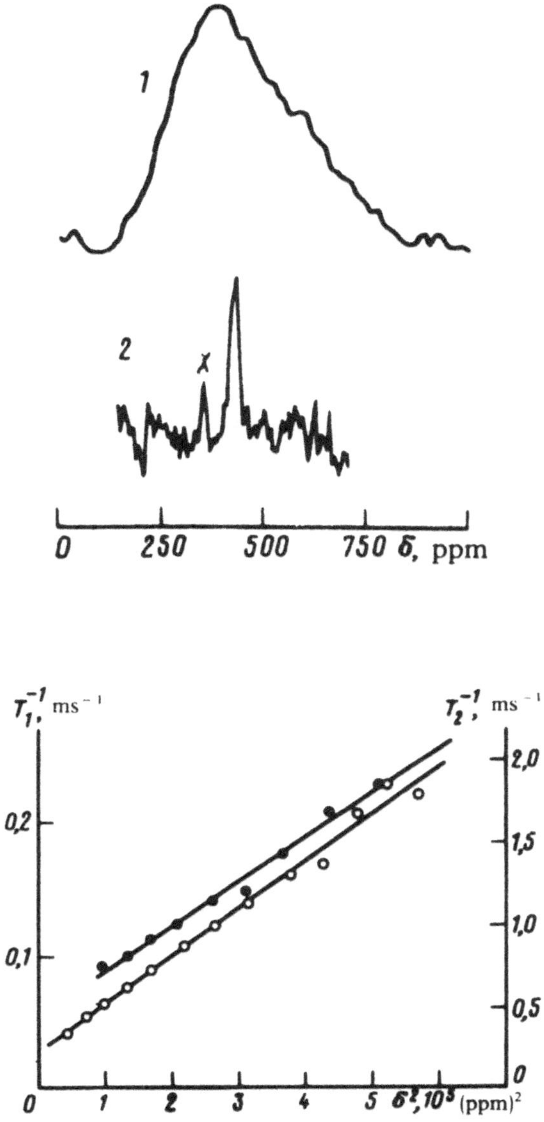

Fig. 10 Top: ^{13}C MAS spectra of ^{13}C labelled CN groups in TTF-TCNQ (2) and in Qn(TCNQ)$_2$ (1) according to Vainrub *et al.* [67]. The broadening due to localization is caused by a Knight shift distribution. This is supported by the dependence of the relaxation rate on the line position (bottom).

It is well known that a separation of charge and spin degrees of freedom can occur in organic conductors. The appearance of a Knight shift alone is no signature for delocalized charges and could result as well from localized charges but delocalized spins. An example of this behaviour has been demonstrated recently in the case of DMTM$(TCNQ)_2$ where a reversed metal-insulator transition occurs at about 290 K [69]. ^{13}C Knight shifts of the cyano group of TCNQ have been observed in the insulating state above 290 K as well as in the conducting state below 290 K as is shown in Fig. 11. The average Knight shift correlates very well with the total spin-susceptibility.

5.4. CHARGE AND SPIN TRANSFER

It has been proposed from X-ray investigations that a redistribution of charge occurs in MEM- and DMTM$(TCNQ)_2$ salts in the conducting phase. NMR can in principle determine this charge redistribution if spin and charge degrees of freedom are not separated. In Fig. 10 it is shown that the ^{13}C Knight shift data in the conducting phase (below 275 K) show a large dispersion of Knight shifts. This has been interpreted as an unequal charge distribution between two different TCNQ chains. Under the assumption that the spin density corresponds to the charge density a charge transfer of 0.62 for the TCNQ molecules on one chain and 0.38 for the other chain was derived from the NMR data. More recent ^1H CRAMPS NMR data by Kolbert *et al.* [70] reach similar conclusions.

The local susceptibility data for different DMe-DCNQI salts were shown in Fig. 9. Two inequivalent DMe-DCNQI stacks were found for (DMe-DCNQI)$_2$Na. From a careful analysis of the hyperfine interaction it was possible to determine the charge (spin) redistribution among these different stacks. A much smaller charge redistribution was found in (DMe-DCNQI)$_2$Cu where all DMe-DCNQI stacks are equivalent, but acquire only about 0.6 of the total charge, whereas the rest is located at the Cu-cyano complex. From MAS NMR experiments at low temperatures it was found that with decreasing temperature the charge decreases on the DMe-DCNQI molecules slightly.

5.5. ONE-DIMENSIONAL TRANSPORT

One-dimensional transport of spins causes a characteristic frequency dependence of the relaxation rate as was already discussed in section 4.9. There have been a number of ^1H-NMR relaxation time measurements on different TCNQ salts and other radical salts. Usually the characteristic magnetic field dependence is utilized in order to obtain some information on the dimensionality of the spin transport. The typical $\omega^{-1/2}$ behaviour where ω is the nuclear Larmor frequency has been verified in a number of cases [71,72,73,5]. It should be noted, however, that NMR relaxation is only sensitive to spin transport and any conclusion concerning charge

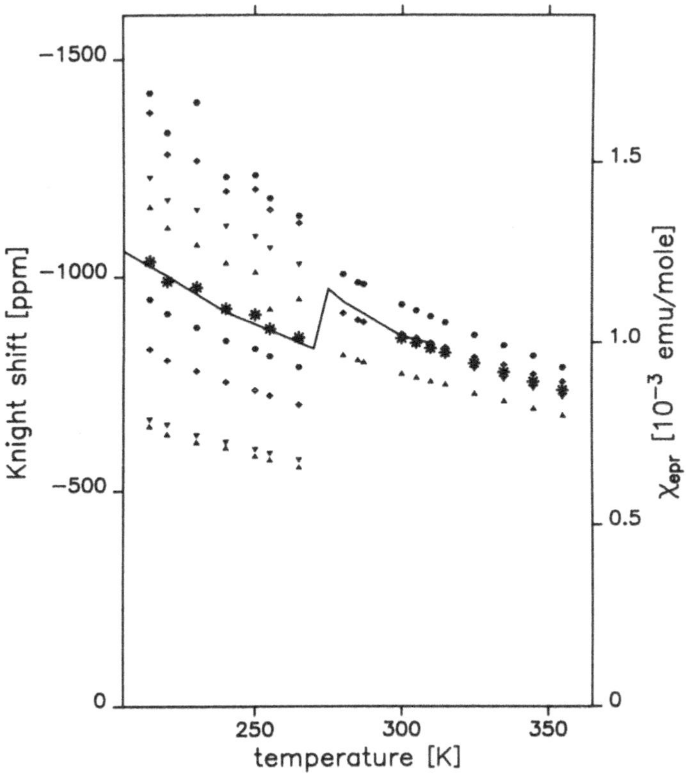

Fig. 11 ^{13}C Knight shifts of ^{13}C labelled CN groups of TCNQ and corresponding spin-susceptibility for DMTM(TCNQ)$_2$ versus temperature [69]. The stars represent the average Knight shift and average spin susceptibility which is compared with the susceptibility observed in ESR (line). Note the change at the inverse Peierls transition at T_c = 270 K. See text for discussion.

transport depends on the direct correlation of spin and charge which has to be proven by other means. The $\omega^{-1/2}$ behaviour levels off usually below a certain cut-off frequency. Below this cut-off frequency interchain hopping destroys the one-dimensional diffusive motion. The interchain hopping rate can be determined directly from this cutoff frequency.

Bourbonnais *et al.* [74,75,76] have analysed in a series of publications the temperature dependence of proton and ^{13}C nuclear spin lattice relaxation in different radical salts. In particular the theory of relaxation mechanisms in this class of systems was advanced by Bourbonnais [77]. Based on a renormalization group approach Bourbonnais could show that a Korringa like relaxation (T_1TK^2 = const) occurs in one-dimensional itinerant anti-ferromagnetically correlated electron systems, whereas a non-Korringa law (T_1TK = const.) holds for three-dimensional systems. These results can be summarized in the following equation

$$T_1^{-1} \simeq C_0T(\chi_s(T))^{(5-d)/2} + C_1T\xi_{AF}^{-\gamma/\nu+z-d} \tag{5.2}$$

where d is the dimensionality of the transport, $\chi_s(T)$ is the temperature-dependent magnetic susceptibility and C_0 and C_1 are positive constants. ξ_{AF} is the AF correlation length and ν is its critical exponent, γ is the critical exponent of the staggered susceptibility and z is the dynamical exponent for the scaling of the frequency of the fluctuations.

This behaviour has been illustrated for the Spin-Peierls system $(TMTTF)_2PF_6$, for the antiferromagnet $(TMTSF)_2PF_6$ and the superconductor $(TMTSF)_2ClO_4$. A particularly interesting case is$(TMTTF)_2Br$, which possesses an antiferromagnetic ground state. The relaxation behaviour near the Neel temperature T_N = 15K follows a power law according to $T_1^{-1}(q \approx Q_0) \propto (T - T_N)^{-1/2}$ that proves the 3D ordering [76].

5.6. Phase Transition in $(FA)_2X$

FA exhibits a metal-insulator phase transition at about 180 K. On the first sight this transition seems to be a textbook example of the Peierls transition in the limit of $4t_{\|} \geq U$, however the dimerization does not increase considerably below T_c [38,78,79]. There is no abrupt change in the lattice constants at all, only certain symmetry elements are lost due to a rotation of the fluoranthene molecules with respect to each other. Pouget, however, pointed out in a private communication that he has observed precursor effects and a Peierls like transition at 180 K.

Definitely, there is an abrupt change in the conductivity and a corresponding decrease in the magnetic susceptibility as seen in NMR and static susceptibility measurements. In Fig. 12 we show the correlation between the ^{13}C Knight shift data obtained from temperature dependent MAS NMR experiments and from the magnetic susceptibility, both above and below the phase transition.

A careful analysis of the paramagnetic susceptibility [80] gave evidence for a real gap in the density of states below T_c and the temperature dependence of χ and the conductivity σ [81] can be fitted by a CDW fluctuation model [82,83]. More recent studies of the crystal structure [78,79] and calorimetric studies [81] showed, however, that in fact there are two transitions, one at T=200 K involving only the counter ions and the second order transition mentioned above which takes place on the FA stack. Recent ^{13}C NMR measurements in our laboratory show substantial broadening in the region where the Peierls transition occurs. This is attributed

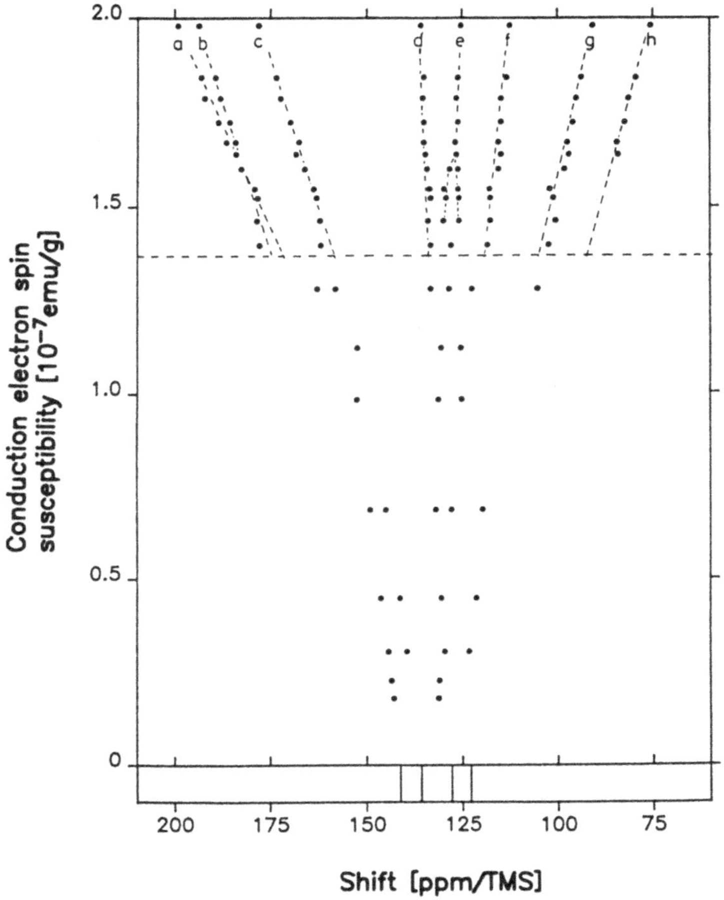

Fig. 12 Correlation between the ^{13}C Knight shifts and the spin susceptibility in $(FA)_2X$ above and below the Peierls transition, obtained from temperature dependent Knight shift and susceptibility measurements.

to the occurrence of a charge density wave (CDW) and the corresponding electron localization. A nonlinear current-voltage characteristic, typical for the CDW, has been observed in conductivity measurements which show nonlinear resistivity in combination with broad band noise above a threshold voltage [84].

5.7. PHASE TRANSITION IN $(TMTSF)_2ReO_4$

A special case of disorder arises in systems with non centro-symmetric counter ions: at higher temperatures the counter ions are either free to rotate or jump between different sites and are therefore disordered on the electronic time scale. When cooling down the sample either an ordering of the counter ions can occur, leading to a new phase with ordered counterions, i.e. sharp NMR lines or the disorder of the counterions is frozen resulting in a disordered phase with localized electrons and corresponding broad NMR lines. Examples are MEM-, DMTM-, $Qn(TCNQ)_2$, NMP-TCNQ, $(TMTSF)_2ReO_4$.

A metal-insulator transition due to anion ordering takes place in $(TMTSF)_2ReO_4$ at 180 K [85], which makes this transition well suited for high resolution studies. The anion ordering leads to a superstructure with a corresponding wave vector $\mathbf{q}=(1/2; 1/2; 1/2)$, and an energy gap of about 1000 K opens up in the density of states at the Fermi level. In the ^{13}C CP/MAS study reported by Stein *et al.* [86] the NMR lines could be followed through this first order transition as is shown in Fig. 13.

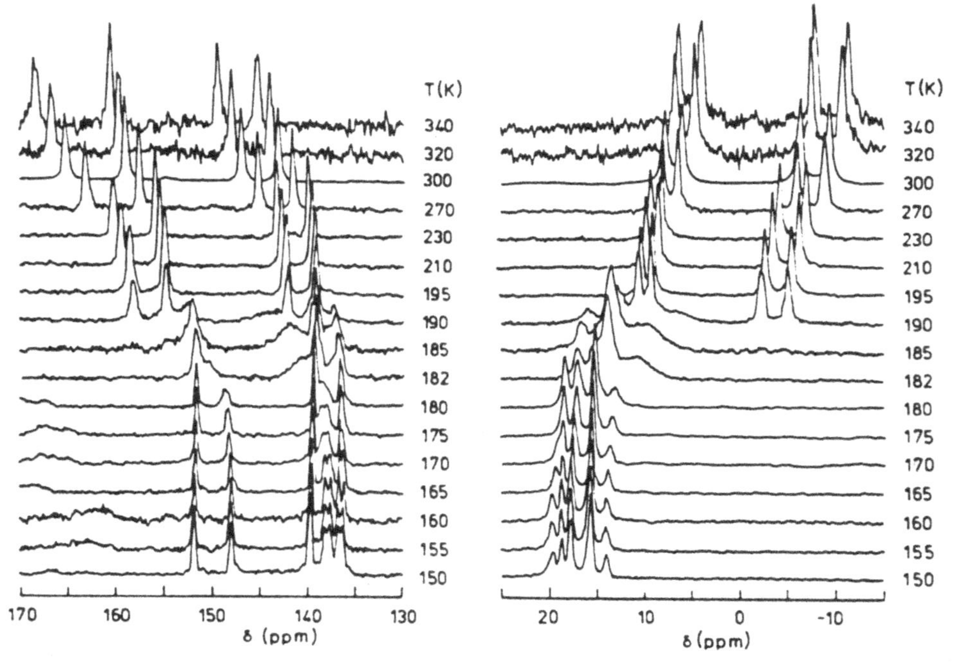

Fig. 13 ^{13}C MAS line shifts versus temperature in $(TMTSF)_2ReO_4$ according to Stein *et al.* [86]. Doubling of the lines is observed below the phase transition at 180 K due to the appearance of a superstructure.

Below T_c one observes a doubling of the carbon lines according to the appearance of the superstructure. The assignment of the lines below T_c concerning the center carbons seems to be doubtful. By comparison with the room temperature spectrum of $(TMTSF)_2ClO_4$ it is much more likely that the center carbons were not observed in both phases. The determination of the Knight shift is difficult in this case because significant chemical shift differences may arise between the metallic phase and the insulating phase due to molecular distortions. Indicative for this is already the observed chemical shift difference between neutral TMTSF in solution and the chemical shift in the radical salt below T_c. The molecular distortions may arise from a redistribution of the charge on the two TMTSF molecules driven by the influence of the ordered counter ions via oxygen-selenium contacts.

5.8. PHASE TRANSITION IN $TTF[Ni(dmit)_2]_2$

$TTF[Ni(dmit)_2]_2$ shows a metallike conductivity at ambient pressure. X-ray investigations [87] detected 1D structural fluctuations with a wavevector $\mathbf{q} = 0.4\ b^*$, which have been attributed to a CDW instability. Despite the development of a 3D CDW ordering at 40 K this compound becomes superconducting below 1.6 K at a pressure of 7 kbar. The coexistence of a metallic state with a 3D ordered CDW is one of the puzzling properties of this compound.

Bourbonnais *et al.* [88] analyzed the proton spin lattice relaxation as a function of temperature. As far as the TTF is concerned Bourbonnais concluded that the electronic properties are 1D. This was in fact predicted by Kobayashi *et al.* [89]. However also a 3D electronic structure has been suggested [90] having in mind the short S $\cdots$ S intermolecular contacts between the TTF molecules.

The first ^{13}C broad line studies revealed a drop in the electronic spin susceptibility below 200 K and proved that the CDW develops on the $Ni(dmit)_2$ stacks [91]. However, due to the limited resolution, a more detailed analysis was impossible. In a subsequent high resolution study Vainrub *et al.* [92] were able to monitor all six carbon resonance lines individually. The different temperature dependence of the ^{13}C Knight shifts within the same Ni(dmit) molecule proves the existence of a multisheet Fermi surface which has been predicted for the Ni(dmit) subsystem according to bandstructure calculations [93]. From a carefull analysis of the locally resolved Knight shifts Vainrub *et al.* succeeded to assign the individual contributions of the HOMO and LUMO to the density of states at the Fermi level across the TTF molecule. It is further suggested that the CDW effects occur in the LUMO band.

5.9. SPIN PEIERLS TRANSITION

Analogous to the Peierls transition a Spin-Peierls (SP) transition occurs in a number of quasi-one-dimensional Heisenberg anti-ferromagnetic $S=1/2$ compounds.

In all these solids the electron-electron correlation represented by the Hubbard U is larger than the transfer integral t. The most interesting property of these systems is the variation of the bandfilling with the magnetic field. This introduces field induced phase transitions with a wavevector which deviates from π/a for the dimerized state. This leads to commensurate-incommensurate phase transitions. For details on the spin Peierls transition see Ref. [94].

Standard proton NMR measurements were performed on MEM$(TCNQ)_2$, and TTF-$AuS_4C_4(CF_3)_4$ [95]. Since we are dealing in this review merely with the application of high resolution NMR techniques we only mention these investigations for further encouragement to apply also high resolution techniques to these solids. The basic results of the proton NMR investigations can be summarized as follows. TTF-$AuS_4C_4(CF_3)_4$ shows a spin Peierls transition at 2.03 K, which becomes field dependent with a field induced phase transition at 2.25 T [96,97]. It is interesting to note that the electronic Zeeman energy is on the same order as the transition temperature. This transition becomes first order at temperatures below 1 K [98]. The low-temperature phase is dimerized at low magnetic fields, whereas at higher fields it becomes incommensurate. By using both spin-lattice relaxation (T_1) and spin-echo relaxation (T_2) on protons and fluorine spins Hijmans *et al.* [95] were able to map out the magnetic field induced phase diagram. Moreover they calculated the relaxation rate for nuclear spins coupled to a spin 1/2 Heisenberg chain based on the Luttinger model.

A spin Peierls transition has also been investigated by ^{1}H-NMR in methyl-ethyl-morpholinium-$(TCNQ)_2$ (MEM$(TCNQ)_2$) [99,100]. Due to the small tight binding bandwith in MEM$(TCNQ)_2$ the conducting electrons are highly correlated. As a consequence the Peierls transition is split into two phase transitions, an electronic Peierls transition occurring at 340 K and a spin-Peierls transition at 18 K. The electronic transition leads to charge localization and a lattice dimerization whereas the spin-Peierls transition affects the spin degrees of freedom. The magnetic susceptibility in this system can be described with the help of a Bonner-Fisher model [101], however, below 20 K the susceptibility drops rapidly to zero. The development of an accompanying lattice tetramerization could be detected by X-ray and neutron scattering [102]. At 10 K the long range order of the superlattice structure exceeds 1000 Å.

5.10. SDW TRANSITION

Some radical salts undergo a metal to antiferromagnetic spin density wave (SDW) transition which can be suppressed under pressure and is then replaced by a superconducting transition [103]. As in the Spin-Peierls state only standard proton NMR measurements have been performed in the spin density wave state.

First NMR results (T_2, T_1 and linewidth) in the SDW state were obtained by Peo *et al.* [104] in $(TMTTF)_2SCN$. Inhomogeneous line broadening occurs below the transition temperature of 8.8 K indicating the onset of a SDW with an amplitude estimated to be about 10%. In addition a λ like increase of the relaxation rate was reported at the phase transition. Proton NMR was also applied to $(TMTSF)_2ClO_4$ and $(TMTSF)_2PF_6$ in the SDW state [105]. By comparing their results with MO calculations these authors were able to estimate the amplitude of the SDW to about 25% for PF_6 and 13% for ClO_4.

A more detailed investigation including an analysis of the wavevector of the SDW in $(TMTSF)_2X$ radical salts was performed by Delrieu *et al.* [106,107] and Takahashi *et al.* [108]. These authors used samples which were quenched rapidly in order to obtain the SDW state. Pronounced linebroadenings, lineshape distortions and satellites appeared reflecting the local field contribution from the SDW. Since these investigations were performed in single crystals an orientational dependence of the magnetic field was observed. It was proven that the SDW in $(TMTSF)_2PF_6$ is incommensurate with an amplitude δ = 8.5 % and $Q = 0.2\ b^*$ where b^* is a reciprocal lattice basis vector [106]. In $(TMTSF)_2ClO_4$ these authors find a different wavevector ($0.5\ a^* + 0.1\ b^*$(or $0.3\ b^*$)) in agreement with predictions that anion ordering should play a role in the SDW transition. Takahashi *et al.* [108] were able to determine the complete wavevector from a detailed analysis of the proton NMR spectra. Their result is (0.5, 0.24, -0.06) in the units of a^*, b^* and c^*. This vector seems to be near the ideal nesting vector for this system.

More recently Kanoda *et al.* reported proton NMR data on $(MDT\text{-}TTF)_2Au(CN)_2$ [109] and $(DMET)_2Au(CN)_2$ [110]. The appearance of the SDW transition showed up in a linebroadening and enhanced relaxation. No attempt was made to investigate the wavevector.

Delrieu [111] tried to find evidence for a moving SDW by observing a line narrowing of the proton NMR line when an electric field above threshold was applied to the sample. Although the data do not seem to be completely convincing this is an interesting proposal which should be followed up.

5.11. CDW TRANSITION

NMR has proven to be extremely successful in the detailed understanding of charge density waves (CDW) in inorganic solids where quadrupole interaction is a very sensitive NMR parameter for observing CDW's. Berthier and Slichter and coworkers have published a number of interesting results in which the structure and dynamic of CDW's was inferred from NMR measurements (see chapters 4 and 5 of this book). A similar investigation is lacking in organic conductors. Although there are a number of possible candidates for showing CDW behaviour, no experimental results are reported so far. Among the possible candidates are: $(FA)_2X$, $(DMe\text{-}DCNQI)_2Ag$, TTF-TCNQ, Ni-dmit, and other TCNQ salts.

There are a number of reasons why it is difficult to detect CDWs in organic conductors in particular. In a commensurate CDW, bond order waves result which give rise to the regular lines because of symmetry. For an incommensurate CDW, however, the changes in the charge redistribution are usually very small and are difficult to resolve. One should remember that most CDWs were detected by quadrupolar interactions in inorganic solids. In organic solids, however, most nuclei have spin 1/2 and the nuclear spin interactions which can be varied by CDW's, namely chemical shift and Knight shift, are rather small in these solids.

Let us now discuss how charge density waves could be detected by NMR. This problem was already treated by Slichter *et al.* [112] in the case of classical metals. The following NMR parameters (for spin 1/2) could be affected by charge density waves.

- chemical shift
- Knight shift

(a) Detection of the CDW by chemical shift.

It is known that the chemical shift of a neutral molecule changes when a charge is placed on the molecule. The dependence of the ^{13}C chemical shift on the partial carbon charge can be approximated in aromatic systems by (see [113] and references therein)

$$\delta_c \approx 160 \text{ ppm} \quad \cdot q_c + \delta_0 \tag{5.3}$$

where q_c is the additional charge in units of the elementary charge and δ_0 is the chemical shift of the neutral molecule. The factor 160 ppm is an empirical number. This equation allows an estimate of the change in chemical shift due to formation of a charge density wave. The CDW instability will mainly affect the SOMO. Along the lines of perturbation treatment one can express the variation in chemical shift

$$\Delta q_c \approx \Delta \rho_{CDW} \cdot q_c^{SOMO} \tag{5.4}$$

where $\Delta \rho_{CDW}$ is the amplitude of the CDW and q_c^{SOMO} is the charge in the SOMO. This latter quantity may be simply approximated by the inverse of the number of atoms over which the single electron orbital extents, i.e. by the extra charge per atom. Taking the amplitude of the CDW modulation equal to 10% for example, one obtains as an estimate for the expected variation in chemical shift

$$\Delta \delta_c \approx 0.1 \cdot \frac{1}{n_{atoms}} \cdot 160 \text{ ppm} \tag{5.5}$$

For a molecular unit with 20 atoms the variation in chemical shift is in the order of 1 ppm. Obviously, even with high resolution techniques it is very difficult to detect such a small change at the CDW transition.

(b) Detection of the CDW by Knight shift

If electron-electron correlation is not too strong charge and spin degrees of freedom are coupled. In this situation one expects a variation in spin density in correspondence with the variation in charge density at the CDW transition. Let us suppose again a 10% change in the charge which would result in Knight shift variation in the order of also 10%. Finally the variation of the Knight shift in the CDW state can be expressed as

$$\Delta K \approx 0.1 \cdot K \approx \frac{1}{\mu_0} \frac{a_0 \Delta q_c}{\hbar \gamma_e \gamma_c} \frac{\chi^{mole}}{N_L} \quad . \tag{5.6}$$

With a typical hyperfine coupling constant of about $a_0 \approx 2\pi \cdot 10\text{MHz}$ the Knight shift variation is $\Delta K \approx 6 \cdot 10^4 \cdot \chi^{mole}[\text{cm}^3/\text{mole}]$. This amounts to Knight shift variations in the range of 1 - 10 ppm in most cases. With current high resolution techniques it should be possible to detect these changes.

Berthier [114] pointed out that an additional difficulty arises which can obscure the observation of a Knight shift variation due to the CDW, namely the fact that the Pauli susceptibility usually drops at the CDW transition causing a corresponding drop in the Knight shift which is not related to the CDW. If electron-electron correlation becomes appreciable spin and charge degrees of freedom can become completely decoupled and the arguments used here do not hold anymore.

6. Summary

In this article we have tried to present a few examples where high resolution NMR techniques have revealed some important information about the electronic structure of radical salts. The main emphasis was placed on the hyperfine interaction, i.e. the Knight shift and its use to detect charge and spin transfer as well as phase transitions. Relaxation was only covered briefly and only one-dimensional diffusive motion of conduction electrons was discussed.

It was shown that the Knight shift provides a unique way to determine locally resolved hyperfine interactions as well as locally resolved magnetic susceptibility. There is hardly any other way to obtain these quantities. Overhauser shift could in principle provide part of this information but is limited in resolution at least for standard microwave frequencies (X band).

We could demonstrate in this article that the locally resolved hyperfine interaction depends sensitively on the molecular orbitals of the conduction band and a detailed spin density map, i.e. the molecular wavefunction of the conduction band can be obtained from the NMR data. Moreover, the locally resolved magnetic susceptibility allows to distinguish different charge transfers to different molecular chains. It could also be demonstrated that pronounced changes in the magnetic susceptibility occur at phase transitions. The advantage of measuring the magnetic susceptibility by NMR with respect to static susceptibility measurements with a magnetometer is the insensitivity of the NMR method to magnetic defects and the local resolution not obtainable by any other method.

There is no doubt that the examples shown here are by no means complete and the field is still in its infancy. There are many more possibilities which have not yet been exploited. It is expected that much more can be done and will be done in order to get a much better understanding of the subtleties involved in the electronic structure of radical salts in different phases.

Abbreviations for Substances

AuBDT	Au-bis-dithiolene
BEDT-TTF	Bisethylenedithiolotetrafulvalene
CuBDT	Cu-bis-dithiolene
DMET	Dimethyl(ethylenedithio)silenadithiafulvalene
(dmit)	1,3-dithia-2-thione-4,5-dithiolato
DMTM	Dimethylthiomorpholinium
DMe-DCNQI	2,5-dimethyl-N,N'-dicyanoquinodiimine
DPPH	α, α-diphenyl-p-picrylhydrazyl
FA	Fluoranthene
MDT-TTF	Methylenedithio-tetrathiafulvalene
MEM	Methylethylmorpholinium
NMP	N-methylphenazine
Qn	Quinolinium
TCNQ	7,7,8,8-Tetracyano-p-quinodimethane
THF	Tetrahydrofurane
TMTSF	Tetramethyltetraselenafulvalene
TMTTF	Tetramethyltetrathiafulvalene
TTF	Tetrathiafulvalene

References

[1] M.E. Anderson, P.J. Zandstra, and T.R. Tuttle, jr., *J. Chem. Phys.* **33**, 1591 (1960).

[2] M.E. Anderson, G.E. Pake, and T.R. Tuttle, jr., *J. Chem. Phys.* **33**, 1581 (1960).

[3] F. Devreux, Cl. Jeandey, and M. Nechtschein, *J. Physique* **40**, 671 (1979).

[4] S. Oostra, B. v. Bodegom, S. Huizinga, and G.A. Sawatzky, *Phys. Rev.* **B24**, 5004 (1981).

[5] J. Wieland, U. Haeberlen, D. Schweitzer, and H.J. Keller, *Synth. Metals* **19**, 393 (1987).

[6] F. Hentsch, M. Helmle, D. Köngeter, and M. Mehring, *Phys. Rev.* **B37**, 7205 (1988).

[7] A. Hackmann, H. Seidel, R.D. Kendrick, P.C. Myhre, and C.S. Yannoni, *J. Mag. Res.* **79**, 148 (1988).

[8] P.C. Myhre, G.G. Webb, and C.S. Yannoni, *J. Am. Chem. Soc.* **112**, 8991 (1990).

[9] P.C. Myhre, G.G. Webb, and C.S. Yannoni, *J. Am. Chem. Soc.* **112**, 8992 (1990).

[10] J. Hubbard, *Proc. Roy. Soc.* **A276**, 238, (1963).

[11] J. Hubbard, *Proc. Roy. Soc.* **A281**, 401 (1965).

[12] S. Mazumdar and A.N. Bloch, *Phys. Rev. Lett.* **50**, 207 (1983).

[13] H. Shiba and P.A. Pincus, *Phys. Rev.* **B5**, 1966 (1972).

[14] H. Shiba, *Phys. Rev.* **B6**, 930 (1972).

[15] J.B. Torrance, Y. Tomkiewicz, and B.D. Silverman, *Phys. Rev.* **B15**, 4738 (1977).

[16] R.E. Peierls, *Quantum Theory of Solids*, Oxford University Press (1955).

[17] G. Beni and P. Pincus, *J. Chem. Phys.* **57**, 3531 (1972).

[18] H. Fröhlich, *Proc. Roy. Soc. London* **A223**, 296 (1954).

[19] A.W. Overhauser, *Adv. Phys.* **27**, 343 (1978).

[20] S. Kagoshima, H. Nagasawa, and T. Sambongi, *One-Dimensional Conductors*, Springer Series in Solid-State Sciences **72**, Springer (1988).

[21] U. Haeberlen, *High Resolution NMR in Solids, Selective Averaging*, in: *Advances in Magnetic Resonance*, **Supplement I**, J.S. Waugh, ed., Academic Press, (1976).

[22] M. Mehring, *Principles of High Resolution NMR in Solids*, Springer (1976).

[23] A. Abragam, *Principles of Nuclear Magnetism*, Oxford University Press (1961).

[24] C.P. Slichter, *Principles of Magnetic Resonance*, Harper and Row, New York (1963).

[25] M. Weissbluth, *Atoms and Molecules*, Academic Press (1978).

[26] A. Carrington and A.D. McLachlan, *Introduction to Magnetic Resonance*, Harper & Row, New York (1969).

[27] L. Salem, *Molecular Orbital Theory of Conjugated Systems*, Benjamin, New York (1966).

[28] J.A. Pople and D.L. Beveridge, *Approximate Molecular Orbital Theory*, McGraw-Hill, New York (1970).

[29] J. Sadlej, *Semi Empirical Methods of Quantum Chemistry*, Ellis Horwood Limited (1985).

[30] G.N. La Mar, W.DeW. Horrocks, and R.H. Holm, *NMR of Paramagnetic Molecules, Principles and Applications*, Academic Press (1973).

[31] T. Cole and C. Heller, *J. Chem. Phys.* **34**, 1085 (1961).

[32] H. McConnell and D.B. Chesnut, *J. Chem. Phys.* **28**, 107 (1958).

[33] M. Karplus and G.K. Fraenkel, *J. Chem. Phys.* **35**, 1312 (1961).

[34] A.D. McLachlan, *Mol. Phys.* **3**, 233 (1960).

[35] M. Mehring in *'Low Dimensional Conductors and Superconductors'*, NATO Advanced Study Institute, Series B, Vol. **155**, D. Jérome and L.G. Caron, eds., Plenum Press, N.Y., (1987), p. 185.

[36] P.K. Kahol, M. Mehring, and X. Wu, *J. Physique* **46**, 1683 (1985).

[37] M. Mehring and J. Spengler, *Phys. Rev. Lett.* **53**, 2441 (1984).

[38] V. Enkelmann, B.S. Morra, Ch. Kröhnke, and G. Wegner, *Chem. Phys.* **66**, 303 (1982).

[39] C.A. Coulson and A. Streitwieser, *Dictionary of π-Electron Calculations*, Pergamon Press (1965).

[40] R.A. Wind, H. Lock, and M. Mehring, *Chem. Phys. Lett.* **141**, 283 (1987).

[41] H. Brunner, K.H. Hausser, H.J. Keller, and D. Schweitzer, *Solid State Commun.* **51**, 107 (1984).

[42] A. Carrington, F. Dravnieks, and M.C.R. Symons, *J. Chem. Soc.*, 947 (1959).

[43] J.P. Colpa and J.R. Bolton, *Mol. Phys.* **6**, 273 (1963).

[44] P. Bernier, M. Audenaert, R.J. Schweizer, P.C. Stein, D. Jérome, K. Bechgaard, and A. Moradpour, *J. Physique Lett.* **46**, L-675 (1985).

[45] R.M. Metzger, *J. Chem. Phys.* **75**, 482 (1981).

[46] N. Thorup, G. Rindorf, H. Soling, and K. Bechgaard, *Acta Cryst.* **B37**, 1236 (1981).

[47] P.C. Stein, P. Bernier, and C. Lenoir, *Physica* **143B**, 491 (1986).

[48] P. Bernier, P.C. Stein, and C. Lenoir, *Physica* **143B**, 494 (1986).

[49] P.C. Stein, and P. Bernier, *Phys. Rev.* **B37**, 10637 (1988).

[50] P.C. Stein, and P. Bernier, *Synth. Metals* **27**, B77 (1988).

[51] D. Köngeter and M. Mehring, *Phys. Rev.* **B39**, 6361 (1989).

[52] A.M. Vainrub, I.A. Heinmaa, and E.B. Yagubskii, *JETP Lett.* **44**, 317 (1986).

[53] J. Herzfeld and A.E. Berger, *J. Chem. Phys.* **73**, 6021 (1980).

[54] I. Hennig, Ph.D. thesis, University of Heidelberg, (1989).

[55] T. Klutz, I. Hennig, U. Haeberlen, and D. Schweitzer, subm. to *Appl. Magn. Resonance* (1992).

[56] T. Takahashi, D. Jérome, F. Masin, J.M. Fabre, and L. Giral, *J. Phys. C* **17**, 3777 (1984).

[57] E.F. Rybaczewski, L.S. Smith, A.F. Garito, A.J. Heeger, and B.G. Silbernagel, *Phys. Rev.* **B 14**, 2746 (1976).

[58] Y. Tomkiewicz, A.R. Taranko, and J.B. Torrance *Phys. Rev. Lett.* **36**, 751 (1976).

[59] Y. Tomkiewicz, A.R. Taranko, and E.M. Engler *Phys. Rev. Lett.* **37**, 1705 (1976).

[60] M. Helmle, J. Reiner, U. Rempel, M. Mehring, J.U. von Schütz, P. Erk, H. Meixner, and S. Hünig, *Synth. Metals* **42**, 1763 (1991).

[61] D. Köngeter, F. Hentsch, H. Seidel, M. Mehring, J.U. von Schütz, H.C. Wolf, P. Erk, and S. Hünig, *Solid State Commun.* **65**, 453 (1988).

[62] U. Langohr, M. Bair, J.U. von Schütz, H.-P. Werner, H.C. Wolf, P. Erk, H. Meixner, and S. Hünig, *Fiz.* **21**, 23 (1989).

[63] A. Kobayashi, R. Kato, H. Kobayashi, T. Mori, and H. Inokuchi, *Solid State Commun.* **64**, 45 (1987).

[64] Y. Suzumura in: *'Low Dimensional Conductors and Superconductors'*, NATO Advanced Study Institute, Series B, Vol. **155**, D. Jérome and L.G. Caron, eds., Plenum Press, N.Y. (1987), p. 343.

[65] J.P. Pouget, *Chemica Scripta* **17**, 85 (1981).

[66] F. Devreux, M. Nechtschein, and G. Grüner *Phys. Rev. Lett.* **45**, 53 (1980).

[67] A.M. Vainrub, I.A. Heinmaa, M.A. Alla, S.Kh. Viya, and E.T. Lippmaa, *JETP Lett.* **41**, 571 (1985).

[68] V.V. Avilov, L.N. Bulaevskii, and O.N. Dorokhov, *JETP Lett.* **42**, 192 (1985).

[69] F. Rachdi, T. Nunes, M. Ribet, P. Bernier, M. Helmle, M. Mehring, and M. Almeida, *Phys. Rev.* **B**, in press (1992).

[70] A.C. Kolbert, G. Zimmer, F. Rachdi, P. Bernier, M. Almeida, and M. Mehring, *Phys. Rev.* **B**, in press (1992).

[71] G. Soda, D. Jérome, M. Weger, L. Alizon, J. Gallice, H. Robert, M. Fabre, and L. Giral, *J. Physique* **38**, 931 (1977).

[72] F. Devreux and M. Nechtschein, *Lecture Notes in Physics* **95**, 145, Springer (1978).

[73] G. Sachs, E. Dormann, and M. Schwoerer, *Solid State Commun.* **53**, 73 (1985).

[74] L.G. Caron, F. Creuzet, P. Butaud, C. Bourbonnais, D. Jérome, and K. Bechgaard, *Synth. Metals* **27**, B123 (1988).

[75] F. Creuzet, C. Bourbonnais, P. Wzietek, H. Nelisse, L.G. Caron, D. Jérome, and K. Bechgaard, *Synth. Metals* **27**, B65 (1988).

[76] P. Wzietek, C. Bourbonnais, F. Creuzet, D. Jérome, and K. Bechgaard, *Europhys. Lett.* **12**, 453 (1990).

[77] C. Bourbonnais, P. Wzietek, F. Creuzet, D. Jérome, P. Batail, and K. Bechgaard, *Phys. Rev. Lett.***62**, 1532 (1989).

[78] V. Enkelmann and K. Göckelmann, *Ber. Bunsenges. Phys. Chem.* **91**, 950 (1987).

[79] V. Enkelmann,*Synth. Metals* **42**, 2547 (1991).

[80] U. Köbler, J. Gmeiner, and E. Dormann, *J. Mag. Mag. Mat.* **69**, 189 (1987).

[81] Th. Schimmel, W. Rieß, G. Denninger, and M. Schwoerer, *Ber. Bunsenges. Phys. Chem.* **91** , 901 (1987).

[82] P.A. Lee, T.M. Rice, and P.W. Anderson, *Phys. Rev. Lett.* **31**, 462 (1973).

[83] D.C. Johnston, *Phys. Rev. Lett.* **52**, 2049 (1984).

[84] W. Rieß, W. Schmid, J. Gmeiner, and M. Schwoerer, *Synth. Metals* **42**, 2261 (1991).

[85] K. Bechgaard, K. Carneiro, F.B. Rasmussen, M. Olsen, G. Rindorf, C.S. Jacobsen, H.J. Pedersen, and J.C. Scott *J. Am. Chem. Soc.* **103**, 2440 (1981).

[86] P.C. Stein, P. Bernier, and C. Lenoir, *Phys. Rev.* **B35**, 4389 (1987).

[87] S. Ravy, G.P. Pouget, L. Valade, and P. Cassoux, *Europhys. Lett.* **9**, 391 (1989).

[88] C. Bourbonnais, P. Wzietek, D. Jérome, F. Creuzet, L. Valade, and P. Cassoux, *Europhys. Lett.* **6**, 177 (1988).

[89] A. Kobayashi, H. Kim, Y. Sasaki, R. Kato, and H. Kobayashi, *Solid State Commun.* **62**, 57 (1987).

[90] M. Bousseau, L. Valade, J.P. Legros, P. Cassoux, P. Garbanska, and L.V. Interrante, *J. Am. Chem. Soc.* **108**, 1908 (1986).

[91] A. Vainrub, D. Jérome, M.-F. Bruniquel, and P. Cassoux, *Europhys. Lett.* **12**, 267 (1990).

[92] A. Vainrub, E. Canadell, D. Jérome, P. Bernier, T. Nunes, M.-F. Bruniquel, and P. Cassoux, *J. Physique* **51**, 2465 (1990).

[93] E. Canadell, I.E.-I. Rachidi, S. Ravy, J.P. Pouget, L. Brossard, and J.P. Legros, *J. Physique* **50**, 2967 (1989).

[94] J.W. Bray, L.V. Interrante, I.S. Jacobs, and J.C. Bonner, *Extended Linear Chain Compounds* **Vol. 3**, J.S. Miller, ed., Plenum Press (1983), p. 353.

[95] T.W. Hijmans, H.B. Brom, and L.J. de Jongh, *J. Phys. C* **19**, 1581 (1986).

[96] J.A. Northby, H.A. Groenendijk, L.J. de Jongh, J.C. Bonner, I.S. Jacobs, and L.V. Interrante, *Phys. Rev.* **B25**, 3215 (1982).

[97] J.A. Northby, F.J.A.M. Greidanus, W.J. Huiskamp, L.J. de Jongh, I.S. Jacobs, and L.V. Interrante, *J. Appl. Phys.* **53**, 8032 (1982).

[98] T.W. Hijmans, H.B. Brom, and L.J. de Jongh, *Phys. Rev. Lett.* **54**, 1714 (1985).

[99] P.I. Kuindersma, G.A. Sawatzky, J. Kommandeur, and C.J. Schinkel, *J. Phys. C: Solid State Physics* **8**, 3016 (1975).

[100] S. Huizinga, J. Kommandeur, G.A. Sawatzky, B.T. Thole, and K. Kopinga, W.J.M. de Jonge, and J. Roos, *Phys. Rev.* **B 19**, 4723 (1979).

[101] J.C. Bonner and M.E. Fisher, *Phys. Rev.* **135**, A640 (1964).

[102] B. van Bodegom, B.C. Larson, and H.A. Mook, *Phys. Rev.* **B24**, 1520 (1981).

[103] D. Jérome and H.J. Schulz, *Adv. Phys.* **31**, 299 (1982).

[104] M. Peo, J.C. Scott, and E.M. Engler, *Mol. Cryst. Liq. Cryst.* **119**, 303 (1985).

[105] Z. Toffano, K. Bechgaard, and A. Moradpour, *Mol. Cryst. Liq. Cryst.* **119**, 151 (1985).

[106] J.M. Delrieu, M. Roger, Z. Toffano, A. Moradpour, and K. Bechgaard, *J. Physique* **47**, 839 (1986).

[107] J.M. Delrieu, M. Rogers, Z. Toffano, E. Wope Mbougue, R. Saint James, and K. Bechgaard, *Synth. Metals* **19**, 283 (1987).

[108] T. Takahashi, Y. Maniwa, H. Kawamura, and G. Saito, *J. Phys. Soc. Japan* **55**, 1364 (1988).

[109] K. Kanoda, Y. Kobayashi, T. Takahashi, T. Inukai, and G. Saito, *Phys. Rev.* **B42**, 8678 (1990).

[110] K. Kanoda, T. Takahashi, T. Tokiwa, K. Kikuchi, K. Saito, I. Ikemoto, and K. Kobayashi, *Phys. Rev.* **B 38**, 39 (1988).

[111] J.M. Delrieu and N. Kinoshita, *Synth. Metals* **43**, 3947 (1991).

[112] D. Follstaedt and C.P. Slichter, *Phys. Rev.* **B13**, 1017 (1976).

[113] H.O. Kalinowski, St. Berger, and S. Braun, *13C-NMR-Spektroskopie*, G.Thieme Verlag, Stuttgart (1984).

[114] C. Berthier and P. Ségransan in: *'Low Dimensional Conductors and Superconductors'*, NATO Advanced Study Institute, Series B, Vol. **155**, D. Jérome and L.G. Caron, eds., Plenum Press, N.Y. (1987), p. 455.

INDEX

Physics and Chemistry of Materials with Low-Dimensional Structures

Previously published under the Series Title:

PHYSICS AND CHEMISTRY OF MATERIALS WITH LAYERED STRUCTURES

1. R.M.A. Lieth (ed.): *Preparation and Crystal Growth of Materials with Layered Structures.* 1977 ISBN 90-277-0638-7
2. F. Lévy (ed.): *Crystallography and Crystal Chemistry of Materials with Layered Structures.* 1976 ISBN 90-277-0586-0
3. T. J. Wieting and M. Schlüter (eds.): *Electrons and Phonons in Layered Crystal Structures.* 1979 ISBN 90-277-0897-5
4. P.A. Lee (ed.): *Optical and Electrical Properties.* 1976 ISBN 90-277-0676-X
5. F. Hulliger: *Structural Chemistry of Layer-Type Phases.* Ed. by F. Lévy. 1976 ISBN 90-277-0714-6
6. F. Lévy (ed.): *Intercalated Layered Materials.* 1979 ISBN 90-277-0967-X

Published under:

PHYSICS AND CHEMISTRY OF MATERIALS WITH LOW-DIMENSIONAL STRUCTURES
SERIES A: LAYERED STRUCTURES

7. V. Grasso (ed.): *Electronic Structure and Electronic Transitions in Layered Materials.* 1986 ISBN 90-277-2102-5
8. K. Motizuki (ed.): *Structural Phase Transitions in Layered Transition Metal Compounds.* 1986 ISBN 90-277-2171-8

PHYSICS AND CHEMISTRY OF MATERIALS WITH LOW-DIMENSIONAL STRUCTURES
SERIES B: QUASI-ONE-DIMENSIONAL STRUCTURES

B1. P. Monceau (ed.): *Electronic Properties of Inorganic Quasi-One-Dimensional Compounds.* Part I: Theoretical. 1985 ISBN 90-277-1789-3
B2. P. Monceau (ed.): *Electronic Properties of Inorganic Quasi-One-Dimensional Compounds.* Part II: Experimental. 1985 ISBN 90-277-1800-8
B3. H. Kamimura (ed.): *Theoretical Aspects of Band Structures and Electronic Properties of Pseudo-One-Dimensional Solids.* 1985 ISBN 90-277-1927-6
B4. J. Rouxel (ed.): *Crystal Chemistry and Properties of Materials with Quasi-One-Dimensional Structures.* A Chemical and Physical Synthetic Approach. 1986 ISBN 90-277-2057-6

Discontinued.

PHYSICS AND CHEMISTRY OF MATERIALS WITH LOW-DIMENSIONAL STRUCTURES
SERIES C: MOLECULAR STRUCTURES

C1. I. Zschokke (ed.): *Optical Spectroscopy of Glasses.* 1986 ISBN 90-277-2231-5
C2. J. Fünfschilling (ed.): *Relaxation Processes in Molecular Excited States.* 1989 ISBN 07923-0001-7

Discontinued.

Physics and Chemistry of Materials with Low-Dimensional Structures

9. L.J. de Jongh (ed.): *Magnetic Properties of Layered Transition Metal Compounds.* 1990
ISBN 0-7923-0238-9
10. E. Doni, R. Girlanda, G. Pastori Parravicini and A. Quattropani (eds.): *Progress in Electron Properties of Solids.* Festschrift in Honour of Franco Bassani. 1989
ISBN 0-7923-0337-7
11. C. Schlenker (ed.): *Low-Dimensional Electronic Properties of Molybdenum Bronzes and Oxides.* 1989 ISBN 0-7923-0085-8
12. R. H. Friend (ed.): *Conducting Polymers.* 1991 (forthcoming)
13. H. Aoki, M. Tsukada, M. Schlüter and F. Lévy (eds.): *New Horizons in Low-Dimensional Electron Systems.* A Festschrift in Honour of Professor H. Kamimura. 1992
ISBN 0-7923-1302-X
14. A. Aruchamy (ed.): *Photoelectrochemistry and Photovoltaics of Layered Semiconductors.* 1992 ISBN 0-7923-1556-1
15. T. Butz (ed.): *Nuclear Spectroscopy on Charge Density Wave Systems.* 1992
ISBN 0-7923-1779-3

KLUWER ACADEMIC PUBLISHERS – DORDRECHT / BOSTON / LONDON

GPSR Compliance

The European Union's (EU) General Product Safety Regulation (GPSR) is a set of rules that requires consumer products to be safe and our obligations to ensure this.

If you have any concerns about our products, you can contact us on ProductSafety@springernature.com

In case Publisher is established outside the EU, the EU authorized representative is:

Springer Nature Customer Service Center GmbH
Europaplatz 3
69115 Heidelberg, Germany

Batch number: 10761599

Printed by Printforce, the Netherlands